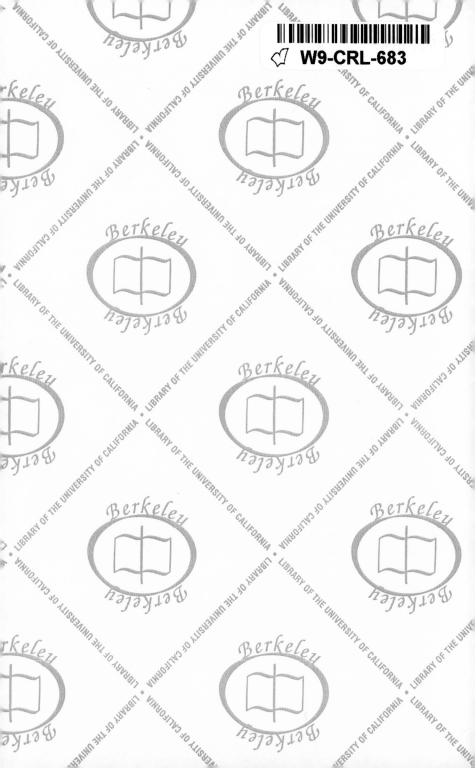

Membrane Processes: A Technology Guide

Membrane Processes:
A Technology Guide

P T Cardew and M S Le

North West Water Ltd

PTG

Chemistry Library

A Specialist Review Commissioned by The Process Technology Group of The Royal
Society of Chemistry, Industrial Affairs Division.

We acknowledge the sponsorship of The Environmental Technology Best Practice
Programme, DTI and the help of the Tony and Angela Fish Bequest.

ISBN 0-85404-454-X

A catalogue record for this book is available from the British Library

Published by The Royal Society of Chemistry,
Thomas Graham House, Science Park, Milton Road,
Cambridge CB4 0WF, UK

For further information see our web site at www.rsc.org

Printed by Athenaeum Press Ltd, Gateshead, Tyne & Wear, UK

PROLOGUE

This book is not meant as a "do it yourself" guide to membrane technology, nor is it a comprehensive treatise. These aims could not be met in a single volume. Instead our objective is to provide a book for those engineers and scientists who are coming across membranes for the first time and are interested in looking under the "bonnet" to see whether or not the membrane engine meets their requirements. Even with this restriction, decisions have had to be made as to what to include, what to assume, what to leave out, what to approximate. Inevitably a balance is achieved which reflects our personal assessment of the key aspects of the technology.

The diversity of membranes is immediately apparent to those who visit different companies or talk to colleagues in different disciplines. A visit to a pharmaceutical company will engender a different response to that of a doctor, an engineer in the water industry, or a process engineer in the chemical industry. This diversity is reflected in the introductory chapters which range over a number of membrane technologies. However, the focus of the book is on the membrane process that has been most widely exploited - the filtration processes. For this reason the examples given in the final four chapters are all examples of filtration processes.

Membranes are undoubtedly a successful technology, but they have failures. Such failures occur because of over selling, over optimistic expectations, inadequate pilot testing, etc. Failures usually occur in new applications where the technical problems have not been fully resolved. This lack of robustness and simplicity of design has meant that certain groups have shied away from membranes. However, environmental pressures on the wastewater and water quality demands on potable water continue to change the separation landscape and drive membranes along the technological and learning curves. After 30 years the technological spotlight is once again on the application of membranes to potable water. The largest, and lowest cost process industries in the world, have traditionally seen membranes as applicable where water costs are high due to the lack of availability of low salinity water. This new wave of plants is being driven by increasing water quality demands, and the increasing scarcity of good quality supplies.

Membrane technology can be readily packaged as another chemical engineering unit operation. To this end the early part of the book focuses on the more basic aspects that impact on the design and operation of such processes. However, to divorce the technology from its applications is like "learning history without the politics". Many of the most successful examples of membrane technology come from a more holistic approach, where the needs of each process is considered in the context of the total process objective. For this reason the latter half of the book illustrates several significant examples of membranes with the full context in which they lie. In this way we hope to provide a greater understanding of the issues than a pure distillate of experience.

ACKNOWLEDGMENTS

The idea for this work was first initiated a number of years ago, just after ICI sold its membrane interests to North West Water. After several years of nurturing, North West Water transformed the business from a research organisation to a membrane company. Eventually, North West Water decided its interest lay elsewhere and sold this fledgling business to US Filter. We are indebted to the many colleagues and consultants, who have worked with us and through the shared experiences have contributed to this work.

Special thanks go to our families who have supported and encouraged us during the years of change.

Contents

Chapter 10 - SURFACE WATER TREATMENT

Chapter 11 - MEMBRANES IN BIOLOGICAL WASTEWATER TREATMENT

Chapter 1

OVERVIEW

Contents

1.1   Membrane Technology - What is it?

Membrane technology is devoted to the separation of the minutiae of particles ranging from bacteria to atoms. To some people the concern is simply the removal of this detrious. To others the recovery of the inhabitants of this sub-microscopic kingdom is the essential goal. In size its constituents span some 4 orders of magnitude, and they are dominated by colloidal/molecular forces, rather than by the gravitational forces of their larger brethren. The various inlet and outlet streams can be all liquids, all gases or combinations. Not surprisingly membrane technology is not one technology but many technologies with one common aspect; the use of a membrane which separates two streams enabling materials to be selectively transported across it. As might be expected there is plenty of commonality between these various membrane processes, but, equally, the diversity and range of applications mean that there are significant differences. In recognition of these differences a classification of membrane processes has developed.

Of the various membrane technologies, the class of membrane filtration is the largest and most diverse. One of the commonest questions is where does conventional filtration end and membrane filtration begin. In a similar vein where does ultrafiltration take over from microfiltration. To answer this sort of question can be likened to defining where does the desert end and arable land begin; the two are clearly different but there is obviously some arbitrariness in defining the boundary. Nevertheless, a semantic definition provides a quick and expedient guide as to what to expect. However, to focus too heavily on the boundary is to miss the point. Customers are not interested in whether something lies on one side or other of a boundary but on what that something can do for them. The purpose of a classification is to convey the potential use.

Membrane technology is generally regarded as addressing the separation needs of sub-micron particles. Selectivity comes through the interaction between the membrane and the surrounding phases. Two factors contribute to selectivity, the partitioning of molecules and or particles between the membrane and the surrounding phase, and the relative diffusion rates of these materials once in the membrane. It is invariably the product of these two factors which contributes to the overall selectivity of the membrane.

One feature that is common to many membrane processes, though not to all, is cross-flow. Cross-flow involves moving fluid tangentially across the membrane surface (see figure 1.1) as well as normal to it. The benefit is that particles/solutes that would otherwise accumulate at the membrane surface are moved along, achieving a steady-state distribution of particles or solutes at the interface, rather than the continually developing one that is seen in conventional filtration. The consequence of cross-flow is that in continuous operation the flux through the membrane tends to a constant while in conventional filtration the flux continues to fall. If higher fluxes are desired then higher cross-flows are required.

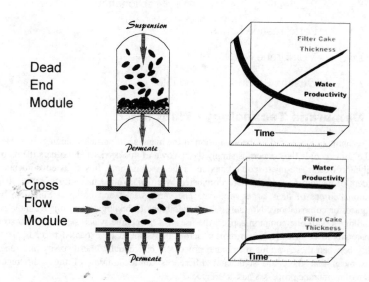

***Figure** 1.1 Schematic illustrating difference between cross-flow and dead-end filtration.*

The benefits of cross-flow do not come without a penalty, which is the energy required to move the fluid across the surface. Fortunately, the additional cost is small compared to that required in conventional filtration to push the fluid through a filter cake. A key factor in this effect is the ratio of the cross-flow to the permeate flow. Not surprisingly, this ratio is a key aspect underlying the design of membrane elements, and selecting optimal operating conditions.

Another consequence of cross-flow is that the system is basically designed to remove only a small proportion of the feed. Thus a feature of most membrane plants is how to design systems to overcome this limitation (see Chapter 8).

In the last few years the boundary between conventional filtration and membrane filtration has been further blurred with the development of hybrid processes. These processes allows some-build up of material at the membrane surface but then the material is dislodged by passing water or air back through the membrane. By repeating this process at frequent intervals (circa 15 min) a reasonable flux through the membrane can be maintained. In this way the deposits on the surface have limited effect and the membrane remains the controlling factor.

1.2   The Development of Membrane Technology

Membrane technology grew out of a 19th century endeavour to investigate a kingdom of particles too small to be seen. With no way of seeing these sub-microscopic constituents, membranes proved to be a useful tool to probe these invisible components. The resulting exploration that ensued provided key ingredients in the development of molecular theory of matter, which burst onto the scene at the start of the 20th century. In contrast it took nearly a 100 years to engineer membranes from a scientific tool to an industrial tool.

The Early Years - A Scientific Tool

A significant contributor in these early years was Thomas Graham, a Scottish chemical physicist and Master of the Mint. In 1861 he discovered that substances like salt and sugar rapidly passed through parchment, whereas material like gum arabic and gelatin would not pass. Materials that permeated he called crystalloids, since these materials could easily be crystallised. Those materials which did not pass, typified by glues, which at the time he believed did not crystallise, he called colloids after the Greek word for glue (Kolla). Graham showed how colloidal material could be purified from crystalloid contamination by putting the colloid in a porous container which was then placed in running water. The crystalloids pass through and the colloids remain. This process he called *dialysis* and the transport through - *osmosis*.

Thomas Graham made another important contribution as a result of studying the diffusion of gases through flat rubber membranes. In explaining his results he regarded the rubber as a liquid in which the gas dissolves and then diffuses due to a concentration gradient. This is the so called solution-diffusion mechanism which is an important element in the molecular theory of transport in some of the membrane technologies.

Another early contributor was Thomas Fick, of Fick's law fame. In 1855 he made a membrane by dissolving collodion (cellulose nitrate) in ether/alcohol solution which he then coated onto a ceramic thimble. This enabled him to dialyse biological fluids.

Membranes Coming of Age - A subject of scientific investigation

The first half of the twentieth century saw membranes themselves become the topic of investigation. Bechold provided the first systematic study, and coined the term "ultrafiltration" [1]. He pointed out that in addition to particle size effects adsorption processes play a role in the degree of separation that is achieved. This was perhaps the first clear recognition that membrane filters frequently involve more than a mechanical basis of separation i.e. one depending purely on size. In 1911 Donnan published his work on the distribution of charged species across a semi-permeable membrane[2]. Teorell[3] and Meyers and Sievers [4] were able to build on this and provide a model for the behaviour of charged membranes which is the basis of much of our understanding of electrodialysis membranes.

By 1927 membranes were in sufficient demand for Sartorius to start selling ultrafiltration and microfiltration membranes. This commercial reality though was largely aimed at those who used membranes as a laboratory tool rather than an industrial tool.

Table 1.2.1 *Some early contributions in the development of membranes*

Development	Contributors	Year
Laws of diffusion	Thomas Fick	1855
Dialysis	Thomas Graham	1861
Solution-diffusion transport mechanism	Thomas Graham	1866
Osmotic pressure	Van't Hoff	1887
Affinity effects in ultrafiltration	Bechold	1906
Distribution of ions	Donnan	1911
Pervaporation	Kober	1916
Membrane potential	Teorell, Meyer and Sievers	1935

Research into the nature of the microporous structure of membranes was severely hampered by a lack of tools to investigate these structural aspects. A significant development came in the 60's with the application of electron microscopy which allowed an understanding of the relationship between manufacturing variables and membrane morphology. At last the science that underpinned the empirical development of membrane manufacturing processes became understood, and meant that new manufacturing processes could be quickly developed the new generation of synthetic polymers such as the polysulphones could be exploited.

The Development of Membrane Technology - Commercialisation

Large scale commercial application of membrane technology started in the 50's, with the development of electrodialysis membranes for the desalination of brackish water[5]. The next major development was by Loeb and Sourirajan who successfully modified an ultrafiltration cellulosic membrane to create a viable reverse osmosis membrane for desalination of brackish water[6]. This opened the door, and by the mid 60's a number of companies had developed systems. Most notably to General Atomic (now Fluid Systems) who by 1965 had manufactured and built the first large scale reverse osmosis plant[7]. This industrialisation catalysed other membrane applications and developments. In particular it spurred on the development of ultrafiltration membranes for industrial usage, with applications like paint recovery in the electrocoat process. A process which is now used throughout the automotive industry. Another development of the 60's was Nafion. As part of a study into fuel cell technology by NASA, DuPont developed a hydrophilic type of PTFE by grafting onto the extremely hydrophobic polyfluoroethylene backbone, side chains with charged groups[8]. It was quickly recognised that this material could be exploited in the extremely challenging application of the production of chlorine and caustic from salt. The 70's and 80's saw a number of chemical companies trying to use their more advanced synthetic polymers and skills to enter the membrane market. One chemical company which had an initial success was Monsanto who developed the Prism membrane, based on polysulphone, for gas separation [9]. The interest of chemical companies waned in the late 80's and many who had entered in the 70's and 80's exited in the 90's as they sought to streamline their businesses.

A major development of the late 70's was the development of the composite membrane by Cadotte et al (see ref [10] for history of development). They had recognised that conventional reverse osmosis membranes were limited because different regions of the membrane had to carry-out the duties of mechanical support, and separation. They reasoned that if the separation layer and the mechanical support could be manufactured from different materials and tuned to the demands of each function, it should be possible to create a higher performing membrane. After many false starts they eventually succeeded in generating a good interfacial composite membrane that surpassed many others and laid the foundations for Film Tec which was later bought up by Dow in the late 80's.

As products have become established, suppliers have tried to open up new markets with varying degrees of success. The 90's brought a new factor into the equation, that of the environment. This has impacted on both the waste and supply side. Perhaps the largest single development has been the developing ultrafiltration and microfiltration technology for use in the municipal production of potable water to deal with cysts, bacteria, and viruses. What characterises many of these developments is not the universality of the technology but how the technology has to be developed for each application segment.

Table *1.2.2 Approximate dates for commercialisation of membrane technology for various applications*

Industrial Application	Commercialisation	Technology
Desalination of brackish water	1952	Electrodialysis
Desalination of brackish/sea water	1965	Reverse Osmosis
Paint recovery (Electrocoat)	1965	Ultrafiltration
Chlorine/caustic production	1972	Electrosynthesis
Hydrogen recovery	1979	Gas separation
Alcohol removal from water	1979	Pervaporation
Softening of hard water	1990	Nanofiltration
Filtration of potable water	1994	Microfiltration

As table 1.2.2 highlights different applications demand different membrane technologies (see table 1.2.3). Sometimes different membrane technologies can be used to solve the same problem. For example both reverse osmosis and electrodialysis can be used to produce potable water from sea water. In the former water is passed through the membrane , while in the latter the salts are removed. A comparison to determine which is best inevitably depends on the customers requirements, and circumstances. Despite the obvious differences in the various membrane technologies there are many common features at a fundamental level (see Chapter 2 and 3).

Table 1.2.3 List of various membrane technologies and abbreviations used

Technology	Abb	Technology	Abb	Technology	Abb
Reverse osmosis	RO	Gas Separation	GS	Gas Contacting	GC
Nanofiltration	NF	Membrane Distillation	MD	Dialysis	D
Ultrafiltration	UF	Pervaporation	PV	Haemodialysis	HD
Microfiltration	MF	Electrodialysis	ED	Haemofiltration	HF
		Electrosynthesis	ES	Membrane Bioreactors	MBR

1.3   The Driving Forces of Separation

For processes like crystallisation, distillation, adsorption, the separation achieved is related to the thermodynamic stability, with kinetics serving to dictate the time and size of plant required. In contrast for membrane processes separation is determined by the relative kinetics of permeation, with thermodynamics providing the time-scale and size of plant required.

Irreversible thermodynamics provide the framework for understanding membrane separations. The driving force for separation comes from gradients in thermodynamic variables. Commercial separation processes are governed by the differences in 1 or more of four thermodynamic factors

- *Pressure*
- *Concentration*
- *Electric Potential*
- *Temperature*

that exist between two phases being separated by the membrane. In response to these forces there are flows of mass, heat, electricity. At a local level the relationship between the forces, X_j, and fluxes, J_j, is a linear one of the general form

$$J_i = \Sigma_j L_{ij} X_j$$

(1.3.1)

where the L_{ij} are phenomenological coefficients to be provided either by experimentation or molecular theories. These coefficients occur in a variety of problems and many have been given names (see table 1.3.1). In the application of these principles to membranes the problem frequently becomes more complex in that the coupled sets of equations have to be solved over regions. Nevertheless the linear nature of the equations means that the fluxes are in general related to the differences in the thermodynamic properties of the two phases on either side of the membrane.

Table 1.3.1 Relationship between thermodynamic driving forces and fluxes

FLOWS	DRIVING FORCE			
	Pressure	**Concentration**	**Potential**	**Temperature**
Volume Flux	Filtration	Osmosis	Electro-osmosis	Thermo-osmosis
Solute Flux	Piezodialysis	Dialysis	Electrodialysis	Thermo-dialysis
Ionic Current	Streaming	Reverse Electrodialysis	Ionic conduction	Thermo-electricity
Heat Flow		Thermal osmosis	Thermo-potential	Thermal conduction

These thermodynamic forces are exploited in a number of different ways to give rise to a wide variety of membrane processes (see Table 1.3.2).

Table 1.3.2 Membrane technologies can be classified by the thermodynamic variables they exploit.

Driving Force	Separation Process
Potential difference (Voltage)	Electrodialysis, Electrosynthesis, Bipolar
Pressure difference	Reverse Osmosis, Nanofiltration, Ultrafiltration, Microfiltration, Gas Separation, Haemofiltration
Temperature difference	Membrane Distillation, Pervaporation,
Concentration difference	Dialysis, Haemodialysis, Gas Contacting

For the most part the different processes arise out of using different membranes to meet the different needs of applications. Some work has gone into membrane processes which used combined fields; most notably the use of potential fields to control fouling in pressure driven processes. To date though these process have not proved sufficiently attractive for large-scale development.

While thermodynamics provides a framework for separation it does not provide a mechanism for separation. Even in filtration processes it was recognised that separation was more than that of sieving. In general the relationship of

- *Size*
- *Charge*
- *Affinity*

between the membrane and the feed all play a role in determining the selectivity. The relationship between the size of particulates in the feed and the pores of a microporous membrane is a basis for separation. However, for large polymeric molecules conformational fluctuations can allow them to slip through pores much smaller than their radius of gyration might suggest. One can express this relationship as a solubility with size exclusion being an extreme example of the relationship. Size and conformation are not the only factors that determine solubility. Most colloidal materials carry some charge. If the charge on the colloid and the membrane are similar then there will be a tendency to exclude the colloidal material. Another fundamental factor that influences the separation is how fast the molecule or particle diffuses. At a fundamental level a recurring theme in membrane separation is that the power to separate two constituents is given by the the ratio, α , of the solubility, s, times diffusivity, D, for each component

$$\alpha_{i/j} \equiv \frac{s_i D_i}{s_j Dj} \qquad (1.3.2)$$

from which it can be seen that the separation power is a product of a thermodynamic factor (relative solubility) and a kinetic factor (relative diffusivity).

A general equation which embodies the issues discussed above is

$$Solute\ Flux = Concentration * Mobility * Force \qquad (1.3.3)$$

One further point which is worthy of attention is that the thermodynamic force is the gradient of the thermodynamic parameter. In many cases this can be approximated by the difference in the thermodynamic variable across the membrane divided by its thickness. Thus, the flux can be increased by decreasing the film thickness or increasing the thermodynamic difference. Films as thin as 500 nm have been manufactured. Such thin films create many practical issues of how to handle and support them, without damaging them. For other reasons flux values are usually set by the application and other operating conditions, so reductions in thickness are usually reflected in reductions in the applied force.

In some applications a lot of the energy supplied is there to overcome the thermodynamic free energy difference between the inlet and outlet phases. Equally there are many cases e.g. MF, UF processes where the energy supplied is principally there so that a separation can be carried out in a short time, or as a continuous separation in a small volume.

1.4   Purification, Concentration, Fractionation

Membranes are used for

- *Purification*
- *Concentration*
- *Fractionation*

Typical examples of purification occur in the fine chemical and pharmaceutical industries where membranes are used to purify products by purging low molecular contaminants through the membrane. An "inverse" example that is carried out in the food industry is to use a membrane to retain the heavier colloidal components and give a "clear" permeate product. The result is a product with a longer shelf life.

In effluent applications membranes are used to concentrate waste products, and hence reduce the volume to be shipped and disposed. In recovery applications such as in the paint industry membranes are used to concentrate the product to a level either sufficient for selling or to allow it to be recycled.

Membranes do not usually provide a sharp separation based on molecular weight. However, the fractionation that can be achieved is sufficient for its wide use in the processing of biological fluids, e.g. blood.

1.5   Performance Limits

Productivity and separation are dependent on the permeation kinetics of the various constituents, and these will vary with the materials and structure of the membrane. Limits to these parameters set a guide to what is feasible and what is unfeasible. They also provide a guide to the structure and design of membranes. A simple illustration is provided through consideration of flow through a homogeneous microfiltration membrane. The maximum flux through such a material occurs if it is made up of a uniform set of parallel pores running across the membrane. The flow through the pores is invariably laminar, so that Hagen-Poiseuille's equation applies. Summing the contributions from all pores gives a simple relation between the flux, J, and the pressure drop across the membrane ΔP:

$$J = \varepsilon \frac{a^2}{8\eta} \frac{\Delta P}{l.}$$

(1.5.1)

where, ε, is the porosity, η, is the viscosity, a is the pore radius, and L is the membrane thickness. The size scale of pores in a microfiltration membrane is some 4 orders in magnitude larger than in reverse osmosis, and as a consequence the flux will be 8 orders in magnitude less if other things are left equal (see table 1.5.1). For a 100 micron thick film with 0.1 micron pores the pressure loss is a modest 0.1 bar. Reduce the pores to 10 A (tight ultrafiltration) and the minimum pressure rises upto 1000 bar to maintain the same flow!

Table 1.5.1 *Maximum water permeability of a membrane (porosity = 0.88) of various pore diameters and thicknesses. Also includes minimum operating pressure that would be required to deliver a flux of 100 L $m^{-2}hr^{-1}$*

Pore Diameter nm	Film thickness μm	Permeability L/m²/bar	Pressure bar
100	100	0.28	0.1
10	100	0.0028	10.0
1	100	0.000028	1,000.0
1	10	0.00028	100.0
1	1	0.0028	10.0
1	0.1	0.028	1.0

This indicates why the smaller the pores the thinner must be the active separating part of the membrane. Thus for reverse osmosis the active layer of the membranes is less than 0.1 microns in thickness.

Making thin films is not difficult but handling them is, and handling films less than 1 micron in thickness in air presents numerous practical issues due to strong electrostatic forces. Practically the problem is overcome by creating membranes with a graded microporous structure (asymmetric membranes), where the critical separation layer is only at one face of the membrane. Thus, in the manufacturing of ultrafiltration and reverse osmosis membranes, the key is to create an asymmetric structure with fine pores on the side facing the feed and more open pores on the other.

In molecular separations transport models are not so simple. However, within any class of materials there does appear to be a trade-off between permeability and selectivity, with more permeable materials having lower selectivity. One of the most significant of these trends is shown in figure 1.5 for oxygen/nitrogen separation through polymeric materials. The lack of any highly selective membrane with high permeability has meant that applications such as oxygen enrichment for combustion still remain a tantalising opportunity, despite considerable research effort and piloting.

However, the limitation observed in polymeric materials appears to be associated with the type of molecular interactions. In materials which have different molecular characteristics quite different permeabilities and selectivities are observed. A noted example of this is the production of high purity hydrogen, which uses a metal membrane composed of palladium. The great selectivity of these membrane provides hydrogen of the highest purity (99.9999 %)[12]. A rare example of a membrane that provides extremely high selectivity.

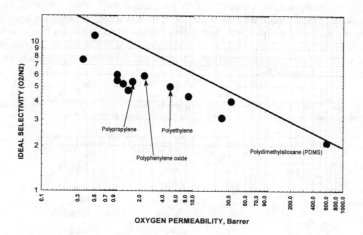

Figure 1.5 Trade-off between selectivity and gas permeability (data from[11]). 10^{10} Barrer corresponds to 1 $cm^3(STP)$ cm^{-2} s^{-1} $(cm Hg)^{-1}$ (cm thickness).

If the membrane material cannot be made any more permeable, it raises the question can it be made thinner? Certainly there does appear to be some scope for this in processes like gas separation. However, the major practical problem is how to make thin films without defects on the scale, and at the cost required.

While improving permeation rate by using more permeable membranes or higher driving forces might seem desirable there is a limit to the benefit that can be obtained. This limitation comes from two directions. The first is a practical one, that as the permeability gets higher it becomes more difficult to get the fluid to the membrane surface without incurring a pressure drop penalty. The second is a phenomenon called concentration polarisation, which creates a selectivity penalty (Chapter 6).

1.6   Membrane Materials and Membrane Structure

1.6.1 Introduction

In reading the research and patent literature one might think that any material can be made into a membrane. Fortunately, most such membranes have academic rather than commercial value. Nevertheless, the number of materials that are available is large, and viable membranes can be made from polymers, ceramics, inorganics, and metals. This variety stems from the fact that there is no such thing has a perfect membrane material. One membrane material might be the best for one application but hopeless at another. For a membrane to be viable it must not only have the necessary performance properties but also statisfy a number of secondary properties (see table 1.6.1)

Table 1.6.1 Secondary features of membranes.

Property	Feature	Property	Feature
Thermal	• Temperature Limit • Activation Energy • Sterilisation	Hydrophilicity	• Bubble Point • Breakthrough pressure • Adsorption properties
Mechanical	• Maximum Operating Pressure • Scratch resistance • Indentation • Yield Strength	Hygiene	• Particle shedding • Leachables • TOC
Chemical Stability	• pH tolerance • Oxidative Degradation • Swelling • Degradation	Sorption	• Organic uptake
Biological	• Biocompatability • Biodegradation	Other	• Radiation Resistance

Membranes come in a wide range of forms, and structures. While many materials come in more than one form there is frequently a processing reason which can make it difficult to obtain a particular material in a desired format. At the macro-scale there are three main forms

- *flat sheet*
- *hollow-fibre*
- *tubular*

Tubular membranes differ from hollow-fibre only in that their larger diameter means that they have to be supported. Other variants, such as reticulated surfaces to encourage mass transfer, have been investigated but remain at the margins of the commercial world with their interest being essentially a scientific one.

Table 1.6.2 Summarises the various manufacturing processes that are used to create membranes

Structure		Membrane Process	Manufacturing Process
Dense	Homogenous	ED, ES, GS	• Cast • Extrusion
	Composite	ED, ES	• Lamination • Reaction
Microporous	Homogenous	MF, UF, HD, MD	• Stretching, • Sintering • Track-etch
	Asymmetric	MF, UF, RO, GS, HD	• Solvent Phase Inversion • Thermal Phase Inversion
	Composite	RO, GS, PV	• Coating • Interfacial polymerisation, • Plasma • Lamination

Close examination of each membrane form reveals a wealth of structural detail and variation. Much of the detail which derives from the manufacturing processes (see table 1.6.2) is optimised to meet both the needs of the application and the mechanical requirements during handling and operation. Some of the features are however incidental.

1.6.2 Dense Films

The simplest structural type of membrane is the dense polymer film. Such materials are nearly exclusively of the ion-exchange type and are used in electrically driven or dialysis type processes.

Electrodialysis uses dense films of anion and cation exchange polymers. These membranes were based on cross-linked styrene-divinyl benzene chemistry with the negative charge being introduced by sulphonic or carboxylic acid groups. Anion membranes are available based on aliphatic chemistry with the positive charge being introduced through quaternary ammonium groups. These materials show good resistance to harsh environments pH 0-10. More recently acrylic based anion exchange membranes which have good oxidative resistance have become available. The large number of ionizable groups make the materials extremely hydrophilic and without the cross-linking there would be a tendency for them to dissolve. Typically membranes are several hundred microns thick, and are sometimes reinforced with a suitable polymer mesh to ease handling. having thinner membranes does not lead to a significant benefit since the impedance is largely associated with the solution between the membranes.

The 1970's saw the development of Nafion. The polymer is essentially PTFE with a perfluorinated side chain carrying a charged sulphonic or carboxylic acid group. The polymer resists swelling by having regions of crystalline PTFE. Some of the unique properties that derive from such materials derives from their microstructure. The hydrophilic groups tend to aggregate and form clusters some 40 A in size. These regions provide conduits for water and charge species to pass through the otherwise hydrophobic polymer. The high local charge density created gives the material its "*superselectivity*". Higher selectivities are achieved with carboxylic materials rather than with sulphonic acid materials. In order to get the correct balance of properties a laminate of more than one of these polymers is sometimes used. As with electrodialysis membranes there is often a reinforcing PTFE mesh to give additional mechanical robustness. Another development during the 1980's was a laminate of a cationic and anionic material. The so called bipolar membrane could be used to split water and hence simultaneously produce acid and base during electrodialysis.

In the late 80' s Dow brought out a hollow-fibre membrane with a dense film of 1,4-methylpentene polymer for gas separation. the membrane wall thickness was around 10 microns in thickness. The thickness meant that the net flux was low, but this was partially compensated by making the membrane in the form of hollow-fibres with an outer diameter of 60 microns.

1.6.3 Homogenous Microporous Membranes

There are a large number of microporous membranes, made by a variety of processing methods. Each product has found a number of niche markets. PTFE membranes are widely used in laboratory work on account of their inertness. On the large industrial scale the stretched polypropylene membranes of Enka and Memcor have found the widest application.

Table *1.6.3 Some of the homogeneous microporous membranes, methods of manufacturers*

Manufacturer	Material	Technology
Nuclepore	Polycarbonate	Track-Etch
Gore	Gortex - PTFE	Stretching
Enka	Accurel - Polypropylene	Stretching
Memcor	Polypropylene	Stretching
Gelmans Science	Repel - Polyacrylic	Photopolymerisation

1.6.4   Asymmetric Membranes

As discussed in section 1.4 asymmetric membranes provide a way of having a thin membrane form that can be handled. UF, NF, RO, GS processes all use asymmetric membranes. Even a number of the microfiltration membranes[1] utilise this structure. In these process the dense layer is presented to the feed solution.

Some degree of asymmetry is absolutely essential to make reverse osmosis work, since the key separating layer has to be extremely thin (typically 500-2000 A). The remainder of the material is there to provide mechanical support and allow one to physically handle the material. If the wrong side of the membrane is presented to the feed there would be severe polarisation due to the stagnant layer of liquid trapped in the membrane. While the support layer is there to provide mechanical support for the separating layer it can have some impact on the permeability. Pressure lost in the sub-structure reduces the permeability of the membrane, and in the case of reverse osmosis it also serves to reduce the separating power. Conversely, if the support structure is very porous it means that defects in the separating layer can become more significant. In other words the sub-structure throttles the effect of defects in separating layers. Thus it can seen that there is a practical element to balancing various features in the membrane structure.

> *Development of the Asymmetric Reverse Osmosis Membrane*
> *Asymmetry also plays a significant role in cross-flow membranes. In the development of the reverse osmosis membrane by Loeb and Sourirajan it was noted that about 50 % of the membranes did not perform properly. It was eventually discovered that this was determined by which way round the membranes were placed in the test cells (see Loeb [6]). Good performance was achieved when the shinier of the two surfaces was presented to the feed. It was then recognised that the reflective properties indicated differences in microstructure; the matt surface was covered with micropores, while the shiny surface was closed. A distinction easily recognised today with modern electron microscopes, but not so easy in the 50's when such devices did not exist.*

Two types of microporous structure are commonly seen. The basic type is known as the honeycomb and consists of an interpenetrating network of pores and polymer (see figure 1.6-1). The other type is the macro-voided structure. In this type of membrane a regular pattern of macro-voids grow down from the top surface. Macro-voided structures tend to have higher permeability and are mechanically weaker than the honeycomb structure.

---

[1]   For some microfiltration cartridge filters the opposite orientation is sometimes used with the large open pores facing the feed and becomes smaller as one moves through membrane. This depth filter configuration is designed to maximise the dirt holding capacity and keep the pressure rise that occurs has solids accumulate to a minimum. The latter arises has a result of solids distributing itself more uniformly across the membrane, with the coarse at the front end and the fines at the tighter end.

Outside the difficulty of making such free standing films with high porosity, there is a major mechanical handling question. The practical way round this is to create membranes with a graded microporous structure (asymmetric films).

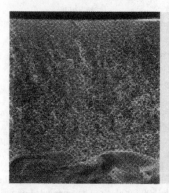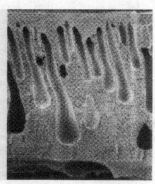

Figure 1.6-1 shows a scanning electron micrograph of sections through two ultrafiltration membranes without backing support. The thickness of the layer is approximately 50 microns in both cases. The left picture shows the honeycomb structure. The right shows the presence of macro-voids. In both cases there is a noticeable asymmetry of the pore size with the more finely porous side acting as the filtering surface, and facing the feed solution.

Manufacturing Asymmtric Membranes - Solution Phase Inversion

The most common method of manufacturing asymmetric membranes is known as solution phase inversion. The process consists of dissolving a polymer in a suitable solvent, and then casting this on to a cloth. After a small time the cloth and the attached polymer solution film are quenched in water. Water diffuses into the cast film which becomes thermodynamically unstable and expels the now non-solvent to create a honyecomb structure. The effect of water is to create an unstable polymer solution that phase separates. It is the nature of this process that creates the detailed microstructure. By carefully selecting the solvents, and adjusting the time the cast membrane sees air before it quenches the asymmetry and structure of the membrane can be controlled.

Typically asymmetric membranes are about 150 microns thick, made up of 95 microns of backing cloth, and 50 microns of membrane.

1.6.5   Composite Membranes

The major problem with asymmetric reverse osmosis membranes is that one material has to carry-out all the functions e.g. separate, support, protect. It therefore seems only natural to try making a membrane from layers of different material designed for each function rather a compromise material for all functions.

In the late 70's Monsanto were developing polysulphone membranes for gas separation. The membrane structure was similar to the cellulose reverse osmosis membrane but made from polysulphone. The major problem though was that to achieve the permeability required meant that the dense film had to be very thin, and in manufacture this led to defects which have a profound effect on performance. Monsanto overcame this problem by overcoating the membrane with a coating of highly permeable polydimethylsiloxane (PDMS). The coating process plugged the defects and it is readily shown that while the coating has only a marginal effect on the permeability of the membrane, it dramatically reduces the effect on the defects, and hence substantially improves the selectivity.

One of the most important composite membranes is the interfacially polymerised membrane developed by Film Tec (now owned by Dow). The membrane was the culmination of more than 10 years of research, and is formed by reacting two monomers on a polysulphone support to form a polyamide coating.

Figure 1.6-2. *Scanning electron micrograph of the surface of an interfacialy polymerised membrane produced by Film Tec (approximately 2 by 5 microns in size). Interfacial membranes exhibit a complex surface structure, unlike more traditional membranes which are totally smooth at this magnification. Some of the high flux characteristics of interfacially polymerised membranes has been attributed to this roughness.*

Another popular method of making composite membranes is by dip coating. This process involves coating an ultrafiltration membrane with a very dilute polymer solution, and then drying off the solvent. The major difficulty with this method is to choose a coating formulation that will not significantly swell the supporting structure during coating. Dip coating is also widely used to apply a protective layer on top of the membrane to protect it during manufacturing. Another layer that is sometimes required is a drainage layer. This consists of a thin coating of a highly permeable polymer, with little intrinsic selectivity. This layer is needed if the top surface of the ultrafiltration layer is of low porosity.

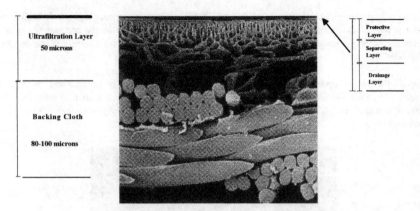

Figure *1.6-3 Structure of a composite RO membrane made on a woven support. The key separating layer is imperceptible at the magnification shown and is supported in this case by a thin drainage layer, and protected by a thin highly permeable/non-selective coating to avoid damage during handling and manufacture.*

Various chemical treatments are sometimes employed to enhance particular physical and chemical characteristics, e.g. enhance the hydrophilicity of the surface by attaching negatively charged surface groups.

1.6.6 Ceramic Membranes

Ceramic membranes grew out of a need for the nuclear power industry to separate the isotopes of uranium. This was achieved by utilising Knudsen diffusion through a ceramic layer with pores in the range 6-40 nm.

Knudsen Diffusion and Uranium Isotope Enrichment [13]
When the mean free path of a gas molecule is larger than the diameter of a pore then it is more likely that the faster smaller molecule will enter the pore. This is known as Knudesn diffusion, and provides a basis for separation on gigantic scale. The theoretical maximum separation between two gases is given by the square root of the molecular weight ratio. Despite the small difference in molecular weights of uranium 235 and 238 Knudsen diffusion has been used effectively to enrich Uranium. The process separates uranium by forming uranium hexafluoride (UF$_6$), a gas, and uses a ceramic membrane. The enrichment factor is extremely small (1.0043), and thus to achieve the modest enrichment from, 0.7 to 3 %, required takes more than a 1000 stages. For economic reasons the plants used for this are enormous (circa 2,000,000 m^2 of membrane). It is reported that despite the aggressive nature of UF$_6$ the Eurodif plant in Southern France is still operating satisfactorily after 20 years with its original membranes!

With the emergence of new methods of separating isotopes, and the limited scope for replacement business, the companies (SFEC, Ceraver, Norton) that had developed these membranes turned their energies to non-nuclear applications. Today a number of companies market ceramic membranes. These membranes are usually of a monolithic structure. By selectively coating the various tubes one can manifold the flows through the device. Ceramic membranes offer good thermal and abrasion resistance. However, per unit surface area they are more expensive than most polymer membranes. As well as the high cost of ceramic membranes they are vunerable to brittle failure. The monolithic structure helps to overcome some of the handling issues, but it does represent an issue when designing a device, e.g. the sealing arrangements. Ceramic membranes are now available in zirconia, alumina, and titania. Zirconia membranes offer the coarest in size (microfiltration) while titania can be made with pores in the nanofiltration range. Like their polymer bretheren as the pores become finer the membranes have to be made more asymmetric. This is achieved through fabricating the membrane in a series of layers with different ceramic materials.

In the late 80's Anotec developed a ceramic membrane (Anopore- now marketed by Whatman) made by anodic oxidation of aluminium. By controlling the conditions the alumina membranes could be made within a wide range of pore sizes, and with a symmetric or asymmetric structure. These membranes are extremely fragile and have to be specially mounted so as to avoid undue mechanical stresses being applied during handling and use. One solution to the mechanical weakness of ceramic membranes was developed by Ceramesh Ltd. Their ingenious solution was to mount a zirconia coating on an inconel wire mesh. The manufacturing process resulted in the mesh acting like reinforcing rods in concrete. As a result a ceramic membrane was made which had sufficient tensile strength and flexibility that it could be wound into a spiral element.

1.7   Quality, Productivity, and Life

This section illustrates the interactions between Quality, Productivity and Life. The quality of a membrane process is determined by the selectivity of the membrane, while the productivity is determined by the amount of membrane and its permeability. However, the membrane only sets the ultimate limit of what can be achieved. The feed, system design, operating conditions all restrict the scope and limit the performance that can be obtained from the membrane system. Each application brings different factors to bear, and thus for design an understanding of these factors and their interaction is required (see figure 1.7).

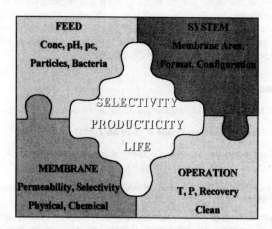

Figure 1.7 Quality parameters depend on 4 different types of factor.

An essential element in new applications is establishing these interactions, which is invariably done through laboratory and pilot testing. From these the basic design parameters and contingencies can be derived. The complexity of these interactions make the design process more than a simple recipe. Some examples of how these various factors interact follow. Further detail is given in the ensuing chapters.

Feed

The initial design question is as to whether or not a membrane can achieve the separation required. However, it is the secondary components that often limit the life expectancy. For example cellulosic membranes will hydrolyse if not operated between a narrow pH range. Particulate fouling in reverse osmosis sets the maximum flux, and hence pressure that the membranes can be run at to avoid undue fouling.

System

The larger the application the more membrane area that is required. This poses the question of how should the membrane be packaged. As one increases the membrane area the more pressure that is going to be lost in getting the feed to and from the membrane. This not only produces a productivity loss but a selectivity loss.

Operation

The structure of the membrane, and how it is package sets limit to the pressures and temperature at which the membrane can be operated at. The build-up of solids on the membrane surface have to be remedied by a physical or chemical cleaning. The frequency of the cleaning process limits the productivity. Cleaning is frequently required to combat the problem of fouling. This is not only a tax on the productivity, but if an aggressive agent is used it can slowly degrade the membrane.

Membrane

While the selectivity of the membrane set the limits of what can be achieved it is frequently its physical and chemical properties that limit the conditions under which it can be used. For example microporous membranes made from polytetrafluoroethylene (PTFE), which is very chemically resistant and very hydrophobic, is mechanically weak. The latter has consequences of how it can be used in an element, and what conditions it can be used under.

1.8   References

1   Ferry, J D, "*Ultrafilter Membranes and Ultrafiltration*" Chemical Reviews, 18 (1936) 373-455

2   Donnan, F G "*The Theory of Membrane Equilbria*" Chem Rev 1 (1924) 73-90

3   Teorell, T "*An attempt to formulate a Quantitative Theory of Membrane Permeability*" Proc Soc Expt Biol & Med 33 (1935) 282-285

4   Meyer, K H and Sievers, J-F "*La Perméabilité des Membranes I Théorie de la Perméabilité Ionique* " Helv Chim Acta 19 (1936) 649-665

5   McRae, W A "*Applications of Ion-exchange Membranes, Current State of Technology After 45 Years*" in "*Effective Membrane Processes - New Perspectives*", Publ.Mechanical Engineering Publications Ltd, 1993.

6   Loeb, S "*The Loeb-Sourirajan Membrane: How it Came About*" ACS Symposium Series 153 Synthetic Membrane Vol1, Ed A F Turbak, (1981)

7   Lonsdale, H K *"The Evolution of Ultrathin Synthetic Membranes"* J Membrane Sci 33 (1987) 121-136

8   Smith, P J *"Perfluorinated Ionomer Membranes for Use in the Production of Chlorine and Caustic"* chpt 7 in "Electrochemical Science and Technology" ed R G Linford, Publ. Elsevier, 1987

9   I W Backhouse *"Recovery and Purification of Industrial Gases Using Prism Separators"* in *"Membrane in Gas Separation and Enrichment"*, 4th BOC Priestley Conference, Royal Society of Chemistry, 1986

10   R J Petersen *"Composite Reverse Osmosis and Nanofiltration Membranes"* J Membrane Science 83 (1993) 81-150

11   U Werner *"Some Technical and Economical Aspects of Gas Separation by Means of Membranes"* in 4th BOC Priestley Conference, 1986, Leeds "Membranes in Gas Separation and Enrichment"

12   Grashoff, G J, Pilkington, C E, Corti, C W *"The Purification of Hydrogen: A Review of the Technology Emphasising the Current Status of Palladium Membrane Diffusion"* Platinum Metal Reviews 4 Oct 1983, 157-169

13   Bhave R R, *"Inorganic Membranes Synthesis, Characteristics and Applications"*, Van Nostrad, 1991

Chapter 2

MEMBRANE TECHNOLOGIES

Contents

2.1 Introduction

Membrane technology is used across a whole range of industries. There is no single membrane type at the root of these applications, nor a single technology. However, specific applications and developments have fuelled its growth and provided the seeds for subsequent opportunities. For example the early 60's saw the emergence of reverse osmosis for the desalination of sea water. Once created it was not long before a vast range of additional applications emerged, from improving the clarity of ice crystals, to dewatering of fruit juices. The 1990's has brought the environment and water quality as key drivers for the advancement of membrane separation technologies, and with it the interest in water re-use and product recovery processess.

2.2 Pressure Driven Processes

Filtration Processes

In value terms, filtration processes are the most important group of membrane processes for industrial users. At the top end of the scale there is microfiltration which deals with particles at the boundary of visibility, such as biological cells. Despite being the oldest of the filtration technologies, microfiltration is perhaps the least mature, and offers great potential. At the other end of the scale there is reverse osmosis which deals with the separation of ions and small molecules from water. Historically, reverse osmosis has been called hyperfiltration, but the term reverse osmosis reflects the fact that there is frequently a significant thermodynamic component to be overcome in the separation. Inbetween there lies the processes of ultrafiltration and nanofiltration. The former is principally concerned with separation of macromolecules ranging from molecular weights of a few thousand to a million, and in market terms has the largest value. The latter is concerned with the separation of small low molecular weight non-volatile organics from water. Over the last ten years a small band of processes that lie between traditional ultrafiltration and reverse osmosis designs has developed. Emphasis of this difference has been created by calling such processes nanofiltration.

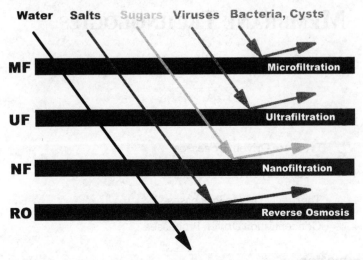

Figure *2.2-1 Filtration processes for separation, clarification, and concentration extend from the rejection of salts to the rejection of bacteria.*

A precise definition for the various technologies is frequently given in terms of size, and or molecular weight (see table 2.2.2). The reality though is that there is no key principle separating one technology from another. However, the terminology is more than just one of semantics. The classification is there to convey a difference in the importance of various features which have significance in the use, design, and application of the technology. Thus, the boundaries should not be regarded as rigidly defined, but as an indicator of technology differences. In keeping with the different uses the various membrane technologies are characterised in different ways. Microfiltration membranes are characterised in terms of pore size, while ultrafiltration membranes are normally described in terms of a molecular weight cut-off (MWCO).

Table *2.2.1 Characteristics of filtration processes.*

Process Technology	Separation Principle	Size Range	MWCO	Rejection Characterisation
Microfiltration	Size	0.1 μm - 1 μm	-	Absolute, nominal, or beta
Ultrafiltration	Size, Charge	1 nm - 100 nm	> 1,000	MWCO
Nanofiltration	Size, Charge, Affinity	~ 1 nm	200 - 1,000	Rejection, MWCO
Reverse Osmosis	Size, Charge, Affinity	< 1 nm	< 200	Rejection

As might be expected the operating parameters for these filtration processes vary drastically, with microfiltration processes offering a high recovery of feed at low pressure, while reverse osmosis offers lower recoveries at much higher pressures.

Table 2.2.2 *Typical operating parameters and membrane materials used for pressure driven processes*

Process Technology	Material	Typical Operating Range		Rejected Species
		Pressure	Recovery	
Microfiltration	Polymers Ceramics Metals	0.5 - 2 bar	90-99.99	Bacteria, Silts, Cysts, Spores
Ultrafiltration	Polymers Ceramics	1 - 5 bar	80-98	Proteins, Viruses, Endotoxins, Pyrogens,
Nanofiltration	Polymers	3 - 15 bar	50-95	Sugars, Pesticides
Reverse Osmosis	Polymers	10 - 60 bar	30-90	Salts, Sugars

Membrane filtration processes are used to concentrate constituents, extract water, and exchange electrolytes. Fractionation is possible, though sharp selection on the basis of molecular weight is not a key characteristic of membranes. However, it is sufficiently sharp to provide a useful separation in the case of blood, (haemofiltration) where blood is retained and low molecular weight materials pass through the membrane.

Along with the change in particle size separation goes a change in basic characteristics of devices. The essential feature in going from RO to MF is that the characteristic dimensions of the devices become larger, as do the cross-flow velocities, and fluxes (see table 2.2.3).

Table 2.2.3 *Typical dimensions and conditions in membrane filtration devices (the RO figure refers to that of a typical spiral wound element, while the UF and MF figures refer to that used in hollow-fibre devices). These figures are derived from devices designed to treat thin waters (i.e. systems with viscosity close to that of water).*

Technology	Distance Between Membranes cm	Cross-flow Velocity cm/s	Flux L/m2/hr
Reverse Osmosis	0.05	10 - 25	40 - 90
Ultrafiltration	0.1	25 - 100	50 - 150
Microfiltration	0.15	100 - 300	100 - 200

Reverse Osmosis

Reverse osmosis has its roots in desalination of sea water, and brackish water, and to this end is now widely used throughout the world. From this starting point, the use of reverse osmosis has extended to process waters, where it can provide a reduction in dissolved solids, or reduce the load on ion-exchange plants. The potentiality of membranes to dewater many food related products was established and plants can be found treating anything from apple juice to soy sauce. One of the attractive features of membranes is that the separation can be done under ambient conditions and hence avoid the denaturing effects of heat. The electronics industry has found that the water quality is one of the critical factors in obtaining high production yields (figure 2.2-1). As integration density has increased, the water quality requirements have steadily risen [1]. A typical water treatment unit for electronics applications may have up to 20 different stages of treatment, and very exacting standards of materials of construction.

Table 2.2.4 Some of the requirements for electronic-grade water Type E-1

Scale of Integration	Resistivity MOhm.cm	TOC ppb	SiO2 ppb	Bacteria count/mL
64K-bit	15	1,000	30	1
256K-bit	17	200	10	0.1
1M-bit	17.5	100	5	0.05
4M-bit	17.5	50	5	0.01

Nanofiltration

During the 60's and 70's the focus of the reverse osmosis manufacturers was very much directed to increasing rejection and or increasing the productivity. However, towards the end of the 70's the idea of using a poorly rejecting RO membrane to soften water as an alternative to lime softening was conceived. As the application developed so did the membrane design. In order to differentiate these membranes from the traditional desalination membranes they were christened nanofilters.

Table 2.2.5 Nanofiltration applications

Application	Feature
Softening	Since multivalent ions like calcium and magnesium are more readily rejected than monovalent ions like sodium, the permeate is relatively softer. While traditional RO also reduces hardness, it requires higher pressures. Also, the recovery limit is higher for an NF process since the concentration factor is lower.
Sulphate removal	Oil extractions from undersea oil fields can be enhanced by pumping water into outlying areas of the oil bearing strata to help displace the oil. Using natural sea water results in precipitation of calcium and barium sulphate, causing blockage. Nanofiltration provides high rejection of sulphate without the high pressures required for normal sea-water RO. By reducing the sulphate concentration, the threat of precipitation is removed.
TOC reduction	Waste water is being used on the West Coast of the USA to inject in aquifers (aquifer recharge) to stop sea water intrusion. Nanofilters provide good reduction of TOC, while allowing the majority of the salts to pass through.
Colour Removal	Natural and synthetic colours are often associated with low molecular weight species. Some examples are removal of colour from dye-stuff effluents, brine regenerant recovery from ion-exchange plants used to decolourise cane sugar, and re-use of glycol anti-freeze by removal of coloured contaminats
Desalting	High BOD wastes from food processing can be desalted, allowing the retained carbohydrates and proteins to be further used in food production.
DBP removal	Disinfection by-products (DBP) are often associated with the low molecular weight components found in surface water supplies (i.e. fulvic, and humic acids)
Pesticide removal	NF is an alternative to using granular activated carbon. Pesticides have molecular weights in the range 200-400 and can be removed by a nanofilter.
Lignin removal	Wash waters from pulp mills contain a high quantity of lignins. Nanofiltration provides an effective method of removal.
Sugar concentration	Nanofilters largely retain sugars, but pass slats. Concentration of sugar solution by reverse osmosis leads to an increase in the salinity. Using a nanofilter means that sugar concentration can be achieved without increasing the salinity.

As understanding and engineering of this new breed of membranes developed so have the applications (see table 2.2.5)[2]. There are several large applications concerned with aquifer recharge to prevent sea water intrusion. These plants use RO/NF to treat low grade waters (tertiary off-take from waste-water treatment works). A related application is in off-shore oil wells where water is used to help abstract oil. If sea water is used the conditions are such that precipitation can occur. By treating water with a nanofilter sufficient TDS can be removed to avoid the danger of precipitation.

Ultrafiltration

One of the first major uses for ultrafiltration on an industrial scale was in the concentration of paint that had been rinsed off cars during initial stages of electrophoretic painting. This technology is now the established method of providing the primary coat of paint on cars. From this base UF is now used in an ever increasing range of food processes. Traditional separation processes like evaporation and freezing denature proteins and hence, alter the flavour and taste of food products. In contrast membrane processes can operate at ambient or any desired temperature at which the membrane is thermally stable. This makes membranes a particularly attractive method of separation in the food and beverage industry. In a single process organisms and large macromolecules, which contribute to haze, can be removed and produce a clear "cold sterilised" product. UF is now widely used in the food and beverage industry (see figure 2.2-2 for examples of its use in the treatment of milk products) [3]. Environmental pressure has focused interest in effluent treatment. Particular interest has been in effluents which contain hazardous or valuable products. Membranes have been used to recover and concentrate products (e.g. paint latex). In some cases the water recovered as part of this process can also be re-used so giving the additional benefit of reducing water demand. Even where the material recovered has no intrinsic value, concentration of it can make other treatment processes more viable (e.g. biological treatment), and or reduce the volume of waste to be transported, and thereby provide a cost saving.

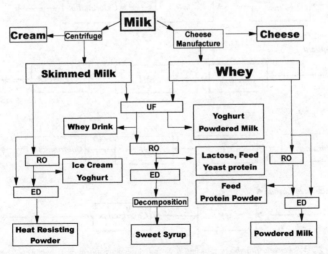

Figure 2.2-2 Membrane processes used or considered in the food industry for milk related products.

Microfiltration

As a laboratory tool microfiltration has a long established history. As a process tool it is far less mature. Its widest use is in clarification processes where a key aspect is frequently the "dirt holding" capacity. Recently though water quality requirements have driven interest in microfiltration for treating water to remove chlorine resistant pathogens (e.g. Cryptosporidium parvuum).

> **Cryptosporidium Parvuum**
> *The spotlight fell on this organism in 1989 when an estimated 400,000 people were infected in Milwaukee. The organism is implicated in about 10 % of all gastro-inestinal illnesses, and can pose a life threat for immuno-depressed people. With an infective dose of less than a 100 and a single infected animal capable of producing 10^{10} cysts per day it is clear that water treatment works have a severe challenge when conditions are "right". Matters are made worse by the fact that in the cyst form the organism is chlorine resistant and suffers only limited deactivation during its passage along the distribution network. The cysts are about 5 microns in size, which means that conventional filtration does provide a substantial reduction. The attraction of membranes is that they should provide an absolute barrier.*

Extensive field trialling has been carried out in the Great Lakes area of the USA and a number of large microfiltration plants are planned. North West Water, which had been trialling membranes in the North West of England, selected microfiltration as the only way in which it could access additional water supplies to meet drought conditions in the UK. In the space of 6 months it was able to go from concept to operation for 80 ML/day plant (the largest of its type to date). The membrane systems used are based on hollow-fibre and operate in dead end mode[1] . Accumulated solids are removed through either air or water backwash. In France, Lyonnaise Des Eaux has decided to go further and employ UF membranes. Again the preferred format was hollow-fibres. These operate in cross-flow, and therefore there is a higher energy requirements.

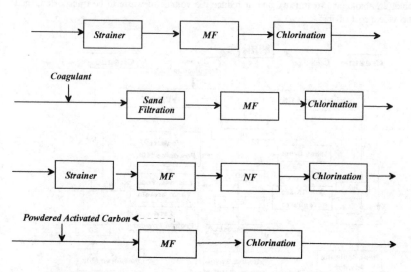

***Figure** 2.2-3 Schematic flowsheets showing in various ways microfiltration is being evaluated for use in potable water production.*

A number of different microfiltration processes are being evaluated for large scale use by the water utilities (see figure 2.2-2). In the case of good quality raw water, microfiltration with a strainer

---

[1]   The companies in this field are Memcor, USF Acumem, Zenon, Koch

guard is sufficient (*direct filtration*). For slightly turbid waters some of the solids loading is being removed and or size enlarged through a pre-treatment of coagulation/sand filtration. TOC is largely untouched by microfiltration. For applications where the TOC is high and could give rise to high levels of triholmethames (THMs), as a result of post chlorination, consideration is being given to post treating the water by nanofiltration. In the US concern over the environmental impact of THMs has led to the Disinfection By-Product Rule. An alternative approach to MF/NF is to add powdered activated carbon before the membrane filtration stage and to use the membrane to recover and recycle the material.

Energy Requirements

For low cost products, such as potable water, the energy demand is a key operating cost issue. A rough estimate of the total power demand is given by the product of the feed flow rate and the feed pressure viz.

$$\text{Energy} = \tfrac{1}{\eta}(\text{Flow Rate}) \, x \, (\text{Feed Pressure})$$

where η is a product of the motor and pump efficiencies. For applications which operate at low recovery and high pressure such as sea water desalination (such systems operate at around 30 % recovery) a significant amount of the energy remains in the reject stream. For such applications it is worth recovering this energy by "recycling" the pressure energy to the feed. Up to 80 % of the energy that might otherwise be lost can be recovered in this way.

Design

Most modern plants are designed by the use of a computer-aided-design tool. A key ingredient in these designs is the operating flux. This is fixed either by past experience or pilot testing. From the operating flux the membrane area required for a plant is readily calculated viz.

$$\textit{Membrane Area} = (\textit{Production Rate}) \, / \, (\textit{Operating Flux})$$

Sometimes designers and operators are attracted by designs based on high operating flux since it reduces the membrane area, and, as a consequence, the capital needed. However, such over optimism leads to rapid fouling and loss in performance, and higher operating costs as a result of the higher operating pressure and additional membrane cleaning required.

Gas Separation [4]

One of the earliest commercial gas separation membranes was that based on silicone rubber. This membrane was developed to produce enhanced oxygen for medical applications. The early 60s saw the development of palladium coated materials for hydrogen recovery. This membrane provides the highest purity hydrogen of any other commercial process [5]. In selectivity terms, this is one of the exceptions to the rule that membranes do not provide high separation factors. The 70s saw Monsanto develop a polysulphone hollow-fibre membrane (Prism). This membrane allows hydrogen to be recovered from various chemical processes, most notably from the purge gas in the ammonia process.

Another development, in the 80's, was the Generon process by Dow. This used the polymer 1,4-methylpentene, which was fabricated into hollow fibres with a 10 micron thick dense wall. The small diameter of the fibres means that a lot of membrane area can be packed into a small volume and this compensates for the relatively low permeability of the fibre. The principle use of this membrane has been to remove oxygen from air. Membranes can readily remove 75 % of the oxygen to provide 95 % nitrogen, which is sufficient for controlled atmosphere packaging and blanketing of flammable chemicals.

The Prism Membrane
The membrane was made from poysulphone using a phase inversion process. This involves dissolving polysulphone in a suitable organic solvent and, after extruding through an appropriate spinneret, quenching into a non-solvent (usually water). By careful control of conditions an asymmetric membrane could be made with one side forming a closed surface. As in reverse osmosis, the surface layer provides the separation through a solution-diffusion mechanism. In order to get high fluxes, conditions are chosen to make the thickness of the membrane skin as thin as possible. It was found that this created defects in a layer of polydimethylsiloxane (PDMS). This plugged the leaks through the defects while only marginally reducing the permeation through the polysulphone.

The majority of industrial membrane gas separations employ membranes with a closed surface that achieves selectivity through the relative solubility and diffusivity of gases in the membrane material. It is generally found that for permanent gases the selectivity order does not change with membrane material. Changing material serves only to trade selectivity for permeability. Only when there are specific interactions between the gas and the material can one obtain significant deviations from the general trend.

Gas separation does not necessarily require a closed surface as is evident from the largest single membrane application; for the last 30 years microporous ceramic membranes have been used to effect the enrichment of uranium. Separation is achieved as a result of the fact that heavier molecules travel more slowly than lighter ones. Thus, in any given time period they are less likely to enter a pore and so separation can be achieved. The criterion for this is the average mean free path of the molecules must be greater than the pore size. It can be shown that if this is the case then the separability of two gases is proportional to the square root of their molecular masses. Other mechanisms that can result in separation of gases are surface flow and capillary condensation in pores.

In contrast with permanent gases, organic vapours generally have very much higher permeabilities than nitrogen through rubbers. Some people have proposed to use this as a method of concentrating organic vapours such as petrol, and cleaning fluids, e.g. trichloroethylene [6].

An estimate of the total power demand can be obtained from the the total molar rate passing through the membrane and the pressure ratio across it, viz.

$$\text{Energy} = \frac{RT}{\eta}(\text{Molar Flow Rate}) \, x \, \ln[\text{Pressure Ratio}],$$

(R= Gas Constant, T = temperature in Kelvins, η =Pump efficiency). It can be seen from this equation that a if there is a large pressure ratio there is a significant cost implication. The pressure ratio also plays a significant role in limiting the selectivity that can be acheived. Hence, there is a trade between energy and selectivity that determines the practical set of operating conditions for gas separation with a given membrane

Some of the processes that have been developed are:-

Nitrogen from air. Membranes can be used to provide 95-98 % N_2, This gas is used to provide controlled atmospheres for storage and shipment of fresh produce, food packaging, and blanketting of flammable chemicals. The main advantages over pressure swing adsorption is that the membrane plant is lighter and smaller.

Oxygen from air. Oxygen enrichment of air is widely used for home oxygen respiratory therapy. The low selectivity of membrane materials has meant that large scale applications in combustion have not been very attractive.

Ammonia purge gas, Refinery hydrogen recovery. The recovery of hydrogen from chemical processes such as that in the ammonia process was one of the pioneering membrane applications developed by Monsanto. The process operates with feed pressures as high as 140 bar.

Carbon dioxide, Enhanced oil recovery, Landfill gas. Carbon dioxide is a contaminant of natural gas, and can be readily removed by using a membrane. In addition, other contaminants such as water vapour and hydrogen sulphide are removed. Also carbon dioxide is pumped into dying oil reservoirs in order to extend the life. This CO_2 eventually appears with the oil. The membrane process can be used to remove the CO_2 and reinject the gas.

Uranium enrichment. Some extremely large membrane plants have been built to create enriched uranium. The basis of this is a ceramic membrane, which can be used to separate the different isotopes of uranium, which are made gaseous by reacting with fluorine to form uranium hexafluoride, which is a gas. Separation is achieved through Knudsen diffusion which gives a separation factor equal to the ratio of the square root of the molecular weights.

Helium recovery. Helium is easy to separate from natural gas by using membranes. A small market exists for helium recovery in applications such as deep sea diving, and for its removal from natural gas.

Water vapour removal. Water vapour passes through membranes extremely quickly, and a water/nitrogen selectivity of 500 is readily achieved. Examples of its use are in the drying of decompressed air and natural gas.

Solvent vapour recovery. During operation of solvent based processes, air frequently becomes contaminated with solvent vapour. With environmental pressure increasing, there is a need to concentrate and recover these contaminants. Some membranes have a high affinity for organics and thus a significant concentration of the organics can be achieved. Membrane Technology and Research (MTR) have piloted a number of trials looking at gasoline recovery, and concentrating chlorinated hydrocarbons. More recently they have developed a process for concentrating monomers such ethylene and propylene from nitrogen in polyolefin plant vents. In a two stage process they can concentrate the monomers from 15 % to 95% which can then be re-used instead of flared.

2.3 Electrically Driven Processes

Electrodialysis [7]

Electrodialysis is a membrane process which removes ionic material from a solution by using ion-selective membranes[2]. This is done by passing the solution between two membranes across which an electric field is applied. The electric field moves the ions towards the respective electrodes. By using an anion permeable membrane on one side and a cation permeable on the other, ions can be removed from one stream and concentrated in another (see figure 2.3-1).

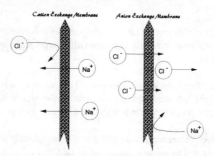

Figure 2.3-1 Diagrammatic cross-section of electrodialysis cell working on sodium chloride

Experimental work on electrodialysis started in the 30's for treatment of food related products, e.g. fruit juice. In the 40's Meyer and Strauss developed the concept from a single to multiple compartments (see figure 2.3-2). However, the separation units consisted only of a single cell, and the membranes were not physically up to the job. In the early 50's MacRae developed suitable membranes, and this opened up the commercial possibility of using electrodialysis. Several hundred pairs of membrane are often placed in series.

Table 2.3.2 Some Manufacturers of Electrodialysis membranes/ technology

Electrodialysis Manufacturers		
Aquatech Systems	Corning (France)	Tokuyama Soda
Asahi Chemical Co	Graver Water	US Filter/IWT
Asahi Glass Co	Ionics	
Christ	Negev Institute	

The first large scale applications were for the production of potable from brackish water. This technology is also widely used in Japan for the recovery of salt from the sea. The major use of electrodialysis is to produce potable water from sea water and brackish waters. Estimates indicate that there is more than 1000 ML/day of installed capacity. ED is widely used to partially desalt food stuffs, e.g. milk, fruit juices, sugar, soy sauce. Industrially, ED is used to recover valuable electrolytes such as silver and nickel.

The membranes used in electrodialysis systems are thick and non-porous, and structurally quite different from that used in filtration processes. A key feature of ion-exchange membranes is that the polymers bear large numbers of charged groups. This makes the membrane hydrophilic and selective. Membranes which pass positive ions (cations) are called cation exchange membranes. Similarly, those that pass negative ions (anions) are called anion exchange membranes.

---

[2] These membranes are made from cation or anion exchange materials. A cation exchange membrane is made of a polymer with negatively charged groups, e.g. sulphonate, carboxylate. The negatively charged groups allow cations to pass but inhibit the flow of anions (see section 2.8). Anion-exchange membranes work in the converse way. An example of a positively charged group is a quaternary ammonium base cation.

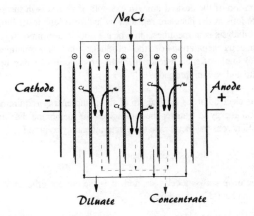

Figure 2.3-2 Schematic of electrodialysis stack. Typically, there are several hundred pairs of membrane

As with other membrane processes fouling is a common occurrence. Fouling problems can be reduced in electrodialysis by a process called electro-dialysis reversal (EDR). The essential feature of this is that the polarity of the stack is reversed. When the field is changed the concentrate and diluate are reversed. The result is that deposits that have formed on the membrane redissolve and get dispersed. The performance improvement usually more than counterbalances the additional cost of the system.

There are four types of cell design

- **Sheet flow** - *flow is introduced, between the two membranes, from one side of a stack and allowed to flow to the other side. Flow velocities are typically 0.05 to 0.1 m/s.*

- **Tortuous path** - *in this design a winding channel with straps to improve mixing is included in the compartment,. This type of design provides a long liquid flow path and high flow velocities (0.1 to 0.5 m/s). As a consequence of the higher velocities and longer path pressure losses are higher*

- **Unit cell** - *in this design the feed water is fed into the concentrate compartment. Thus only water which transfers with the ions across the membranes is withdrawn from this compartment.*

- **Spiral Wound** - *a relatively recent design which involves electrodes at the centre and edge of a cylinder. These produce a radial electric field.*

The major applications of electrodialysis are in demineralisation/concentrations of salts. The largest applications are for water purification where electricity is cheap, concentration of sea water (Japan), and cheese whey purification. Other smaller industrial applications include nickel recovery from plating solutions, drug desalting, acid recovery from metal finishing solutions, and HF recovery from glass etching solutions. As an example, a dilute acid stream of 1-3 % can be used to generate a 10-15 % acid stream by ED. Electrodialysis can be used to replace continuously an anion or a cation species by using a stack madeup of all anion or all cation membranes respectively. Each alternate compartment is fed with a solution to be processed, while the other compartments are fed with a relatively high concentration of the desired ion. An example of its use is in the sweetening of citrus juices, where the citrate ions in the juice are replaced by hydroxyl ions from caustic. The third type of application of electrodialysis is in metathesis in which a double decomposition reaction is carried out with the simultaneous separation of the products; this is sometimes referred to as electrometathesis. An example of such a reaction is the production of silver bromide and sodium nitrate from silver nitrate and sodium bromide.

The energy required for electrodialysis is made up of the pump energy requirements of the various streams and the electrical energy to transfer the ions across the membrane. For one gram equivalent of electrolyte transported across the membrane the electrical energy required is

$$Energy = \frac{\Delta V * F}{\eta} \qquad \text{Joules/g. equivalent.}$$

where ΔV is the voltage drop across a cell pair, and η is the current efficiency, and F is Faraday's constant. On a metric ton (1000 Kg) this gives

$$Energy\ Consumed = \frac{0.278 * \Delta V * F}{\eta * EW} \qquad \text{KW.h/tonne}$$

where EW is the equivalent weight of the salt being processed. The voltage drop across cell pairs is typically 0.5 to 2 V, while the current density is of the order of 500 A/m^2.

The key economic issues are the operating current density, and recovery targets. A phenomenon called polarisation (see Chapter 6) sets an upper limit to the current density at which an ED unit can operate effectively. For brackish water the limiting current density is of the order of 1000 A/m^2, but this falls as the concentration is lowered. The degree of concentration that can be achieved is limited by the amount of water which is transferred with the ions by osmosis, and electro-osmosis (typically, about 100-400 mL/Faraday). Very low concentrations can be achieved in the diluate stream. However, for water the economic lower limit is about 200 ppm, since electrical losses become too severe on account of the high electrical resistance of dilute solutions. Other complications are the management of the rinse stream of the electrodes, and fouling.

Chlor-Alkali

Sodium hydroxide (chlorine) is one of the major commodity chemicals of the world. The majority of this is produced electrolytically from salt. Two processes which have been used for a long time in the manufacture of this chemical are the mercury cell, and diaphragm cell. The idea of using a membrane cell (see figure 2.3-3) developed in the early 50's but failed due to the lack of a suitable membrane material to withstand the harsh chemical and thermal conditions needed.

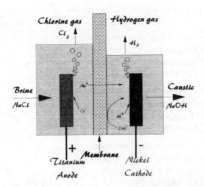

Figure 2.3-3 Schematic of membrane cell for production of chlorine and caustic.

As part of a NASA programme on fuel cells DuPont developed a separator based on PTFE which was made hydrophilic by the attachement of a fluorinated side chain with a charged end group (sulphonic acid[3]). The potential of this material for chlorine/caustic production was recognised and prototype cells were evaluated in the early 70's. By the end of the 70's improvements in the material meant that membranes provided a viable alternative. Since then a large proportion of newly installed capacity has used membrane technology from one of a small range of companies (see table 2.3.3).

Table 2.3.3 Manufacturers of chlor-alkali systems

Asahi Chemical Co	Hoechst-Uhde	Orinzio DeNora
Asahi Glass Co	Ionics	OxyTech
Chlorine Engineers	ICI plc	The Electrosynthesis Co

The products of each of the three types of cell are very different (see table 2.3.4), and this has economic implications beyond the manufacturing process. The membrane process is more efficient than mercury amalgam cells but it produces caustic of lower strength. If the user requires 50 % caustic then an evaporator will have to be installed, which will add to its cost, and increase the energy demand. In terms of capital the diaphragm process is cheaper than the membrane process, but requires the customer to be able to use low caustic concentration contaminated with salt. It can thus be seen that a commercial decision to change has significant implications for customers.

While the development of membrane technology for the chlor-alkali market is a major success story the nature of the process and its requirements mean that its use in other areas has so far been fairly limited beyond the production of similar products, e.g. KOH.

---

[3]   Later, Asahi Glass developed a material with side chains that terminated in carboxylic acid. This gave high current efficiency, but required high voltages to operate. at the thickness required. The solution was to form a laminate of the sulphonic and carboxylic acid materials.

Table 2.3.4 Comparative characteristics of cell systems

Cell type	Features	Product Quality
Mercury	• Expensive to construct • Mercury needs to be reclaimed from waste • Operates at 4.5 V • 3.55 MWh/tonne Cl_2	50 % caustic
Diaphragm	• Simple and inexpensive to construct • Caustic contaminated with salt • Operates at 3.8 V • 3.0 MWh/tonne Cl_2	10 % caustic
Membrane	• Cheap to construct and install • Requires high purity brine • Operates at 3.1 V • 2.5 MWh/tonne Cl_2	35% caustic

Bipolar [8]

The early 80's saw the development and commercialisation of a process based on the bipolar membrane. This membrane is a laminate of cation and anion exchange membranes. Its structure allows the simultaneous production of acid and base, essentially by using electricity to split water into hydroxyl, and hydrogen ions. Once produced at the interface between the anion and cation exchange parts of the membrane the electric field conveys them away to separate compartments (see figure 2.3-5). One consequence of this process is that water has to be supplied at a sufficient rate as it is consumed. The tendency is for the membrane at the interior interface to become dehydrated. This imposes a limit on the operational current density. The process has particular attractions in locations where acid or base are difficult to ship in. In commercial arrangements some 200 sets of exchange membranes are used.

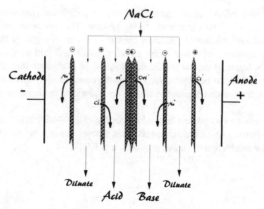

Figure 2.3-5 Schematic of 3 compartment cell used with bipolar membrane

Continuous Deionisation (CDI) [9]

One of the most recent process developments in electrically driven membrane technologies is continuous deionisation (CDI) or electro-deionisation as it is sometime known. Commercial reality started in 1987, although the basic idea had been around a long time. As concentrations fall, conventional electrodialysis becomes uneconomic since the ohmic resistance of water leads to significant power demands. The CDI system combines electro-dialysis with ion-exchange to act as a water polishing stage for RO or ED permeate. The essential idea is to fill the space between the two ion-exchange membranes with a mixture of cation and anion exchange beads (see figure 2.3-6). This reduces the ohmic losses. Its major use is in the production of high purity waters, and is finding wide use in those industries which require such water, e.g. electronic, pharmaceutical industries.

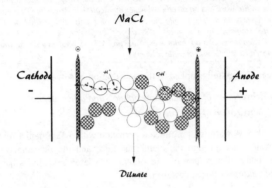

Figure 2.3-6 Schematic of mechanism in EDI

2.4 Thermally Driven Processes

Pervaporation [10]

Pervaporation is unusual among membrane processes in that transport across the membrane is accompanied by a phase change. The process involves contacting a liquid mixture on one side of a membrane and then removing the permeate by a sweep gas or evacuation. The membranes used for this are of the asymmetric variety like those used in reverse osmosis. Invariably, permeation through the thin surface layer of the membrane is rate limiting, and hence, selectivity derives from the relative solubility-diffusivity of the components through this layer. The evaporation at the downstream side creates a substantial heat demand. This has to be provided by the feed through thermal conduction across the film. Within limits, the selectivity is not significantly affected by the upstream or downstream pressure [10]. Also, the upstream pressure has no effect on the transport rate. The composition of the permeate gas is quite different from that predicted by equilibrium thermodynamics. In the case of azeotropic systems, pervaporation allows one to break the azeotrope without recourse to adding additional components (extractive distillation), or by lowering the pressure (vacuum distillation).

The commercial processes developed by GFT in the early 80s are based on poly vinyl alcohol. In the case of ethanol/water systems there is no azeotropic point, and the pervaporate and the gas phase composition is water rich. This illustrates one of the important applications for pervaporation, which is in the dehydration of alcohols and other solvents. Such capability stems from a high affinity and diffusivity of water through such polymers. If a hydrophobic polymer such as silicone rubber is used then the opposite effect is achieved. i.e. the vapour composition is richer in the organics than water. Thus, by selecting the membrane material it is theoretically possible to select the composition of the gas phase permeating.

> **Short History of Pervaporation** [11]
> Research work on pervaporation dates back to the early part of the 20th century. The term pervaporation was coined by Kober in 1917. Research by Binning et al in the late 50's using cellulose acetate membranes and the development of commercial reverse osmosis membranes stimulated further investigation. However, it was not until the early 80's that a pilot trial was carried out. The initial work was on an ethanol/water application in Brazil. GFT provided the lead for much of this early work. The membrane consisted of a polyvinyl alcohol dense film on a polyacrylonitrile support. In the late 80's other companies developed membranes for pervporation. Kalsep developed a tubular membrane and MTR developed a spiral wound element for work on concentrating volatile organic solvents.

Pervaporation has been successfully applied to the dehydration of solvents

- *dehydration of ethanol, isopropanol, and other alcohols*
- *dehydration of ethylene glycol*

Typically a stream with 5 % water will produce a permeate with 5 % organic. As the concentration of water falls, the water flux through the membrane falls, and the percentage of organic in the permeate rises. The ability of pervaporation to obtain such large selectivties has led to its use in separating azeotropic mixtures of organic solvents:

- *alcohols (ethanol, isopropanol, butanol)*
- *ketones*
- *esters (ethyl and butylacetate)*
- *pyridines*

In general, such azeotrope breaking separations are done in combination with distillation.

The use of pervaporation to remove **volatile organic compounds** (VOCs), which occur in a variety of waste waters has been considered. The principal advantage of the technology is that it can recover VOCs in a more concentrated form than obtained by air stripping or carbon sorption. The economic viability of VOC pervaporation depends on the feed concentration.. As concentrations fall so does the flux. Consequently, the membrane area requirements increase making the process unattractive. An ideal feed has between 0.1 and 5 % organics.

Membrane Distillation [12]

Like pervaporation, membrane distillation uses heat to drive the separation, but in this case the membrane is microporous and is not wetted by the process liquid. The function of the membrane is solely to keep the two streams apart and does not affect the vapour-liquid equilibrium of the components. The most common form of membrane distillation is shown in figure 2.4-1 in which liquid is in contact with both sides of the membrane.

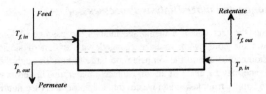

Figure 2.4-1 Schematic of direct contact membrane distillation

Another variant is "gap-gap" membrane distillation in which an air gap exists on the permeate side and the vapour is condensed on an internal surface which is maintained cool. Some other variants involve removing the vapour by sucking (low pressure membrane distillation) or by using a carrier gas (sweeping gas membrane distillation) and condensing the fluid outside the module.

2.5 Concentration Driven Processes - Dialysis

Haemodialysis

In financial terms the medical application of membranes is the largest sector. The major requirement is for small elements (typically with about 1 m² of membrane surface) to dialyse or filter blood, to remove low molecular weight metabolite wastes such as urea, and creatine. The treatment was developed during the 40's by Kolff [13] and became accepted practice for dealing with chronic renal diseases in the 60's. The dialysis unit is either a hollow fibre module or a plate and frame cassette and contains about 1 m² of membrane. In the case of hollow fibre devices[4] blood is passed along the fibre lumen while a dialysate (typically a hypotonic acetate solution with a low potassium content) is passed in counter current direction on the shell side (see figure 2.5-1). The units are either single use or multiple use/single patient. In modern treatment, dialysis is frequently combined with ultrafiltration (haemofiltration) at the start of dialysis which not only reduces the blood volume but removes larger metabolites.

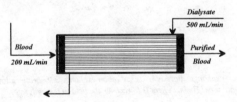

Figure 2.5-1. Schematic of haemodialysis unit. Blood is passed along fibre lumen, and solutes are exchanged with diafiltrate

There are an increasing range of other membrane processes being used/developed in the medical community. Further details can be found in [14].

---

4   The majority of devices come in the hollow-fibre format. A small number of plate and frame devices are also made.

Donnan, Ion-exchange and other Dialysis Processes

Donnan dialysis exploits the fact that if a membrane inhibits the passage of ions of a particular charge, then this allows the counter-ion to be exchanged with an ion of similar type across a membrane. The exclusion can be due to the ion being too large to pass through the membrane (e.g. dyestuff) or the membrane may be made of an ion-exchange material. This process can be used to extract and concentrate copper from low grade wastes into sulphuric acid (see figure 2.5-2) [15].

Anion-exchange membranes do not provide good inhibition to the passage of hydrogen ions. This feature can be exploited to extract acids from metal contaminated solutions (see figure 2.5-3 and table 2.5.1)

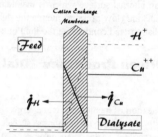

Figure 2.5-2 *Schematic of the concentration profiles occurring in Donnan dialysis using a cation exchange membrane which inhibits an anion from passing. Copper flows to the dialysate despite the negative concentration gradient because there is a much larger electric field acting on it due to the accumulated charge at the interface with the membrane.*

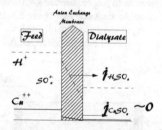

Figure 2.5-3 *In ion-exchange dialysis a weak anion-exchange membrane allows hydrogen ions to pass while inhibiting metal ions. The net effect is to recover the acid in the dialysate.*

Table 2.5.1 *Applications in ion-exchange dialysis identified in [16]*

Applications	Acids
Waste acids from pickling steel	H_2SO_4, HCl, HNO_3, HF
Refining of waste acid from batteries	H_2SO_4
Separation and recovery in mining processes	H_2SO_4, HCl, H_3PO_4
Treatment of acid waste from Alumilite processing	H_2SO_4, HNO_3
Waste acid from etching Al, Ti	HCl
Waste acid from surface treatment in plating line	H_2SO_4, HCl, HNO_3, HF
Treatment of waste acid in deacidification, organic synthesis	H_2SO_4, HCl

Dialysis has also been used to remove alcohol from beer. The dialysate is water saturated with carbon dioxide under 3 bar pressure [17]. The alcohol flows through the membrane faster than water resulting in a reduction in alcohol. Some of the other low molecular weight species are also lost.

Gas Contacting [18]

A particular variant of dialysis is the transfer of volatile gases from one liquid to another. There are many similarities with membrane distillation in that the membrane is hydrophobic and this leads an air gap between the feed and dialysate. Since molecules diffuse significantly more slowly in liquids than in gases such devices are invariably mass transfer limited in the liquid phase[19]. Applications include deoxygenation, and decarbonation. Another application that has been proposed is the removal of gases such as bromine from dilute salt solutions.

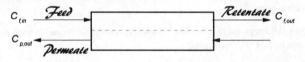

Figure 2.5-4. *Schematic of gas transfer*

Most commercial devices utilise hollow-fibres due to their high surface area and simple design. As well as removing gases these devices have been used to aerate systems without the injection of bubbles which can introduce frothing [20].

2.6 References

1   *"Reverse Osmosis Technology: Application for High Purity Water Production"* ed B S Parekh, Pub Marcel Dekker 1988

2   Conlon, W J, and McClellan S A, *"Membrane Softening: A Treatment Process Comes of Age"* JAWWA Nov (1989) 47-51

3   C M Mohr, D E Engelgau, S A Leeper and B L Charboneau *"Membrane Processing and Research in Food Processing"* Publ. Noyes Data Corp

4   W J Koros and G K Fleming, *"Membrane-based Gas Separation"* J Membrane Science 83(1993) 1-80

5   G J Grashoff, C E Pilkington and C W Corti *"The Purification of Hydrogen"*, Platinum Metal Reviews 1983, 157-169

6   R W Baker, N Toshioka, J A Mohr and A J Khan *"Separation of Organic Vapours from Air"* J Membrane Sci 31 (1987) 259-271

7   H Strathmann *"Electrodialysis and Its Application in the Chemical Process Industry"* Sep & Purification Methods 14(1) (1985) 41-66

8   K N Mani and F P Chlanda *"Recovery of Spent Mineral Acids via Electrodialytic Water Splitting"* Int Tech Conf on Membran Separation Processes, Brighton UK, May 1989

9   C S Griffin *"Advancements in the Use of Continuous Deionization in Production of Ultrapure Water"* Conf on High Purity Water, April 1991, Philadelphia, USA

10   J Neel *"Introduction to Pervaporation"* in *"Pervaporation Membrane Separation Processes"* ed R M Y Huang, Publ Elsevier, 1991

11   C S Slater and P J Hickey *"Pervaporation R&D: A Chronological and Geographic Perspective" Proceedings of 4th International Conference on Pervaporation Processes in the Chemical Industry"*, (1989). Publ Bakish Material Corp, Ed R Bakish

12   S-I Andersson, N Kjellander and B Rodesjo *"Design and Field Tests of a New Membrane Distillation Desalination Process"* 56 (1985) 345-354

13   W J Kolff, and B Watschinger *"Further Development of a Coil Kidney"* J of Laboratory & Clinical Medicine 47 (6) 969-977

14   M S Lysaght, D R Boggs, and M H Taimisto *"Membranes in Artificial Organs"* in Synthetic Membranes ed M B Chenoweth, Publ MMI Press (1986)

15   G Jonsson *"Dialysis"* in *"Synthetic Membranes: Science, Engineering and Applications"* Ed P M Bungay, H K Lonsdale and M N Pinho NATO ASI series C Vol 181

16   Y Kobuch, H Motomura, Y Noma, F Hanada *"Applications of the Ion-exchange Membranes: Acids Recovered by Diffusion Dialysis"* at *"Membranes and Membrane Processes"* Jun 84, Stressa, Italy

17   H Moonen and N J Niefind *"Alcohol Reduction in Beer by Means of Dialysis"* Desalination 41 (1982) 327-355

18   Jansen, A E, Klaasen, R,and Feron, P H M *"Membrane Gas Absorption - A new tool in sustainable technology development"* Process Int for Chem Ind (6 Dec 1995)

19   Z Qi and E L Cussler *"Microporous Hollow Fibers for Gas Absorption II Mass transfer across the membrane"* J Membrane Sci 23 (1985) 333-45

20   T Ahmed and M J Semmens *"Use of Sealed End Hollow Fibres for Bubbleless Membrane Aeration: Experimental Studies"* J Membrane Science 69 (1992) 1- 10

Chapter 3

PACKAGING MEMBRANES

Contents

3.1 Introduction

As applications evolve, a balance has to be maintained between optimising the product, which improves the operational costs at the expense of specialisation, versus the benefits of utilising value engineering which lowers the capital costs. In membrane systems this optimisation process takes place at several levels:

- *membrane*
- *packaging of membrane*
- *system*
- *process*

This chapter looks at the second of these.

In much of the early laboratory work membranes were used in the form of flat sheets, but for industrial applications of any scale the membrane has to be suitably housed in a manageable unit. The structure of such a device has been a continual challenge, which requires balancing the economic costs of a unit versus performance factors and operational factors, e.g. cleaning, replacement There are four major types of packaging of membranes

- *plate and frame*
- *spiral wound*
- *hollow-fibre*
- *tubular*

3.2 Plate and Frame

In the early days membranes were made as simple flat sheets. The needs for large scale automated the process and membrane is commonly manufactured as continuous flat sheets. Typically these production lines are about 1 metre wide and lengths of several hundred metres are produced in a single run. For electrically driven processes the plate and frame geometry reflects the needs of the electrical field. Such design are also seem very natural extension of more conventional low pressure "conventional" filtration designs. However, the high pressures that are required present considerable engineering in order to avoid leakage. Companies like DDS (now part of Dow) and Dorr-Oliver were at the vanguard of the development of plate and frame designs[1]. As many as 100 membrane sheets of membrane can be stacked into one element (see figure 3.2-1). The ability to separately deal/monitor with each permeate line provided scientific and process options.

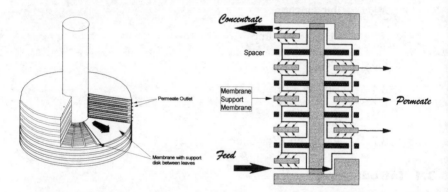

Figure *3.2-1 Schematic diagram of plate and frame devices.*

One of the attraction of these devices are that they are easy to disassemble, sanitise, and replace the membrane sheets. A particular advantage if membranes were damaged or fouling problems. The down-side is that disassembling and re-assembling is a very labour intensive and time-consuming. A practical difficulty was ensuring that during re-assembly all parts were positioned correctly so that water tight seals are made. Despite these problems, in small scale high added value applications, such as pharmaceuticals, and development work, these plate and frame devices are a useful tool. The plate and frame design have also been found to be a useful vehicle in research and food application. A more recent variant is the provision of the membrane in cassettes. It is important to recognise that the cost of the membrane is minimal compared to the cost of a unit (< 5 %). For high pressure applications such as sea water desalination the whole unit is placed in side a pressure vessel.

Table *3.2.1 Commercially available plate and frames modules*

Companies	Technology	Trade Name	Basic Element	Spacing mm	Area m² Element	Area m² Device
Alpha-Laval,	UF		Membrane			
DDS (Dow),	RO, UF		Membrane	0.7		35 - 42
Dorr-Oliver		IOPLATE	Cartridge	2-3	1.3	16
Gambro	HD		Cassette		1	
Millipore	UF, MF	PELLICON	Cassette		0.5	
Rochem	MF		Membrane			
Rhone-Poulenc	UF		Membrane	1.5	0.4	50
Sartorius	UF	SARTOCON	Membrane	0.1		

3.3 Spiral

The spiral wound design is one of the most favoured for reverse osmosis applications. This preference stems from its practical convenience. Despite its complex internal design (see figure 3.3-1), it is easy to use, and is available in a large number of formats with different materials.

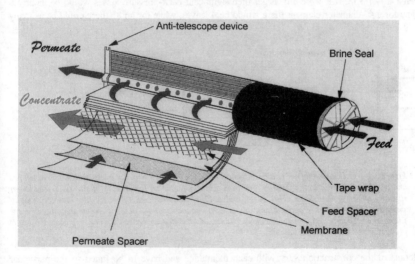

Figure *3.3-1 Schematic of a spiral wound element showing cut-away section of internals.*

In addition spirals have a reasonably high membrane area per unit volume which means the size of the pressure vessels and associated pipework and frames are less, giving an additional economic and practical benefit. Spirals are also used in gas separation, nanofiltration, and ultrafiltration.

Spirals - A history of rediscovery
One of the most striking things about spiral elements is their complexity of construction. Despite this various inventors in different fields have come up with the spiral as a solution to the technical problems that faced them. One of the earliest documented cases dates back to 1938 when William Groves [2] patented a spiral wound dialysis device for use in cleaning dirty sodium lye created in the textile industry. In the 40's Bauer and Curtis patented a disposable spiral filter for use in the automotive industry[3]. An interesting feature of this was the use of a corrugated spacer design. In the early 50's Tom Arden and George Solt patented a spiral wound electro-dialysis unit[4]. However, over 40 years had to pass before the concept was commercialised. In the early 1960's, as companies sought to commercialise the development of the reverse osmosis membrane, Gulf Atomic (now Fluid Systems) produced a number of patents[5]-[7]. The convenience of the spiral was quickly recognised and this opened up a wide range of applications. In the 80's spiral wound elements were developed for ultrafiltration applications involving low solids concentrations, such as pyrogen removal. Most recently, a microfiltration spiral- wound ceramic membrane, based on Ceramesh, has been trialed for recovery of caustic.

Spirals can be viewed as a collection of plastics bags containing permeate spacers on the inside and the active membrane surface on the outside, wound round a hollow mandrel (product tube), and separated from each other by a feed mesh. On large RO elements many sets of leaves might have to be attached and wound round the central product tube. The length of each bag (or leaf) is usually between 1 and 1.5 m. If it were any larger then significant back-pressure losses would be incurred. Typically for a 4" diameter element there would be 4 leaves, while for an 8" there could be as many as 18.

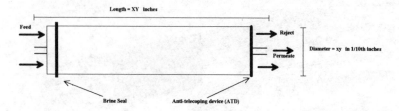

Figure 3.3-2 A size naming convention of spiral elements based on imperial measurement has developed and is still widely used in the industry. Shown is a schematic for an RO element of size xyXY. The larger sizes are 8040 (8" diameter, and 40" long), and 4040 (4" diameter, and 40" long). Elements are made as small as 1512 for domestic applications, and at least one manufacturer makes elements that are 60" long. Elements are slightly smaller than the quoted values, and the measurements are there to indicate the size of pressure vessel that would be suitable.

The details of the construction vary with each membrane and have to be tuned to the particular application e.g. the permeate spacer required for a high pressure application is different from that for a low pressure application. Interfacial membranes can be made which can withstand temperatures as high as 80 C. However, to utilise them requires ensuring that all the components parts are also compatible with such temperatures. Desal have developed a module to operate at 80 C (their Durasan range), by replacing the materials with low temperature stability with high performance polymers such as polysulphone. Another feature of spirals is the outer wrap. Small elements are usually wrapped with polyethylene or PVC tape. Large elements tend to have an additional wrapping of fibre-reinforced plastic (FRP). More recently, manufacturers have introduced a "sanitary" design. This consists of an outer wrap of a mesh, and the brine seal has been removed. The elements fit tightly into pressure vessels and only a small percentage of flow can by-pass the element by flowing on the outside. However, the design does mean that there are no dead spots, and all parts of the system can be chemically sanitised.

Compared to other membrane packages, the most noticeable difference is the complexity of the design. A typical element is made from up to 11 different materials. Until fairly recently each element had to be hand made. Despite these disadvantages the spiral wound element provides a convenient way of packaging large amounts of membrane surface area into a single element, which can easily be managed by one man. Its modular nature is that it is easy to assemble and refit the system.

In use the water is supplied to the device off-axis, and enters the feed channel. The membrane layers area kept apart by a feed mesh, which is usually made from polypropylene.

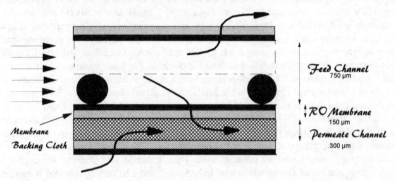

Figure 3.3-3 Schematic of section through a reverse osmosis spiral highlighting channels and size scales.

The orientation of the mesh is important; and is usually positioned to make the water spiral its way through. The narrowness of the spacing means that despite high velocities, the flow is not turbulent. However, the presence of the mesh means that there are inertial pressure losses, and the pressure drop is a non-linear function of the flow-rate, though usually over the small working flow-range a linear approximation suffices for most purposes. RO elements are usually designed to take a maximum pressure loss of 12 psi, which is small compared to the feed pressure. However, for large systems with several stages and elements this can amount to a substantial pressure loss. For RO spirals the feed mesh is typically 500 to 750 microns in thickness. In UF applications thicker spacers, typically 2000 microns, are used for more concentrated solutions, e.g. food. Increasing the thickness of the feed spacer does not change the cross-sectional area of the feed channel, and hence it has no impact on the cross-flow velocity. However, it does reduce the viscous drag. Increased thickness means that it is easier to handle feeds containing large particulates, and more viscous materials. The major disadvantage of a thicker feed spacer is that for a membrane of a given size there is a reduction in membrane area.

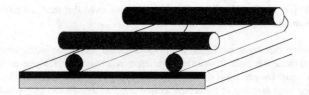

***Figure** 3.3-4 Schematic of feed mesh on membrane surface. Typical spacing of strands is about 3 per cm.*

A particular problem with the mesh type feed spacer is that material can build up around the fibres where the shear velocities are low[1]. The applied pressure forces some of the water to pass through the membrane and into the permeate channel. Once there, it spirals in towards the product tube and exits through holes drilled along the product tube into a central hollow. The permeate channel is kept open by a spacer which is usually a woven material, with deep channels. The permeate spacer is usually a tightly woven material, since it has to withstand the hydraulic load and provide sufficient support to prevent the membrane deforming into the channels. For water to flow along the permeate channels requires a pressure. This "back-pressure" reduces the available head across the membrane and hence produces a small reduction in the productivity and rejection. The resultant pressure loss means that for most designs the length of the permeate channel is kept between 1 and 1.5 m. If it is made too short then there will be a consequential reduction in available membrane area owing to the width of the glue-lines. High pressure applications use a more tightly woven spacer to resist the mechanical load, but this incurs a greater pressure loss for a given flow. Consequentially, for high pressure designs shorter leaves are normally used. Thus, it can be seen that the design of a spiral involves a large number of factors which have to be traded, and a balance sought that is appropriate to the cost-performance equation.

To utilise a spiral requires it to be sealed in a pressure vessel. As the feed moves through the element, it will have a tendency to telescope it. To avoid this an anti-telescoping device (ATD) is placed at the reject end of the element. Designs differ but all are essentially a solid plastic wheel with holes for the reject to pass through. In order to avoid the water by-passing around the element a seal, consisting of a rubber ring, is attached to the outside of the element at the feed end. In some designs ATDs are fixed at both ends, differing in the fact that there is a groove in the outer rim of the ATD to take a sealing O-ring.

The major cost of pressure vessels is in the end fittings. Thus, if more than one element is required, then it makes economic sense to put them in the same pressure vessel. In this way up to 6 elements can be packaged in a single pressure vessel (in a few cases 8 elements are used). Another advantage of this arrangement is that when elements are to be replaced only one end cap needs to be removed to enable access to the elements. The limit to the number of elements that can be used in series is set by the requirement of having sufficient cross-flow for the last element in the series.

---

[1]   Osmonics have tried to overcome some of the problems of the feed mesh by using a corrugated spacer which channels the flow directly through the device. This device reduces the problem of solids accumulation but increases the polarisation problem.

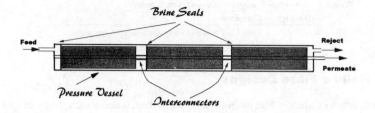

Figure 3.3-5 Schematic of 3 spiral elements in a single pressure vessel.

In addition to any operational limits on the membrane there are a number of limits determined by the nature and form of the module element unit construction. They include

· *Maximum feed flow-rate*
· *Maximum pressure drop*
· *Maximum hydraulic pressure*
· *Maximum temperature*

The maximum feed flow-rate is set to give an acceptable pressure drop along the module at start-up. During operation this pressure differential can increase due to fouling, and a maximum pressure drop is set at which cleaning becomes necessary. Exceeding this limit can damage the element through failure of the outer wrap. It can also cause damage to the ATDs which are there to stop the spiral telescoping. As already indicated the hydraulic limit on a module is set by more than the membrane. In particular the choice of permeate spacer, and product tube design determine the operational limits.

The dimensions and flow rates in spirals mean that the flow is not turbulent. This however does not mean that the pressure loss is a linear function of the feed flow-rate since there are inertial effects associated with the path the water takes through the module. In general the pressure loss is related to the feed flow rate by

$$\Delta P = aQ + bQ^2$$

where the first term is the viscous term, which varies with temperature, and the second term is an inertial term. Usually, the range of flow-rates of interest are sufficiently narrow that a linearised expression with an intercept can be used. A consequence of lowering the temperature is to increase the pressure drop. In operation some of the fluid is removed as permeate while the feed flows along the element, which consequently reduces the pressure loss. To obtain a slightly more accurate estimate, which takes this into account, one can use the average feed side flow rate. Fortunately, in actual designs such effects as this are accounted for in the design programmes provided by suppliers.

One important feature of spiral elements is that they are designed for a specific duty. Examples of this can be found in the trade literature from spiral manufacturers (see table 3.3.1).

Table 3.3.1 Some of the leading spiral wound manufacturers

Technology	Companies
RO	Dow, Fluid Systems,Toray, Osmonics, Hydranautics
UF	Membrex, Amicon, Osmonics, Hoechst

3.4 Hollow Fibre Designs

The basic design of a hollow fibre module is a bundle of fibres sealed at each end by an epoxy resin plug, and the whole element encased in a PVC, or acrylic tube or a fibre re-inforced plastic. The large commercial gas separation units use small hollow-fibres less than 100 microns in thickness, and extending over several metres in length. RO elements also use small fibres but are usually shorter (1 to 2 m). In contrast, haemodialysis units use slightly larger fibres (200 microns) typically, about 25 cm long containing some 10,000 fibres. Ultrafiltration and microfiltration use even larger fibres to meet the hydraulic and fouling demands.

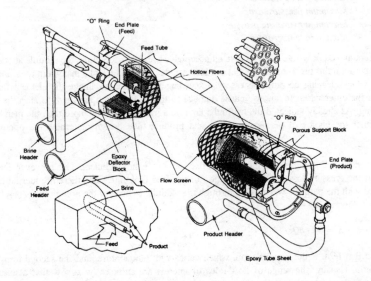

Figure 3.4-1Schematic of Permasep® hollow-fibre RO element. A particular feature of the device is that the flow is from the outside into the lumen of the fibres where it is conducted to one of the ends. The large elements are about 200 mm in diameter, and 1.5 long. (Courtesy of DuPont)

In concept the hollow fibre format is probably the most appealing, providing a very high surface area per unit volume, with no spacers or meshes needed to support the membrane. A further advantage is that, having no support fabric, the membrane can be backwashed. Not surprisingly, this design is widely used with devices being available for reverse osmosis, gas separation, ultrafiltration, microfiltration, and haemodialysis. Within this large range there are a variety of element designs. One of the earliest hollow-fibre designs was DuPont's Permasep B9 and B10 modules for the desalination of brackish and sea water respectively. The fibres used in these elements have an outer diameter of 85 microns, and a wall thickness of 21 microns, giving. a surface area per unit volume of the order of 10000 m^2/m^3(see figure 3.4-1).

A single Permasep element contains 2.5 million fibres [8]. Flow is introduced along a central mandrel and radiates outwards through the fibre bundle. At the outside the reject stream is channelled along the element to exit from the same end that it enters. The flow through the fibre bundle is laminar. The membrane permeability is significantly lower than that used in spiral or plate and frame designs. This is because of difficulties in making defect free fibres and, even if better quality fibres were available, there would be an increase in the amount of polarisation (see chapter 6). Thus, the operating fluxes are significantly lower than that used in spirals, and this off-sets some of the advantage that derives from the high specific surface area. The major practical difficulty with these modules is the fineness of the spaces between the fibres which means that they are vulnerable to fouling.

In contrast with RO, most UF/MF devices the feed flow passes along the fibre lumen and is collected on the shell side of the device. Typically, for such applications fibres are 0.5 to 1.5 mm in diameter. An exception to this is the microfiltration system developed by USF Acumem for municipal water treatment works. In this design the flux is from the outside to inside the fibres just as in RO modules.

Table 3.4.1 *Approximate dimensions of various hollow-fibre membranes. Some manufacturers provide information on the dimensions of the fibres while others prefer to quote the available membrane area in a module.*

Membrane Process	Manufacturer	Product Name	Material	ID µm	OD µm	Wall Thickness µm
GS	Dow	Generon	PM	60	80	10
HD	Baxter-Travenol	CF	Cuprophan	200	222	11
	Gambro	Lundia-10	Cuprophan	200	220	10
	Kuraray	KF	EVAL	200	264	32
RO	DuPont	Permasep B-10	PA	42	85	22
	Toybo	Hollosep	CTA		165	
UF	Amicon		PS			
	A/G Technology		PS	1,000		
	Membrex		PAN			
MF	Memtec		PP	1,500	2,500	625
	Acumem	Megamodule	PS	1,000	1,500	250

Hollow-fibre units dominate the haemodialysis market. Typically, the fibres used are some 200 microns in internal diameter, and have a wall thickness of between 8 and 20 microns. The elements are about 25 cms long, and about 3 cms in diameter, and contain 1 - 1.5 m^2 of membrane area. Blood passes down the central lumen while the dialysate passes through to the shell side in counter-current direction.

The Prism gas separator provides yet another design of hollow-fibre element with the feed gas being applied from the outside of the shell, and the product gases being removed from the lumens at the ends of the element.

One of the key factors that govern the design of these units is hydraulic pressure losses that occur in getting the various streams into and out of an element. Losses not only reduce the flow-rate but also the selectivity. Decreasing fibre sizes results in greater surface area per unit volume, but this is at the expense of greater pressure losses. Irrespective of any fouling effects, it can be seen that there is a minimum fibre internal diameter below which the system as a whole becomes less efficient. Theoretical models have been constructed which look at optimal designs (HD - [9], [10], RO - [11],[12],[13], GS-[14],MF -[15]).

3.5 Tubular Designs

Tubular modules have the largest dimensions of all membrane devices. To some extent they are an extension of the hollow-fibre type with diameters ranging up to 25 mm. However, tubular modules differ from hollow-fibre devices in that the membrane is invariably supported along its length and packed in an array which might contain from 1 to 100 tubes. In these devices the membrane is usually cast on the inside of a rigid porous tube. The feed flows down the centre of the tube, and permeate flows through the walls where it is channelled to a side-port. The flow-rates used in these tubes means that the flow is invariably turbulent.

Tubular membranes are widely used in food applications because of their capability of handling difficult rheological systems, or fluids which contain large suspended solids (which might be inconvenient to remove), or present difficult fouling problems. Although tubular membranes have relatively low surface area per unit volume their simplicity means that they are easy to clean. Many of the available systems are also designed with the capability of being cleaned at high temperatures. The large diameter of the tubes means that the liquid hold up is high. A recent application of tubular membranes has been in the removal of TOC from the supply of potable water in remote areas of Scotland. The tubular membrane format means that no prefiltration is required despite the feed turbidity. However, a mechanical cleaning method involving a foam ball has to be applied frequently.

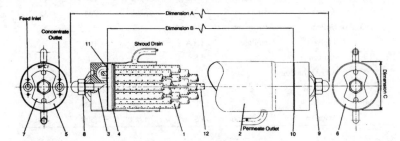

Figure 3.5-1 Illustrated is a tubular system from PCI. The assembly consist of 18 stainless steel tubes which houses tubular membranes of 12.5 mm diameter. The tubes are either 1.2 or 3.66 mm in length. The total membrane surface area is 0.9 and 2.6 m². Depending on end fitting the flow can be in series or in parallel. (Courtesy of PCI Membrane Systems Ltd)

Another examples of a tubular membrane is shown in figures 3.5-2.

Table 3.5.1. Suppliers tubular membranes

Manufacturer	Membrane Process	Manufacturer	Membrane Process
Koch	UF/MF	Nitto-Denko	UF/MF
PCI	RO/NF/UF		

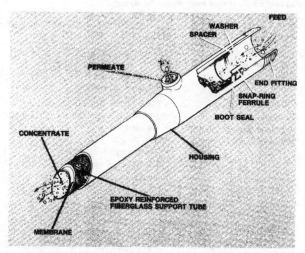

Figure 3.5-2 Illustrated is an element for a single tubular ultrafiltration membrane with 1 inch diameter made by Koch. (Courtesy of Koch Membrane Systems)

A particular variant of the tubular design is the ceramic monolith. In this case a bundle of tubes are co-extruded to form one solid block of tubes. Selected tubes in the block can then be coated so that a particular tube will act as a feed, or permeate line. The tubes are not necessarily circular in cross-section, and can be square, circular or star shaped.

3.6 Packaging Selection

The use of the various formats varies with application and technology. Some of the more important factors that drive selection are summarised in table 3.6.1.

Table 3.6.1 Factors that influence the choice of membrane module

- *membrane availability*
- *fouling*
- *suspended solids concentration*
- *dissolved solids concentration*
- *polarisation*

- *hold-up*
- *ease of cleaning*
- *hygiene*
- *feed viscosity*
- *pressure drop*

- *flexibility*
- *maintenance*
- *serviceability*
- *cost*
- *replacement cost*
- *experience*

The different packaging types have different merits (see table 3.6.2). While inevitably the problem can be reduced to one of cost ther is frequently a question of risk management with new processes.

In practice one or other formats tend to dominate because of the demonstrated benefits that have been proven. In many cases the end user is to some degree reliant on the system designer and past experience.

Table 3.6.2 *Advantages and disadvantages of membrane packaging. The last column indicates the technology which uses the various packaging types (technologies in brackets indicate its availability in the specified format but it is not a major part of the market).*

PACKAGE TYPE	ADVANTAGES	DISADVANTAGES	TECHNOLOGY
Flat Sheet	• Wide choice of membranes • Can be cleaned by dissassembly • Low energy requirement	• High cost • Replacing membrane is time consuming • Can have seal problems	D,ED MF, UF (NF), (RO), (HD)
Hollow-Fibre	• Very compact system • Low liquid hold-up • Low capital cost • Backflushable	• Easily fouled with particulates • Not suitable for viscous systems • Limited range of products available	GS, HD UF, MF, RO
Spiral Wound	• Low hold-up • Compact system • Wide range of materials • Wide range of sizes • Low capital cost	• Can have dead spots • Cannot be backflushed	RO, NF, (UF), (ED)
Tubular	• Can tolerate feeds with high suspended solids • Can work with viscous and non-Newtowian fluids • Easy to clean mechanically	• High energy requirement • High capital cost • Large space demand • Dissassembly long • High hold-up	MF, UF, NF, RO

Usually one format dominates a particular application sector. However, there are exceptions such as that for the electrocoat process in the automotive industry. Here both hollow-fibre, and plate and frame designs are available.

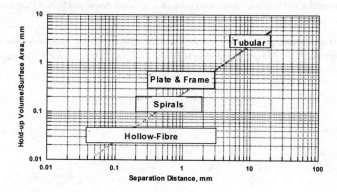

Figure 3.6-1 Hold-up volume for various membrane packages.

Hold-up is particularly important in high value added products such as drugs. The different packaging types cover hold-ups which range two orders of magnitude. Equally in such applications the ability to clean and rinse such device is also important.

3.7 References

1   R F Madsen "Hyperfiltration amd Ultrafiltration in Plate-and-Frame Systems" Publ. Elsevier, 1977

2   W W Groves *"Process and Apparatus for Conducting Dialytic Operations"* BP 489,654. 30 Jan 1937

3   J V Bauer and R H Curtis *"Filter Element"*USP 2,599,604. 13 July 1949

4   T V Arden and G S Solt, BP Patent 759275, 1953

5   J C Westmoreland *"Spirally Wrapped Reverse Osmosis Membrane Cell"* USP 3,368504, 21 Dec 1964

6   D T Bray *"Reverse Osmosis Purification Apparatus"* USP 3,417,870, 22 Mar 1965

7   R D Hancock and D T Bray *"Spiral Reverse Osmosis Device"* USP 3,542,203. 29 Aug 1967

8   A M Moch and I Moch Jr *"Seawater Reverse Osmosis - A study in use"* in *"Desalination and Water Re-use"* 2 (1991) 3-13

9   J E Sigdell *"Operating Characteristics of Hollow-Fibre Dialyzers"*

10 J E Sigdell *"A Mathematical Theory of the Artificial Kidney"* Publ Hippocrates Verlag, Stuttgart 91974)

11 W T Hanbury, A Yuceer, M Tzimopoulos, and C Byabagambi *"Pressure Drops along the Bores of Hollow-Fibre Membranes - Their Measurement, Prediction, and Effects on Fibre Bundle Performance"* Desalination 38 (1981) 301-

12 W N Gill and B Bansal *"Hollow Fiber Reverse Osmosis Systems Analysis and Design"*, AIChE Journal 19 (1973) 823-831

13 R Rautenbach and W Dahm *"Design and Optimisation of Spiral Wound and Hollow Fiber RO-Modules"* Desalination 65 (1987) 259-275

14 H Gorissen *"Studies on Transport Phenomena in Hollow-fibre Membranes by means of Dimensionless Groups"* in *"Future Industrial Prospects of Membrane Processes"* ed L Cecille and J-C Toussaint, Publ Elsevier 1989.

Chapter 4

PROCESS CHARACTERISATION

Contents

4.1 Introduction

A membrane separation process can be characterised as having at least one input stream producing at least two output streams of different compositions. Most commonly there is one input stream, as shown in figure 4.1-1, but dialysis and sweep gas processes involve two input streams.

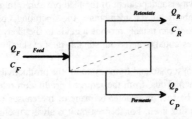

Figure 4.1-1 Schematic outline of membrane separation process (Q - flow rate, c-concentration).

Usually, one of the output streams is regarded as the product and the other as the waste. Which is which depends on whether one is interested in recovering something in the feed, or removing something from the feed. The output stream that passes through the membrane is known as the permeate, while the residual stream is referred to as the retentate, or concentrate, or reject stream..

Two key measures of performance are the productivity of a plant and the quality of the product, (i.e. how good is the separation). Ideally, these performance factors are constant, but in practice they change due to a variety of causes. This is one of the most critical and difficult problems in designing a successful system.

Just how well a membrane system separates is expressed in a variety of ways depending on the technology and application of interest:

Table 4.1 Measures of separation used in various membrane processes. The definitions of these measures of selectivity are provided in section 4.3.

Separation Measure	Membrane Technology
Rejection	Reverse Osmosis, Nanofiltration, (Ultrafiltration)
MWCO	Ultrafiltration, (Nanofiltration)
Log Reduction	Microfiltration, (Ultrafiltration)
Nominal and Absolute Size	Cartridge Filtration, Microfiltration
Beta values	Cartridge Filtration
Clearance	Haemodialysis
Selectivity	Gas Separation

While the composition and mass flows are sufficient to describe the mass balance, additional information is required to provide a full thermodynamic description, e.g. pressure, and temperature.

4.2 Productivity

One of the key economic questions regarding a plant is its production rate. This rate is made up of the sum of the productivities of the individual elements. The productivity performance of an element is usually quoted by a supplier at a particular benchmark set of conditions. Such productivity however, should be considered as a quality control measure, rather than as an operating performance measure, since a number of factors can serve to reduce the productivity. Indeed, in some pressure driven applications, the productivity is independent of the feed pressure. In general productivity is a function of the intrinsic permeability of the membrane, the operating conditions and the feed. Experience from similar applications can usually provide a guide to the likely productivity, but in the absence of such information it is essential to carry-out suitable pilot evaluations.

In some cases, (e.g. a flat sheet for a plate and frame system), the productivity is usually quoted as a flux (production rate per unit area). Thus, from the required production rate one can calculate the membrane area needed. As will be seen in chapter 6, many of the causes of the reduction in flux show a direct dependence on the flux itself. For this reason it is always useful to know the flux rate proposed. A high initial flux might lead to a long term fouling penalty. One particular issue that needs to be taken into account is the difference between total membrane area and the active membrane area. In all applications there will be some membrane that will not be available, (e.g. boundary glue lines, end pots, spacer masking). These boundary areas can occupy up to 40 % of the total area. Thus, if one is estimating the flux from the performance of an element it is important to include only the active membrane area, otherwise, the predicted operating flux will be too low.

Conversely, if the total membrane area is used to estimate the production rate then there will be a significant overestimate of performance.

4.3 Product Quality/Selectivity

Membranes are rarely perfect and in processes which are concerned with rejecting material some contamination of the permeate will occur, or in the case of recovering solutes, some material will be lost in the permeate. Whether it is the concentration of a waste, the concentration of a drug, or the production of pure water for cleaning electronic components, invariably, it is the concentration of some component or components in the permeate which is used as a measure of product quality. A feature of membrane processes is that the quality of the permeate strongly depends on the feed concentration. Indeed, in most cases there is a linear relationship between the permeate concentration and the feed concentration. For this reason, most measures of performance are expressed as some ratio between the permeate and feed concentration.

The measure of the quality of separation is thus its selectivity of the process or membrane. A variety of methods are used to connect the selectivities of membranes/elements with the applications for which they are suitable. The relationship between these is provided either by models, or, more often than not, by testing. In which case the benchmark values are a quality control measure of a membrane. In an ideal world membranes designed for a similar duty would be tested under the same bench mark conditions. That would allow the user to select on the basis of performance, cost etc. However, the premise which lies behind this is that the benchmark performance of a membrane/element will translate to the application conditions in the same way. The reality is far more complex. Benchmark conditions are usually chosen so that they are relevant to the operation of a particular element and provide a quality measure so that benchmark values can be linked to an application. The extent to which one can compare membrane elements varies with the technology field.

Rejection, Retention, Transmittance

In reverse osmosis and nanofiltration the membrane selectivity is usually quoted as a rejection (or retention), and is defined in terms of the fraction of the solute in the feed that appears in the permeate:

$$R = \left(1 - \frac{c_p}{c_f}\right)$$

(4.3.1)

where c_p is the concentration of the permeate and c_f is the concentration. of the feed rejections are usually quoted on a percentage basis. In benchmarking for RO sodium chloride is the most commonly used solution, while in nanofiltration magnesium sulphate is preferred. For solution mixtures the various solutes/ions have different rejections. Most ions have positive rejections, but in a number of cases certain species (usually those involved with weak acids) can show a strong negative rejection. In practice the rejecting properties of a membranes can be characterised through any relevant measure of the composition. Common measures are :-

- *conductivity*
- *UV*
- *refractive index*

- *TOC*

The most widely used measure in RO is conductivity, while UV and TOC are more commonly applied for waste applications. Refractive index is widely used in food applications.

While in operational terms the rejection is usually the measure of key concern, it is the transmittance, T, that is of greatest significance at a fundamental level:

$$T \equiv 1 - R = \frac{c_p}{c_f}$$

(4.3.2)

Molecular Weight Cut-Off (MWCO)

In ultrafiltration a popular measure of selectivity is the molecular weight cut-off. This is done by measuring the rejection characteristics of either a mixture of molecules of differing molecular weights (e.g. proteins) or a material with a range of molecular weights (e.g. dextrans). The feed and permeate are then characterised using, for example, gel permeation chromatography (see figure 4.3-1).

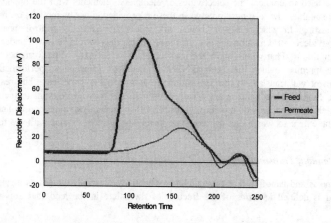

Figure 4.3-1 Gel permeation chromatography (GPC) profile of dextran solution in the range 10,000 to 150,000.

From the curve a rejection as a function of molecular weight. Making some assumptions as to the relationship between molecular weight and retention, one can calculate the molecular weight cut-off (MWCO) which is defined as the point at which 90 % rejection occurs. An alternative approach is to challenge the membrane with a blend of polymers with a tight molecular weight range. One of the most popular species for this are PEGs (polyethylene glycols)

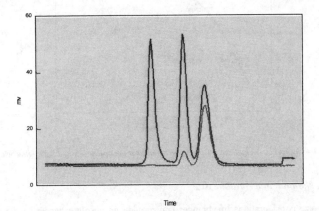

Figure 4.3-2 GPC profiles for a hollow-fibre with 2,000, 20,000, 400,000 polyethylene glycols. The time axis can be equated to molecular weight through reference to the molecular weight of the polyethylene glycols.

Figure 4.3-2 shows a schematic of a molecular weight cut-off curve. Membranes are sometimes characterised as having diffuse or sharp cut-off depending on the width of the cut-off curve. The ideal cut-off is often identified as a discontinuity. For reasons discussed in chapter 6 this ideal appears unreal and will not be observed even if the membrane contained narrow distribution of pores.

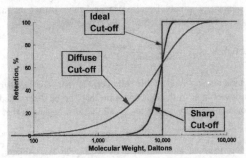

Figure 4.3-3 Different membranes have different molecular weight cut-off curves.

For nanofiltration, or tight ultrafiltration, simple molecules are sought for carrying out molecular weight cut-offs. One of the commonest group of materials chosen for this purpose is the polysaccharides (e.g. sucrose, glucose, raffinose). These materials have exact molecular weights and thus a molecular weight cut-off can be determined. It is tempting to think of molecular weight cut-off as a physical property of the membrane independent of the measurement method. However, as with salt rejection the rejection properties are sensitive to pressure and cross-flow. Thus, it is essential to carry all measurements out under the same operational conditions.

4 - 5

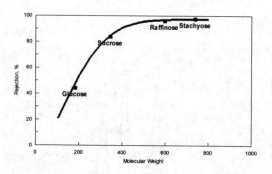

***Figure** 4.3-4 Molecular weight cut-off for some nanofiltrations measured by challenging with a range of sugars.*

Log Reduction, Nominal and Absolute Size

Based on pore size considerations, membranes can provide an absolute barrier to the passage of particles or organisms. An absolute membrane is therefore one which allows no passage of the selected particles. If a specific size of particle is used then this allows one to give an absolute pore size rating to a membrane. This nomenclature is popular in cartridge filtration. A slightly weaker rating is based on nominal size. This is simply the size rating at which only 10 % of the particles at the size rating in the feed pass through the membrane. It is clear that such a definition will be specific to a particular challenge and method of evaluation.

In some cases membranes have an exceedingly high rejection e.g. greater than 99.99 % and yet the exact value of the membrane's performance is still an issue. The integrity of the barrier therefore is a critical question for producing a stream free from a particular pathogen. In such examples the performance of a membrane is expressed in terms of *log reduction*.

$$\log \text{ reduction} \equiv \log_{10}\left(\frac{\text{concentration in permeate}}{\text{concentration in feed}}\right) \qquad (4.3.3)$$

One of the issues in characterising filters with pathogens is that it is based on a number count. Consequently, there is a lower limit to the concentration that can be determined, which depends on the practical limit of the sample volume. As a result, all that can be said is that the log reduction is better than a value.

Table 4.3.1 highlights log removal data obtained by Jacangelo and Adham [1], on 6 different membranes (3 microfiltration, and 3 ultrafiltration) for a range of biological challenges. The log removal values for the cysts (*Giardia muris* and *Cryptosporidium parvuum*) are set by the detection limit for the particular challenge. *E. Coli* is much smaller than *Cryptosporidium parvuum* and yet its reduction is orders of magnitude better, which is a consequence of the measurement method and not an indication that *E.Coli* is better rejected than *Cryptosporidium parvuum*. With the exception of UF-B (this membrane is believed to have a defect) the log removal values were limited by the measurement method. As expected, the ultrafilters performed markedly better than the microfiltration membranes when subjected to a virus challenge.

Table 4.3.1 Shows log reduction measured on 6 different membranes (from Jacangelo and Adham,[1]).

Membrane	Specific Flux Rate, GFD/psi	Cut-off	Log removal				
			Giardia muris	Cryptospodium parvum	Escherichia Coli	Pseudomonas Aeruginosa	Virus
MF-A	28	0.2 μm	> 5.2	> 4.9	> 7.8	> 8.2	< 1
MF-B	45	0.2 μm	> 5.2	> 4.9	> 7.8	> 8.2	< 1
MF-C	13	0.1 μm	> 5.0	> 4.8	> 7.8	> 8.2	< 1
UF-A	26	500 kD	> 5.2	> 4.9	> 9.0	> 8.2	1-2
UF-B	17	300 kD	> 5.0	> 4.8	5.6		4
UF-C	9	100 kD	> 5.2	> 4.9	> 7.8	> 8.2	> 6

Beta value

A characterisation method that is widely used for cartridge filters is the beta value

$$\beta \equiv \frac{\text{Total number of particles larger than a given size in the influent}}{\text{Total number of particles larger than a given size in the effluent}}$$

(4.3.4)

Removal efficiency is also used to measure membrane performance. It is defined by.

$$\% \text{ Removal efficiency} = \left(\frac{\beta - 1}{\beta}\right) * 100$$

(4.3.4)

The relationship between beta, the removal efficiency and log reduction is illustrated in table 4.3.2

Table 4.3.2 Shows the relationship between three different measures of separation

β	% Removal	Log Reduction
1	0	0
10	90	1
100	99	2
1,000	99.9	3
10,000	99.99	4
100,000	99.999	5

At a practical level cartridge filter manufacturers call the particle size at which the beta value is about 10,000 (i.e. log reduction of 4) the absolute rating of the filter.

For these highly selective membranes a minor defect such as that caused by a trapped air bubble or particle of dust during the manufacturing process can significantly reduce performance. As yet it is not viable to validate all of the membrane surface, but instead the focus is on quality control procedures to reduce the problem to a minimum. In addition to defects through the membrane there are leakage paths which by-pass the membrane through the glue lines or around defective seals. Based on size a reverse osmosis membrane should be an absolute barrier to a pathogen. However, an extremely minor leak through the glue lines, which would not significantly effect the salt rejection could be extremely significant when one focuses on the transmittance of pathogens.

Selectivity

In the case of gas separation the separation is defined in terms of selectivity between components i and j with respect to the molar composition $(c_1, c_2, c_3, ...)$

$$\alpha_{i/j} \equiv \frac{(c_i/c_j)_P}{(c_i/c_j)_F} \qquad (4.3.6)$$

Clearance

Clearance is a term that has come to be used in the medical application of membranes, particularly haemodialysis, and is defined by

$$Q_{Cl} = Q_F\left(\frac{c_f - c_r}{c_f}\right) \qquad (4.3.7)$$

where Q_F is the feed flow-rate, c is the concentration of the constituent of interest, and the subscripts f and r refer to the feed and reject respectively.

If the exit concentration for a certain solute were zero, then the clearance would be equal to the feed flow rate. If the exit concentration were half that of the inlet, then the clearance would be half the feed flow-rate. In essence, the clearance indicates the virtual amount of feed that is being treated per unit time.

4.4 Membrane Life

The initial productivity of a membrane is one thing, it is another to maintain that performance. In heavy duty applications membranes may last as little as a few weeks, while in clean applications membranes may last 5 to 10 years. In practice, there is always a compromise between the degree of pretreatment, how intensively the membrane process is operated, the frequency of cleaning and the membrane life. Nearly all membrane processes decline in performance over time-scales that vary from a few minutes to several months. The causes of this decline are various and include amongst others

- *temperature changes*
- *compaction*
- *chemical degradation*
- *accumulation of foulants on the membrane surface*
- *precipitation on the membrane surface*

In carrying out an economic assessment some allowance for fouling has to be taken into account. It is the integrated productivity that is important.

4.5 Recovery, Stage Cut, Yield

The fraction of the feed which is treated is known as the recovery, Y (usually quoted as a percentage), and is defined as the fraction of the feed that passes through the membrane,

$$Y = \frac{Q_P}{Q_F} \qquad (4.5.1)$$

While there is a frequent desire to treat all the feed, there are practical reasons why this cannot be done. These include

- *Thermodynamic/Energy requirements*
- *Phase separation (e.g. precipitation)*
- *Rheological properties of retentate*
- *Fouling*

The energy/thermodynamic issues are exemplified in the reverse osmosis of sea water. The osmotic pressure of sea water relative to pure water is of the order of 350 psi. If a system was run to give a recovery of 67 %, the concentration would have risen by nearly three fold and so would the required feed pressure. Given the usually low cost of sea water, and practical issues due to the high pressure required, sea water RO is rarely run above 30 % recovery.

As water is removed from the feed stream, solutes concentrate, and this can cause some to become supersaturated with a consequent precipitation. For solutes such as calcium carbonate, calcium sulphate, and silica the operating limits are readily predicted by the system design programmes. For more application specific inorganics and organics, one is usually reliant on pilot trials to identify the potential problems. The key objective of calculations based on water analyses or pilot testing is to determine the maximum recovery at which a system can be operated.

Particularly in food/biological systems the effect of removing water is to change the rheology of the feed. Invariably materials get more difficult to pump as they progress through a membrane system. At very high viscosities a number of problems arise

- *generation of heat*
- *degradation of material from shear stresses*
- *increasing concentration polarisation*

Inevitably, there is a practical limit when the viscosity is too high, and the performance too low for membranes to be used.

4.6 Volume Reduction, Concentration Factor

In waste applications, customers are most interested in the quality of the permeate and the degree of volume reduction achieved. In the pharmaceutical industry the focus is usually on the recovery of an ingredient, the concentration factor achieved is often a critical measure of performance. These factors are not independent, but linked to how much water is recovered from the feed (recovery). Indeed, for batch operation

$$VR \equiv \frac{V(final)}{V(initial)} = Y \qquad (4.6.1)$$

Thus, a system with 80 % recovery represents an 80 % reduction in feed volume. As water is removed from a feed the rejected components become concentrated. A mass balance shows that the concentration factor, the rejection, and the recovery are linked via

$$CF \equiv \frac{C(final)}{C(initial)} = 1 + R\left(\frac{Y}{1-Y}\right) \qquad (4.6.2)$$

If the membrane is ideal (i.e. $R=1$) this reduces to (see figure 4.6-1)

4 - 9

$$CF = \frac{1}{1-Y} \qquad (4.6.3)$$

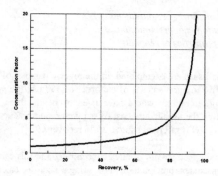

Figure 4.6-1 Plot of concentration factor versus recovery for ideal membrane.

Thus, a 90 % reduction in volume (i.e. 90 % recovery) results in a concentration factor of 10, and for a 99 % reduction in volume there is 100 fold concentration increase (see table 4.6.1) . In many cases these levels of concentration cannot be sustained without inducing a phase separation. In batch processing high volume reduction targets mean that the membrane will see extreme conditions in terms of concentration, and fouling at the end of a batch cycle.

Table 4.6.1 Effect of recovery on concentration factor.

Recovery or Volume Reduction, %	Concentration Factor (Max)
50	2
90	10
95	20
99	100

One further complication is that the system rejection is less than the element rejection which in turn is less than the membrane rejection. For a membrane with intrinsic rejection, r, that is independent of concentration, the concentration factor for a single pass system is

$$CF = \left(\frac{1}{1-Y}\right)^r \qquad (4.6.4)$$

These results can be used in equation 4.6.4 to compute the system rejection and the result of this is shown in figure 4.6-2 for several different membrane rejections. The graph shows that the penalty for increasing recovery is a fall in system rejection as a result of concentration increases within the syste,. Thus, a membrane with an intrinsic membrane rejection of 90 % will only achieve 73 % system rejection if a recovery of 95 % is used.

4 - 10

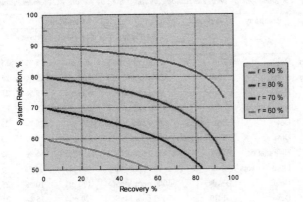

***Figure** 4.6-2 Calculated effect of recovery on system rejection in a single pass system for membranes with various intrinsic rejections.*

4.7 Membrane/Element Characterisation

The performance of an element depends on a large number of characteristics and most manufacturers carry out a large number of tests during the development stage in order to optimise the design. Packaging of the membrane inevitably leads to some degradation in performance (loss of flux, loss of rejection). The art of the designer is to achieve the best possible balance of factors for the intended use. Once made, elements go through a quality control procedure which involves testing the element under a standard set of test conditions. Test conditions are generally chosen to be indicative of those to be used in practice. A glance at the catalogues will reveal the test conditions used by different manufacturers (see table 4.7).

While there is a degree of individuality about this, manufacturers use conditions that are indicative of the end use e.g. a cellulosic RO membrane operates under quite different conditions from an interfacial RO membrane. Not only do they operate under different conditions but the pre- and post-treatment requirements are different. Thus, if comparison has to be done, it should not be whether one membrane out performs another membrane under a chosen set of conditions, but whether the final plant meets the performance requirements and whether or not it gives a better economic return. Most RO manufacturers provide computer aided design tools which means that various designs can be rapidly constructed and costs estimated. One important aspect of the performance of membranes is the tolerance. In the case of RO the productivity tolerances can be as large as +/- 15 %. Such variability has to be taken account of during the design process.

Table 4.7 Some typical characterisation conditions used by various RO element manufacturers.

Type	Company	Product	Conditions			
			Challenge	Pressure	Recovery	Feed Temp, pH
RO/Sea Water	Dow	FT30	32000 ppm NaCl	800 psi	15 %	25 C, pH=8
RO/Sea Water	Desal	Desal-3	32000 ppm NaCl	800 psi	15 %	25 C, pH=6.5
RO/Brackish	Desal	Desal-3	2000 ppm NaCl	425 psi	15 %	25 C, pH=6.5
RO/Brackish	Hydranautics	CPA2	1500 ppm NaCl	225 psi	15 %	25 C, pH=7
RO/Brackish	Dow	Flux, Rejection	2000 ppm NaCl	225 psi	15 %	25 C, pH=8
RO/Brackish	Dow	Flux, Rejection	2000 ppm NaCl	225 psi	15 %	25 C, pH=8
NF	Hydranautics	PVD1	1500 ppm NaCl	150 psi	15 %	25 C, pH=7
NF	Fluid Systems		700 ppm tap	109 psi	15 %	25 C, pH=7
NF	Desal	Desal-5	1000 ppm MgSO4	100 psi	10 %	25 C
NF	Dow	NF 70	2000 ppm MgSO4	70 psi	15 %	25 C

4.8 References

1 J G Jacangelo and S S Adham, *"Comparison of Microfiltration and Ultrafiltration for Microbial Removal"* in "Microfiltration for Water Treatment Symposium" Aug 1994, Irvine, California, USA

Chapter 5

FUNDAMENTALS

Contents

5.1   Introduction

Underlying every technology there is some basic science. With its extensive colloidal roots membrane technology is no exception, and this chapter provides an appreciation of this. There is inevitably a dilemma as to what should and should not be included and the decision must be a personal one. We have been guided by those factors that impact significantly on the process of separation. Wherever possible the various length scales, time-scales and forces that arise will be quantified. The final part of the chapter provides an introduction to models of the transport of water and solutes through membranes and explains how this gives rise to the selectivity that we exploit. Through the large number of approximations that are apparent in the analysis it emphasises the complexity of real systems, and hence provides an appreciation of the care that is needed in extrapolating from such models without also taking into account supportive experimentation.

5.2   Osmosis

Osmotic phenomena plays a significant role in those membrane processes involving the selection of low molecular weight solutes, i.e.

- *reverse osmosis,*

- *nanofiltration,*
- *electrodialysis,*
- *dialysis,*
- *membrane distillation.*

The presence of low molecular weight solutes in water depresses its vapour pressure. Thus, when pure water is brought in to close proximity to a salt or sugar solution, water preferentially evaporates from the pure water and condenses on to the solution. This transfer reflects a purely thermodynamic difference between water in the two solutions. By using a membrane these solutions can be brought very close together, and then the membrane provides a conduit for molecules of water to pass between them. If the membrane is **ideal** (i.e. only water migrates across it) then exactly the same result is achieved; water crosses the membrane to dilute the solution. This flow can be made to stop by applying sufficient pressure to the salt solution (see figure 5.2-1). This pressure equals the osmstic pressure generated by the solutes. If additional pressure is applied to the salt solution then water can be made to flow in the opposite direction, hence the term reverse osmosis[1].

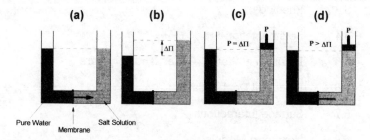

Figure 5.2-1 *Schematic illustration of the processes of osmosis ((a) and (b)) in which water flows through the membrane from pure water to salty water until a pressure head is created (the osmtic pressure difference). If a pressure is applied to the salt solution (figure (c)) the flow of water diluting the salt solution will cease. Applying greater pressure results in reverse osmosis in which water flows from the slat solution into the pure water (figure (d)).*

For dilute solutions the osmotic pressure is expected to be proportional to the solute concentration, and Van't Hoff showed that for an ideal solution

$$\Pi = cRT\frac{1}{MW},$$ *(Van't Hoffs equation)*   (5.2.1)

where the concentration is in g/mL and MW is the molecular weight. This equation shows that osmotic pressure

- *increases linearly with concentration,*
- *decreases with increasing molecular weight,*
- *increases with temperature.*

---

[1] It is due to this that the term reverse osmosis originates despite the fact that many applications to which it is applied the osmotic pressure is not a major factor. For this reason some authors prefer the term hyperfiltration.

The Van't Hoff equation is obeyed well for salts at concentrations typical of tap water. For typical tap waters the osmotic pressure is usually less than 0.5 bar, while for sea water the osmotic pressure rises to about 25 bar[2]. In this case feed pressures of the order of 55 bar are required to overcome the osmotic pressure[3] and provide reasonable fluxes. For feeds with a high osmotic pressure special consideration must be given to what happens when the system is switched off. When the feed pressure is turned off, the osmotic pressure across the membrane can drive a substantial flow of water across it in the reverse direction. If this aspect is not designed out, then delamination of the membrane can occur in spiral and plate and frame systems. Such osmotic pressure effects can also trigger syphoning action if precautions have not been taken to prevent it.

The osmotic pressure of an aqueous solution is usually referenced to pure water. In a membrane process it is the difference in osmotic pressure difference between both sides of a membrane that is of significance. If a loose RO membrane (i.e. a nanofilter) is used, then a substantial amount of the low molecular weight material passes through the membrane. In this case the relevant osmotic pressure is the difference between the feed and the permeate. This is considerably less than that of the feed relative to pure water. If van't Hoff's equation provides a good representation for the osmotic pressure then the osmotic pressure difference is simply related to the feed by the rejection i.e.

$$\Delta\Pi = \Pi_f - \Pi_p \cong R\,\Pi_f\,. \qquad (5.2.2)$$

More generally for a multi-component system

$$\Delta\Pi = \Sigma_j\,R_j\Pi_j\,. \qquad (5.2.3)$$

For food and biological applications where concentrations can be high, significant positive deviations are observed (see Figure 5.2-2). Such deviations are frequently modelled with a virial type expansion:

$$\Pi = \frac{cRT}{MW_s}\left(1 + \frac{B_2}{MW_w}c + ...\right). \qquad (5.2.4)$$

The first virial coefficient can be related to entropic and enthalpic interactions, and can be either positive or negative. Macromolecules can show strong deviations. Deviations from ideality are more marked for long chained polymers than for compact proteins. These deviations are frequently dependent on pH and ionic strength due to the charged state of the anionic, and cationic groups on the protein. The non-linear characteristics of osmotic pressure can be important in food and biological applications where concentrations can be substantial.

The rejection of solutes at the membrane surface results in a locally higher solute concentration. As a result the osmotic pressure at the surface is higher than in the bulk. This serves to reduce the net operating pressure across the membrane (pressure difference less osmotic pressure).

In batch applications of reverse osmosis the osmotic pressure ultimately limits the extent to which a feed can be dewatered. In electrodialysis the osmotic pressure drives water across the membrane

---

[2]   In imperial units 100 ppm of dissolved solids produces an osmotic pressure of about 1 psi.

[3]   The key driver for flux is the net pressure drop = pressure drop across membrane-osmotic pressure drop across membrane

from the dilute to the concentration stream and provides a practical limit to the level of concentration enhancement that can be obtained.

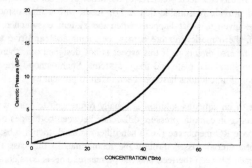

Figure 5.2-2 Osmotic pressure of orange juice versus concentration as measured in°Brix [1]

5.3   Rheology

At its simplest level the rheology of the feed dictates how much energy is lost in getting the feed and product to and from the membrane surface. An aspect which provides an additional level of complexity in many membrane applications is that the rheological properties changes as it passes through the membrane system as a result of increases in the bulk concentration as a result of dewatering. This aspect not only appears along an element, but also between the bulk fluid and the membrane surface, with an inevitable impact on the permeability and rejection characteristics. Thus, assessing the rheological properties of a feed are a vital aspect in the process design. Fortunately, in established applications the rheological properties are known and have been taken into account in bothe the elements used and the system design.

Gases and dilute aqueous solutions behave as Newtonian fluids, for which the shear stress, τ, is proportional to the shear rate, σ:

$$\tau = \eta\sigma,$$

(5.3.1)

where the constant of proportionality is the viscosity, η. Viscosity is sensitive to temperature (see table 5.3.1). For water a decrease of one degree Centigrade gives an approximate decrease of 2.5 % in viscosity.

Table 5.3.1 The effect of temperature on the viscosity of water and air

Temperature, C	Viscosity, (N s m^{-2})	
	Air	Water
0	$1.72 * 10^{-5}$	$1.79 * 10^{-3}$
20	$1.82 * 10^{-5}$	$1.01 * 10^{-3}$
40	$1.91 * 10^{-5}$	$0.65 * 10^{-3}$

Variations in feed temperature have more impact than a change in pumping costs. Changes in viscosity result in a change in membrane permeability. Thus, if the feed temperature varies between 10 and 20 C this could reflect in a 30 % difference in output! This benefit is usually recovered in lower operating pressures since membrane plants are designed around an operating flux rather than an operating pressure, for fouling reasons. Thus, a ten degree fall in temperature must be accommodated by a 30 % increase in net pressure, if constant output is required. For this reason reverse osmosis systems are designed around the minimum operating temperature rather than the average. Where cheap heat is available it can pay to heat the feed prior to treatment

The presence of macromolecules in a solution leads to an increase in viscosity. At low concentrations the viscosity is modelled with a virial type equation

$$\eta(c) = \eta(0) * (1 + [\eta] \, c + \ldots) .$$ (5.3.2)

Higher concentrations give rise to a more rapid increase that is commonly modelled with an exponential expression:

$$\eta(c) = \eta(0) \, \exp\{\alpha \, c\} .$$ (5.3.3)

This concentration dependence has practical implications for the design of ultrafiltration and microfiltration applications dealing with concentrated dispersions.

Concentrated suspensions like paint latex, and fermentation broths have complex relationships between the shear stress and shear rate (see Blanch and Clark [2], Coulson and Richardson[3]). A widely used equation to characterise such non-Newtonian fluids is the power law model in which the shear stress is proportional to some power of the shear rate:

$$\tau = \eta \sigma^n$$ (5.3.4)

where n is an empirical constant. When n <1 the fluid is shear thinning. In a shear thinning system the ratio of the shear stress over the shear rate decreases with increasing shear rate, i.e. the apparent viscosity decreases with shear-rate. This behaviour is a consequence of the shear breaking up the structure between particles and or macromolecules in the fluid. Once the structure has been ruptured it takes time to reform. As it does so the viscosity increases. Fluids like this are called thixotropic.

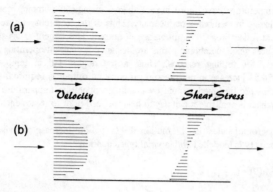

Velocity   *Shear Stress*

Figure *5.3-1 shows the velocity and shear stress in the flow down a pipe for (a) Newtonian fluid and for (b) a power law fluid with n=0.33.*

The exponent in equation 5.3.4 has a significant effect on the velocity profile created when the fluid passes down a duct lined with membrane (see figure 5.3-1). However, while the velocity profile changes, the shear stress remains uniform across the duct, being maximum at the membrane surface and decreasing linearly away from it till it reaches zero at the centre of the pipe.

One further complication that occurs in many colloidal concentrates is the appearance of a yield stress

$$\tau = \tau_o(c) + \eta\sigma^n, \qquad (5.3.5)$$

where τ_o is a yield stress. The yield stress reflects that for some fluids when the concentration reaches a critical value there is sufficient structure to resist a simple mechanical load, but will yield once a sufficient shear stress has been applied (sometimes referred to as a Bingham plastic - see figure 5.3-2). In operation the concentration varies rapidly near the membrane surface and this can have significant impact on gel polarisation(see chapter 6).

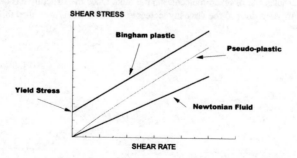

Figure *5.3-2 Schematic of relationship between shear stress and shear rate for different fluid types.*

An example of a thixotropic fluid is illustrated in figure 5.3-3. As can be seen the viscosity is extremely sensitive to the solids loading. These are significant factors in the design of both batch and continuous plants.

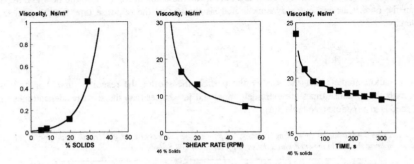

Figure *5.3-.3 Shows the rheological properties of a concentrated latex solution (40 %) as a function of solids, "shear" rate and time.*

In the context of the time-scale it should be noted that the residence time within a membrane element rarely extends to minutes, and in many cases seconds provide a more meaningful measure. The time-scale and the concentration gradients within devices means that care should be taken to obtain rheological characteristics pertinent to the application and the membrane device. Also, consideration of the type of pump and its impact on the fluid prior to processing need to be given.

The rheological properties impact on basic design issues such as

- *the sizing of pipe work,*
- *selection of a pump,*
- *selection of element type, i.e. spiral, tubular,*
- *provision of cooling to remove the generated heat.*

To do this requires an understanding of the rheological properties over the whole concentration range of interest, and under the shear conditions employed.

5.4   Sedimentation

Gravitational forces are frequently exploited to separate macroscopic particles. In membrane devices settlement of particles can cause problems like blockages of the spacers. With the exception of tubular membranes some pre-treatment is necessary to remove coarse materials e.g. screens for microfiltration, media filter for reverse osmosis.

Settlement is governed by the balance between the gravitational forces acting on the particles and hydraulic drag resisting movement. The hydraulic drag on a spherical particle of radius, a, and with a density difference of $\Delta\rho$, is given by Stoke's law. At steady state the force balance equations predict

$$v_t = \frac{2a^2 \Delta\rho\, g}{9\eta}.$$

(5.4.1)

This equation shows that the settlement velocity gets smaller as the particles get smaller. Typical values for these parameter (see table 5.4) show that the terminal velocity for most colloidal materials is very small. As a consequence sub-microscopic particles move with the fluid. However, when the fluid moves across an obstruction such as a fibre its inertia could move it out of the flow and impact it on the fibre. Carrying out a dynamical analysis shows that the response time for the particle is given by

$$\tau = \frac{2a^2\rho}{9\eta} \;\ldots$$

(5.4.2)

The equation predicts that the smaller the particle the shorter the response time. For colloidal materials the response times are extremely rapid and far shorter than the short residence times that the feed is in a module (see table 5.4).

Table 5.4 *Variation of settling velocity, response time, and characteristic length scale for a colloidal spherical particle in water with a density difference[4] of 300 Kg/m³*

Particle Diameter	Settling Velocity	Response Time	Length Scale
μm	μm/s	s	μm
0.1	0.0016	7.2E-10	1.2E-12
1	0.16	7.2E-8	1.2E-8
10	16.35	7.2E-6	1.2E-4
100	1635	7.2E-4	1.2

The product of the terminal velocity and the response time provides a characteristic length scale, which varies as the fourth power of the particle size. For colloidal materials this length scale is sub-atomic indicating that other interactions such as electrostatics will dominate in the colloidal domain.

This analysis suggests that provided adequate pre-treatment has been carried out to remove macroscopic particles, deposition by sedimentation is rarely a significant process in membrane units unless coagulation takes place within it.

A consequence of processes like sedimentation is to produce regions of high solids concentration. Over time these regions consolidate, i.e. they develop mechanical properties such as a yield stress (see equation 5.3.5) as the result of molecular interactions. Once such a structure has developed, it is more difficult to remove, and provides the nucleus for further fouling. This is a reason why frequent cleaning is more effective in maintaing performance, and enables easier removal of such deposits.

---

[4]   The density difference is between that of the particle and the average density of the system. In dilute systems the average density is that of the fluid. Another correction to the settling formula is the effect of neighbouring particles which modify the flow-field around the particle. This is frequently modelled by replacing the viscosity by an apparent viscosity.

5.5 Diffusion in Water, Gases, and Polymers

5.5.1 Brownian Motion

In the early part of the 19th century Thomas Brown studied a "strange" phenomenon where small pollen grains where observed to translate and rotate randomly when placed in a fluid. This motion, now known as Brownian motion, is common to all small particles. It was explained by Einstein in 1905 on the basis of thermal fluctuations resulting from the random collisions of the fluid with the particle momentarily being imbalanced. These fluctuations become more significant as particles become smaller. Brownian motion allows particles to migrate when there is a concentration imbalance. It can be shown that for a spherical particle of radius a (Stokes-Einstein relationship)

$$D = \frac{k_B T}{6 \pi a \eta} .$$

(5.5.1)

The measured values agree well with that predicted by the Stokes-Einstein equation except when molecules are non-spherical (see figure 5.5-1). The important implication of Brownian motion is that molecules can move significant distances in a second. For example a sugar molecule on average diffuses about 20 microns in a second.

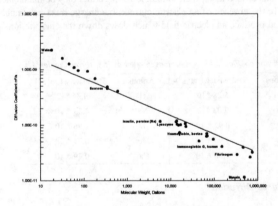

Figure 5.5-1 shows diffusion coefficient for various molecules at 20 C. The line is that predicted by the Stokes-Einstein equation assuming a density of 1300 Kg/m³ .

The Stokes-Einstein equation fails when the particles are not gloubular or there is a large number of internal degrees of freedom.More generally it is found that

$$D = k \, [MW]^{-\beta},$$

(5.5.2)

where

$\beta = 1/3$,   for globular particles,

$\beta = 1/2$,   for random coils,

$\beta = 1$,   *for rigid coils.*

In addition to the random translations particles also undergo random rotations. For non-rigid particles there is an internal Brownian motion which causes it to continually change its shape/conformation. This mechanism allows molecules to wriggle through microporous materials which have pores smaller than the "effective" size of the molecule.

Table 5.5.2 *Diffusion coefficient for various solutes in water at 25 C*

Solutes	MW	Diffusivity ($m^2 s^{-1}$)	Solutes	MW	Diffusivity ($m^2 s^{-1}$)
NH_3	17	$1.64 * 10^{-9}$	Ethanol	46	$0.84 * 10^{-9}$
H_2O	18	$2.25 * 10^{-9}$	Urea	60	$1.38 * 10^{-9}$
O_2	32	$2.38 * 10^{-9}$	Cl_2	70	$1.25 * 10^{-9}$
CO_2	44	$1.92 * 10^{-9}$	Sucrose	342	$0.53 * 10^{-9}$

5.5.2   Diffusion of Ions

Ions diffuse at different speeds (see table 5.5.3) and this has significant implications for reverse osmosis and electrodialysis. Consider the diffusion of salt across a microporous membrane into pure water. Since chloride ions diffuse faster than sodium ions, this could only be maintained if an electric current was allowed. In the absence of an electric current, the charge must accumulate close to the membrane interface and produce an electric field which slows down the chloride ions and speeds up the sodium ions.

Table 5.5.3   *Diffusion coefficient for various ions in water at 25 C (From Cussler[4])*

Cations	Diffusivity ($m^2 s^{-1}$)	Anions	Diffusivity ($m^2 s^{-1}$)
OH	$5.28 * 10^{-9}$	H^+	$9.31 * 10^{-9}$
Cl^-	$2.03 * 10^{-9}$	Na^+	$1.33 * 10^{-9}$
CO_3^{2-}	$0.92 * 10^{-9}$	K^+	$1.96 * 10^{-9}$
NO_3^-	$1.90 * 10^{-9}$	Mg^{2+}	$0.71 * 10^{-9}$
SO_4^{2-}	$1.06 * 10^{-9}$	Ca^{2+}	$0.79 * 10^{-9}$

5.5.3   Shear Induced Diffusion

When particles are passed down a tube they are observed to migrate away from the walls to some equilibrium distribution[5]. This movement has been attributed to the action of shear on the particle creating a lift through the Magnus effect. From this mechanism and a balance of forces it has been shown that the shear induced diffusivity for a 15 % solids suspension is given by

$$D_{sh} = 0.02 * a^2 * \left| \frac{du}{dx} \right|, \qquad (5.5.2)$$

where du/dx is the shear, and a is the particle radius. This equation shows that the dependency on particle size is opposite to that of Brownian motion in that it increases with particle size. Consequentially, there is going to be a critical size above which shear induced diffusion becomes more significant than Brownian diffusion (see figure 5.5-2).

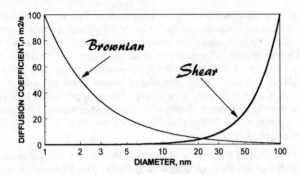

Figure 5.5-2 Shows the Relative effects of Brownian and shear induced diffusion in a shear field.

The consequence of shear induced diffusion is that fouling is likely to be worst for feeds with particles in the mid colloidal range.

5.5.4   Diffusion of Gases

Gases diffuse much faster in air than in water. Thus, in processes like gas contacting it is diffusion in the liquid phase that is rate limiting.

Table 5.5.4 Diffusivity of oxygen, and water in air.

Temperature, C	Diffusivity ($m^2 \, s^{-1}$)	
	O_2	H_2O
0	$17.9 * 10^{-6}$	$20.9 * 10^{-6}$
20	$20.3 * 10^{-6}$	$24.2 * 10^{-6}$
40	$22.7 * 10^{-6}$	$27.7 * 10^{-6}$

As in water, molecules diffuse faster as the temperature increases (see table 5.5.4), and also the heavier they are the slower the diffuse. The kinetic theory of gases predicts that the relative diffusivity of two gases is given by

$$D \propto T^{3/2} \sqrt{\frac{1}{MW_1} + \frac{1}{MW_2}} \, .$$

The reason for this is that at room temperature and above each degree of freedom has the same amount of thermal energy (equipartition theorem), and this means that the average velocity is inversely proportional to the square root of the molecular mass. This can be used to separate gases if the pore size is less than the mean free path (the average distance between collisions) - Knudsen's diffusion, and is the basis of isotopic separation of uranium.

5.5.5 Diffusion in Polymers

For reverse osmosis, gas separation, pervaporation, and electrodialysis separation is more than a size affair. The dimensions of the small molecules and the "holes" in the polymer matrix are comparable. However, the "holes" in the polymer matrix are dynamic. The transport of molecules across a membrane is governed by the solubility of a component in the membrane, and the diffusivity across it. This simple solution-diffusion mechanism was first expounded by Graham over 100 years ago and works well for rubbery polymers. In the 50's Meares observed that the solution-diffusion mechanism could not explain some of the data being obtained for glassy polymers[6]. Work on this has led to the idea that there are two types of sites within the polymer matrix. One type has the traditional Henry law adsorption and is freely mobile, while the other has a Langmuir type adsorption characteristic and no mobility.

In some case there is a high affinity between an organic and the polymer with the result that the organic uptake is extremely high and the polymer noticeably swells. This promotes the transport of the organic. The details of the interaction and consequences in any system are complex. The graph of Hopfenburg and Frisch [7] provides a useful aid to understanding to the various diffusion phenomenon that can occur, and the physical changes that result in the polymer (see figure 5.5-3). Further details and methods of estimating the permeability of gases through polymers can be found in Krevelen and Hofzser [8]

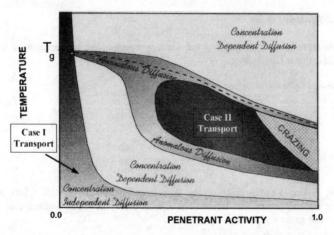

Figure 5.5-3 Schematic showing diffusional processes in various domains of temperature-penetrant activity affects diffusional processes (adapted from [7]) .

Substantial concentration enhancements have been demonstrated for trichloromethane and silicon rubber and these have been used to achieve substantial concentration enhancements.

5.6 Electrical Double-Layer

When small particles are placed in an electric field they invariably move towards one or other of the electrodes. This process, known as electrophoresis, is caused by the particles carrying a surface

5 - 12

charge. This charge can originate in a number of ways. For example if the material carries a fixed bound charge (e.g. sulphonated polystyrene), or the material is a metal oxide, hydroxide which can ionise at the surface. Sometimes the largest contributor comes from adsorption of an ionic species (e.g. a cationic surfactant) which has one end with a high affinity for the particle and the other contains an ionisable group. If amphoteric materials are present then the original charged nature of the materials can be completely masked. These effects also occur on a membrane surface and can have considerable effect on the selectivity of charged species, particularly in the processes of electrodialysis, reverse osmosis, and ultrafiltration.

If one analyses the potential field in the close vicinity of the membrane it is found that it is made of a number of effects (see figure 5.6-1)

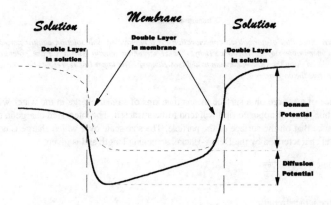

Figure 5.6-1 Schematic graph showing potential fields for a charged membrane in equilibrium with solution (dotted line) and with solutes passing through (solid line).

These potentials are caused by variations in the space charge density around the membrane surface. The strength of electrostatic charges is such that differences in charge cannot be sustained over large distances, and neutrality has to be maintained over macroscopic distance (see figure 5.6-2).

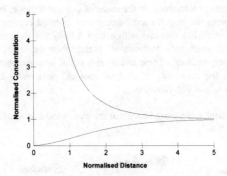

Figure *5.6.2 Plot of normalised concentration for a monovalent salt solution against normalised distance from membrane surface (Distance/Debye-Hückel screening length) with a surface potential of 125 mV, calculated using the Gouy-Chapman model (see Hunter)[5]. The upper line represents the counterions, and the lower line the co-ions.*

The presence of a charge on a surface means that ions of similar charge in the water will tend to be repelled, while those of opposite sign will tend to be attracted. The integrated charge in the layer will counterbalance that on the surface of the particle. The size scale over which charge is non-uniformly distributed is characterised by the Debye-Hückel screening length and is given

$$\lambda_{DH} = \sqrt{\frac{\varepsilon RT}{F^2 I}},$$ (5.6.1)

where I is the ionic strength:

$$I \equiv \tfrac{1}{2} \Sigma_j c_j z_j^2,$$ (5.6.2)

and provides a measure of the concentration of ionic species which emphasises the role of the higher valency components. Increasing ionic strength improves screening of charges and thus reduces the natural length scale over which charge is imbalanced (see figure 5.6-3). For a 2000 mg/L salt solution (i.e. a brackish water) has an ionic strength of about 0.035 mol/L and this gives us a screening length of about 2 nm. This has to be contrasted with an ultra-pure water with about 1 mg/L of dissolved solids for which the screening length is 90 nm.

---

[5]   Close to the membrane surface the finite size of the charged molecule, short ranges forces, and ordering of the water molecules can result in a breakdown of the simple Gouy-Chapman model.

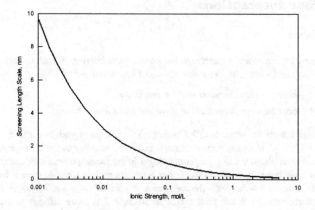

Figure 5.6.3 shows the effect of ionic strength on the Debye-Hückel screening length.

For low surface potentials the local variation in co and counter ions contribute equally to balance the surface charge. For high potentials though it is the counter ion that makes the dominant contribution (see figure 5.6-2).

What this theory indicates is that increasing ionic strength improves the screening of the surface and bound charges. As a result membranes which utilise charge effects for rejection will in general show greater selectivity for charged species at low ionic strength.

Just as the double-layer exists outside the membrane, it also exists inside the membrane. The range however, is much shorter since the dielectric constant is much lower there than in water. A further complication occurs in ion-exchange membranes which are used in electrodialysis and reverse osmosis membranes. The fixed bound charge of the ions produces a potential difference (Donnan potential) between the inside of the membrane and the solution outside. If the bound charge density is small it can be shown that

$$\phi_{DP} \approx \frac{RT}{F} \frac{\rho_X}{2\,F\,I},$$ (5.6.3)

where ρ_X is the bound charge density. Thus, if there is a negative bound charge density, there will be a negative potential. This potential would serve to attract positive charges and repel negative charges. Hence the selectivity of ion-exchange membranes, which are used in electrodialysis and reverse osmosis. As the concentration increases the ions screen out the bound charge making the membrane more transparent to negative ions. For a more in depth analysis the reader is referred to the texts of Hellfrisch [8], Schultz [9], and Hunter[10].

5.7 Surface Interactions

5.7.1 Introduction

The interactions between a feed and a membrane has considerable bearing on both the potential of a membrane to be used, and its life. The two major classes of interaction are

- *those between the water and the membrane.*
- *those between the solutes in the water and the membrane.*

A surface with a high affinity for water is called hydrophilic, while those with low affinity are called hydrophobic (e.g. PTFE). Contact angle measurements provide a measure of the hydrophilicty of the membrane surface. For applications like gas, contacting a hydrophobic surface is desirable, since it will give rise to a high break through pressure. For applications like ultrafiltration a hydrophilic surface is desirable. Solutes interact with the membrane in various ways e.g. hydrogen bounding, coulombic interactions, van der Waals forces. Not surprisingly it is more difficult to characterise fully all these interactions. One component which can be characterised, is the surface charge, which can be inferred from measurements of the zeta potential. These molecular concepts provide a guide to what might be happening, but the complexity that occurs in real systems, where a cocktail of different organic molecules are present, means that for most part the assessment is qualitative.

5.7.2 Hydrophilicity/Hydrophobicity

At a thermodynamic level a microporous membrane would be preferred to be wetted if there is a free energy gain. The free energy gain is given by the difference in the surface energy between the wetted and unwetted surface:

$$\Delta G = \gamma_{sl} S - \gamma_s S = (\gamma_{sl} - \gamma_s)S, \tag{5.7.1}$$

where γ_{sl} is the surface tension between the polymer and the liquid, γ_s is the surface tension between the solid and the air, and S is the internal surface area of the membrane. Now Young's equation relates this surface tension difference to the surface tension of the liquid, γ_l, and the observed contact angle θ_c between the liquid and the polymer surface viz.

$$\Delta G = \gamma_l \cos(\theta_c) S. \tag{5.7.2}$$

Thus, when the contact angle is less than 90° the membrane would be prefer to be wetted, and not if it is greater than 90° This free energy requirement can be translated into a pressure form

$$\Delta P = -\left(\frac{S}{V}\right)\gamma_l \cos\theta_c = -\frac{4\gamma_l}{d_h}\cos\theta_c, \tag{5.7.3}$$

where V is the pore volume and d_h is the hydraulic diameter (see Appendix-H). For a membrane with a simple parallel pore type structure this yields the equation

$$\Delta P = -\frac{2\gamma}{r}\cos\theta_c, \tag{5.7.4}$$

5 - 16

which can be obtained from a force balance approach. It follows that for water to pass through a hydrophobic material a critical pressure has to be exceeded. This pressure increases as the pore size decreases (see figure 5.7-1).

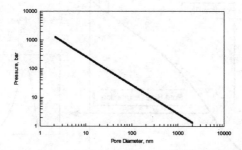

Figure *5.7-1 Pressure required to water to be forced through a highly hydrophobic (i.e. contact angle =180°) membrane with parallel pores.*

This analysis predicts that there is a critical contact angle which if exceeded means that pressure will be required to push water through the membrane. In the case of the simple parallel pore type this angle is 90°, but for the convoluted porous structures that occur in most membranes this critical contact angle is significantly less. Only if the contact angle is less than zero will the membrane surface be spontaneously wetted. In general the more hydrophobic a material is the greater is the difficulty for water to penetrate the pores.

Table *5.7.1 Contact angles for various polymers.*

Polymer	Contact Angle
Polypropylene	83
Polysulphone	73
Cellulose acetate	60
Polyacrylonitrile	53

Most membranes are not monodisperse parallel pores. The surface porous structure can be characterised by coating the membrane surface with a low surface a tension liquid, and then steadily increasing the pressure to the other side. If one observes the mass flow rate of gas then initially it increases uniformly with pressure. At a certain low pressures the mass flow rate increases more rapidly. This is the point of breakthrough on the largest pores. As the pressure increases the smaller pores open up and contribute to pathways. This continues till in theory the pressure is sufficient to open up all the small pores. In practice the pressure that can be used restricts this method to microfiltration and the top end of the ultrafiltration range.

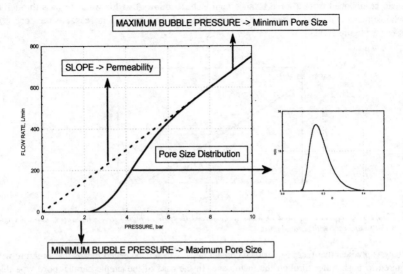

Figure 5.7-2 Graph showing the hysteresis effect that occurs when water is forced through a hydrophobic material. The flow-pressure curve can be used to obtain a pore "size" distribution (porometry).

As a general guide the tighter the membrane in terms of its molecular weight cut-off the more hydrophilic it is required to be. In order to utilise the more hydrophobic material manufacturers use various methods to overcome these problems:

- *Wetting agents are added to change the surface tension of the water e.g. ethanol*
- *Surfactants are added to change the surface tension of the solid*
- *Chemically modifying the polymer surface to change its surface tension characteristics*
- *Over pressurisation*

Over pressurisation is a technique sometimes used in RO type membranes if a wetting out issue is suspected. An example of this is seen in figure 5.7-3 where the permeability of the membrane increases as pressure is raised but stays approximately the same when pressure is reduced. This hysteresis is associated not with the active layer of the membrane but the microporous support it sits on. In practice the scope for using pressure spikes is limited by design of system. Also there are element design issues such as the permeate spacers, product tubes which set mechanical limits for the element. Occasionally, an element which is in use and under-performing will spontaneously increase in flux. This increase is associated with a significant piece of the microporous structure wetting out.

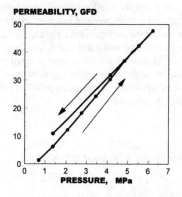

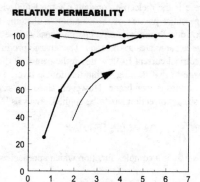

Figure *5.7.3 Shows the variation in permeability of an RO membrane as the pressure is increased in increments from 100 to 900 psi and then dropped back to 200 psi. The challenge solution was water with 500 ppm of common salt. Hysteresis is clearly evident, and is highlighted by plotting the data as permeability against feed pressure (right graph). This highlights that the permeability of the membrane at 100 psi was some 25 % of its maximum value. There is some evidence that the permeability of the membrane is pressure dependent. This is assigned to reversible compression of pores within the ultrafiltration base and hence decreasing the permeability.*

The sensitivity of the pressures to the microporous structure has spurrned a technique for characterising membranes, the bubble point method. The technique consists of wetting the surface of the membrane with a non-reactive low surface tension liquid, and then steadily increasing the pressure. At low pressures there is steady flow of gas due to diffusion. At a pressure called the bubble point the gas flow rate increases more rapidly with pressure. This pressure is the pressure at which gas displaces the fluid and is a measure of the maximum pore size. Further increase in pressure open up more pores. The flow pressure curve can be used to get "pore size" distribution (***porometry***). Some manufacturers use the characteristics of this pore size distribution as a method of determining whether their production process is under control, and thereby maintain consistency between different batches of membranes.

5.7.3   Surface Charge and Zeta Potentials

As discussed in section 5.6 the membrane appears as if it carries a surface charge. The potential field created by this charge can attract or repel charged species in water. The nature of this surface charge has been implicated in the fouling tendency of different membranes and its dependence on pH. Since many biological fragments carry negative charges it is generally believed that having a membrane with a negative charge is preferable.

The surface charge of a membrane can be inferred from measuring the zeta potential,ζ. This is done by measuring the (streaming) potential, $\Delta\phi$, generated when flow is forced between two sheets of membrane separated by a distance $2r$, by a pressure ΔP. If the sheets are sufficiently far apart, then

$$\Delta\phi = \zeta\left[\frac{\varepsilon}{\eta\kappa}\right]\Delta P, \qquad\qquad (5.7.5)$$

where ε is the dielectric constant, η is the liquid viscosity, and κ is the conductivity of the solution. The essential physics of the situation is that the flow induced by the pressure gradient creates a current due to the non zero charge density close to the membrane surface. In an open circuit the charge accumulates at the ends. This charge produces an electrical potential and this in turn induces an electrical current to flow down the centre of the device. At steady state this current is equal and opposite to that flowing within the double layer. The same concepts can be applied to flow through a microporous membrane. However, since the size scale of the double layer is comparable to the pore size a correction has to be applied. Analysis [11] yields the equation

$$\Delta\phi = \zeta \left[\frac{\varepsilon}{\eta\kappa} \right] F(\kappa r)\Delta P \qquad (5.7.6)$$

where $F(x)$ is a complex function which approaches 1 when the gap is large and zero when the gap is small.

Most membrane materials carry a negative charge. This might arise due to charge groups within the material or weakly dissociating groups. For ceramic membranes surface reactions can occur resulting in the material having a net charge. Such interactions are usually sensitive to the pH due to hydrolysis of the various solution species. Most membrane materials carry a negative charge (see figure 5.7-4). Negative charge is often seen as a good thing since many of the naturally occurring foulants carry a negative charge as well as many types of bacteria. The negative charge gives rise to a negative zeta potential. The equilibria are sensitive to pH, and hence so is the zeta potential. As the pH falls the surface charge falls to zero. Typically, the zero point of charge is on the acid side. The surface charge of the membrane however can be changed by adsorption of complex molecules from the solution. Through such adsorption processes the characteristics of the membrane can be radically altered.

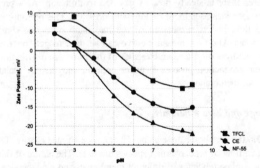

Figure 5.7-4 Zeta potential versus pH (adapted from [16]) for three commercial RO/NF membranes

5.7.4   Adsorption

A number of scientific investigations have looked at the adsorption of macromolecules onto membrane surfaces and their role in membrane fouling. Of particular interest has been the adsorption of proteins and surfactants. Adsorption of proteins usually involves conformational changes which eventually lead to the protein unfolding and irreversibly adsorbing on the surface. The adsorption of these materials can change the surface properties of the membrane. This can be used to advantage in that the membrane can be deliberately pre-treated to change the surface character of the membrane and prevent subsequent fouling. The general view is that adsorption is worst on hydrophobic surfaces. In practice, the problems of adsorption are resolved through selection of membrane materials, cleaning frequency, and operation.

Bacteria readily adsorb on surfaces. If suitable conditions prevail then the surface of the membrane provides a useful breading ground. This is a major concern for cellulosic membranes since the material represents a natural food source. Another problem created by bacteria is that the sugary residues that they excude help bind other particulate materials, creating an increase in both the amount of fouling and the ease to which it can be removed.

5.8   Transport Models

5.8.1   Introduction

At the heart of membranes are the transport processes. These govern the separation and determine their productivity. The theory of irreversible thermodynamics provides the framework for modelling these processes. However, as in traditional thermodynamics one has to provide constitutive relationships which relate to a molecular understanding of what occurs inside the membrane.

Two simple models are presented here, the first is for ultrafiltration of dilute systems, and the second is for reverse osmosis. These models show how the selectivity is goverened by equilibrium factors such as solubility and kinetic factors like diffusivity. Another observation is that the maximum separation is achieved by driving the separation process as hard as possible; a result which is common to all membrane processes. However, as will be seen in chapter 6 other processes (polarisation) come into play, that overide this result, and limit both the productivity and selectivity that can be achieved. At the end some experimental results are shown. These illustrate that the models are insufficient to predict the detail that is frequently encountered, and the importance of obtaining experimental data from field trials.

5.8.2   Filtration through Microporous Structures

In the absence of any concentration gradients the mass flow of water and solute (J_w and J_s respectively) are simply related to the volumetric flow rate, J_v, by

$$J_w = c_w J_v, \tag{5.8.1}$$

$$J_s = c_s J_v, \tag{5.8.2}$$

where c_w and c_s are the concentrations of the water and solute respectively.These equations also apply to flow in a microporous medium. However, the solute concentration in the pores will not equal that in the external environment. The first and most obvious reason for this is a size exclusion (i.e. only particles less than the pore size can pass). Another factor is the so called excluded volume effect[6]. As the size scale becomes smaller molecular effects will dominante. While in general the details are complex it is clear that the ratio of the particle size to pore size will play a significant role.

---

[6]   For rigid particle this simply says that the particle centre cannot be closer to the surface than the particle radius.

At low concentrations the solute will partition itself between the surrounding medium and the pores according to a linear relationship viz.

$$c_s^{(m)} = \beta c_s,$$

(5.8.2)

where β is a partition coefficient for the solute between the external solution and the pores in the medium. Size exclusion predicts that no particle above a certain size can enter a pore. A simple excluded volume calculation predicts that

$$\beta \approx \left(1 - \frac{a}{R}\right)^2, \qquad a < R,$$

(5.8.3)

where a is the particle radius and R is the pore radius. Thus, the partition coefficient is expected to give a much smoother characteristic than that expected on size exclusion alone. This has important implications for polydispersivity of the pores in the membrane. It follows from Hagen-Poiseuille's equation that the larger pores contribute much more to the flow than the smaller ones. Thus, if one can idealise the structure by a distribution of pore sizes an average partition coefficient needs to be used which is given by an expression of the form

$$\overline{\beta} = \frac{\int dR \, R^4 n(R) \, \beta(a/R)}{\int dR \, R^4 n(R)},$$

(5.8.4)

where $n(r)$ is the number density of pores of radius r and the intregation spans all physical pore sizes. It is often thought that engineering a membrane with a monodisperse porous structure will provide a membrane with the best possible selectivity. What this analysis shows is that the benefits from narrowing the pore size distribution is fairly limited due the broadness of the partition coefficients dependence on pore size.

The effect of passing a solution through the medium by pressure will be to create a concentration gradient on account of the partitioning. This concentration gradient will provide an additional force to encourage solute particles to move through the membrane. Thus, the solute flux is made up of a convective term and a diffusive term

$$J_s = \beta c_s J_v - D_s \frac{dc_s^{(m)}}{dz}.$$

(5.8.5)

The concentration difference across the membrane produces an osmotic pressure difference which will reduce the water flow. Hence the water flow equation is of the form

$$J_w = A\left(\frac{dP}{dz} - \frac{d\Pi}{dz}\right).$$

(5.8.6)

Together with boundary conditions these equations can be integrated, and predict that the rejection will satisfy

$$R \equiv 1 - \frac{c_{s,p}}{c_{s,f}} = 1 - \frac{\exp\left(+\frac{J_v}{D}L\right)}{\left[1 - \frac{1}{\beta}\left(1 - \exp\left(+\frac{J_v}{D}L\right)\right)\right]}.$$

(5.8.7)

As expected the equation predicts that in the limit of high volume flow the rejection tends to the ideal value viz.

$$\lim_{J_v \to \infty} R = (1 - \beta) \equiv R_\infty, \qquad\qquad (5.8.8)$$

and at low flow rates the diffusive term drives the system to zero rejection (provided $\beta \neq 0$). As required the equation predicts the rejection is ideal if there is total exclusion. In equation 5.8.7 the feature that governs the rejection characteristics is the ratio of the convective flow to the diffusive flow, which is a Peclet number:

$$Pe \equiv \frac{J_v L}{D} \qquad\qquad (5.8.9)$$

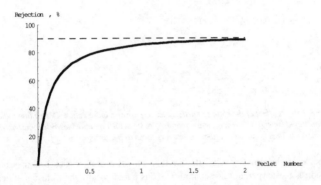

***Figure** 5.8-1 Shows the effect of Peclet number on rejection for a membrane with a limiting rejection of 90%.*

This analysis shows that rejection can be increased to its maximum value by increasing the flux, which in practice means increasing the operating pressure. However, this ignores what happened above the membrane. As will be shown in the next chapter, driving the membrane hard leads to concentration polarisation which reduces the selectivity of the membrane.

Some further insight can be obtained by using the Stokes-Einstein approximation for the diffusion of colloidal particles (equation 5.5.1), and the parallel pore model for the flux (equation 1.4.1) in equation 5.8.9. This gives

$$Pe \equiv \frac{3}{4}\pi \, \varepsilon \, r^2 a \, \frac{\Delta P}{k_B T} \qquad\qquad (5.8.10)$$

This Peclet number is the product of three types of factor. The first is that associated with the porous structure, the second with the particle, and the third with a thermodynamic factor. From the thermodynamic factor it follows that increasing pressure will lead to an increase in rejection. A number of conclusions follow from equation 5.8.10 which have wider generality. Firstly, the Peclet number is independent of membrane thickness, i.e. the diffusional effects are not dependent on film thickness[7]. A consequence of this is that the rejection is independent of thickness. Secondly, dramatic difference is expected for a membrane dealing with particles and pores 0.1 microns in size and these which are 0.001 microns in size. Inserting some typical values[8] into equation 5.8.9 for these two cases gives Peclet numbers of the order of 1300 and 0.001. Thus, as we descend from the top of the colloidal range to the bottom, diffusion processes become more significant and will modify the rejection behaviour predicted on a purely thermodynamic basis. This effect is illustrated in figure 5.8-2

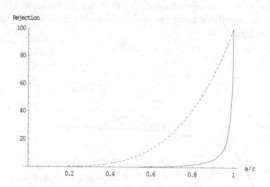

Figure 5.8-2 Shows effect of the ratio of particle size to pore size on rejection. Dotted line indicates limiting separation based on partition coefficient(equation 5.8.3). Solid line indicates rejection based on equation 5.8.7 for pores 0.2 nm in diameter (other parameters taken as per example in text).

Now, as the pore size becomes smaller, the limiting cut-off characteristics becomes broader. The effect of diffusion is to sharpen up the separation. Thus paradoxically the diffusive effects lead to a sharper separation. More complex models which include hydrodynamic effects, wall interactions arising from electrical double layer effects etc., boundary layer effects, etc. have been developed [12][13].

This analysis applies to very dilute systems. As one drives the membrane harder the rejection of solutes modifies the solution above the membrane producing a concentration polarisation. The harder the membrane is driven the more significant is polarisation, and this is the subject of the next chapter.

5.8.3   *Transport Through Dense Films - RO*

As the pores become finer and finer there comes a point when there is no continuous water conduit through the membrane. Instead, we have a picture of water winding its way through a fluctuating

---

[7]   It might be expected that a thicker film would be a better barrier. The argument for this stems from a model of the membrane as a series of layers each with a finite probability of trapping a particle i.e. the membrane acts more like a depth filter. Particles and membrane are seen as rigid objects. On this basis all particles below a certain size would pass, and those above would not. The selectivity that is observed being derived from the polydispersivity of the pores. While this percolation argument is sufficient for large particles, diffusional effects dominate in the colloidal regime. As a result even if the pores were monodisperse they would not have a discontinuous rejection behaviour. This feature becomes more significant as the size of the particles and pores becomes smaller.

[8]   Porosity=0.6 , Pressure difference= 1 bar, Tempertaure =25 C

polymer matrix. This is the domain of reverse osmosis. The concentration of water in the membrane material depends very slightly on the hydrostatic pressure. This difference produces an internal concentration gradient for the water to diffuse across. Similarly the solute molecules adsorb into the membrane and diffuse across. The analysis given here is based on the work of Lonsdale, Merten and Riley [16] and makes the basic assumption that the polymer can be modelled like a fluid in which the water and solute adsorb (solution-diffusion model).

The basic equations are similar to that in ultrafiltration. The water flux and solute flux are respectively given by

$$J_w = A(\Delta P - \Delta \Pi),$$ (5.8.11)

$$J_s = B \Delta c .$$ (5.8.12)

Equation 5.8.11 shows that the water flux increases linearly with feed pressure. In contrast the solute flux tends to a maximum value determined by the maximum concentration difference that can be generated across the membrane (i.e. assume that the permeate contains no sugar). Formally, the equations can be rearranged to give a simple expression for the solute rejection,

$$R \equiv 1 - \frac{c_p}{c_f} = 1 / \left[1 + \frac{P_c}{\Delta P - \Delta \Pi} \right],$$ (5.8.13)

where the characteristic pressure has been introduced:

$$P_c \equiv \frac{c_f B}{A} .$$ (5.8.14)

The characteristic pressure is a measure of the intrinsic selectivity being the ratio of the solute to water permeability. The lower the characteristic pressure the better the rejection, which is consistent with the relative effects of the salt and water permeability.

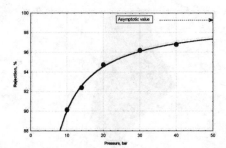

***Figure** 5.8-3 Measured data on an Optimem™ membrane tested with 2000 ppm NaCl solution at 25 C over a range of pressures. The data has been fitted to a modified form of the LMR model which takes account of leakage through micro defects in the coating [17]. In this case 0.7 % of the flow passes through the defects. The calculated characteristic pressure for this data is 0.98 bar.*

If the osmotic pressure is negligible it can be seen that equation 5.8.13 provides a simple scaling formula for the rejection. Thus, once the rejection is known at one pressure it can be used to calculate the rejection at any other pressure. When the osmotic pressure is not negligible and the rejection is not close to the ideal value then account has to be made for the osmotic pressure of the permeate. Manipulation of equation 5.8.13 yields a quadratic equation with the solution

$$R_c = \frac{\left(\Pi_f + \Delta P + P_c\right) - \sqrt{\left(\Pi_f + \Delta P + P_c\right)^2 - 4\Pi_f \Delta P}}{2\Pi_f}. \qquad (5.8.15)$$

The predicted behaviour of rejection and flux is shown in figure 5.8-4, 5.8-5

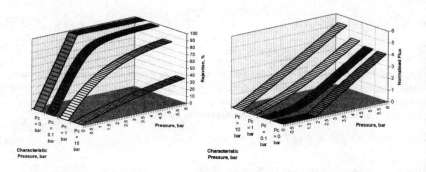

Figure 5.8-4 Plots of rejection and flux versus pressure for membranes with various characteristic pressures. In the flux graph it should be noted that even at pressures below the osmotic pressure of the feed (relative to pure water) there is a flux due to the leakage of solute across the membrane.

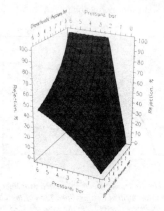

Figure 5.8-5 Variation of rejection with pressure and characteristic pressure for a feed with osmotic pressure of 2 bar relative to pure water, based on the LMR model.

The graphs clearly show that only for an ideal membrane ($P_c = 0$) will there be no flow below the osmotic pressure, but for real membranes there is some flow below it. The model also predicts that for pressures below the osmotic pressure the maximum rejection will be less than that predicted for an ideal membrane. These issues become significant in nanofiltration application where the osmotic pressure of the feed is high relative to the operating pressure.

By applying irreversible thermodynamics to the flows within the membrane and matching those to the external environment relationships between the macroscopic permeabiulity coefficients given for can be related to the microscopic diffusion coefficients, viz.

$$A = D_w \, \tilde{c}_w \, \frac{v_w}{RT} \, \frac{1}{d}, \tag{5.8.16}$$

$$B = \kappa_c \, D_c \, \frac{1}{d}, \tag{5.8.17}$$

where κ_c is the partition coefficient of the solute between the membrane and the water, d is the thickness of the membrane (only the active layer), D_w is the diffusivity of water in the membrane, D_c is the diffusivity of the solute in the membrane, v_w is the partial molar volume of water in the polymer, and $\tilde{c}_w$ is the average concentration of water in the polymer. Thus

$$P_c = \frac{D_c}{D_w} \, \frac{c_f}{\tilde{c}_w} \, \frac{\kappa_c}{v_w} \, \frac{RT}{}, \tag{5.8.18}$$

is a product of the relative diffusivity of the solute to that of water, the relative solubility of the solute and the water, and a thermodynamic factor. The key features predicted by this analysis are that

- *Flux increases uniformly with the net pressure difference across the membrane*
- *Rejection increases with pressure*
- *Rejection depends on concentration only through osmotic pressure effects*
- *There is no interaction between solutes*

The pressure equation 5.8.15 allows a single measurement to predict the rejection behaviour at all pressures, and over limited pressure ranges the equation provides an excellent basis for extrapolation. The concentration predictions of the model are less satisfactory. Interfacial RO membrane shows little concentration dependence over the concentration range 100 to 30,000 ppm, other than that an osmotic pressure effect, and can be modelled reasonablly well with the LMR model. On the other hand cellulosic and other charged membranes show substantial decreases in rejection with concentration over and above that predicted. Deviations are most noticeable when rejections are very much less than ideal which is often the case for NF membranes, and when reverse osmosis is carried out with mixed inorganic solutes. This complexity stems from various effects, bound charges on the polymer, different diffusivities of ions, ion-pairing etc. Figure 5.8-5 shows how complex these effects become in a three component system.

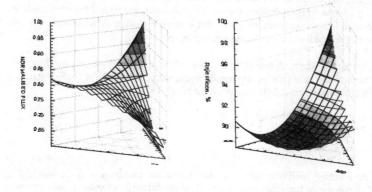

Figure *5.8-5. Graphs of flux, and rejection versus composition (Na, Mg, Ca) for an Optimem-C membrane. The right hand corner is Mg. A notable feature is the sensitivity of the flux and rejection to small addition of hard ions. Another interesting feature is that while calcium depresses the rejection the most, magnesium depresses the flux the most.*

More fundamental formulations have been developed, but their non-linearity and complexity means that computational methods have to be employed. Most leading membrane manufacturers supply this software to assist consultants and equipment manufacturers (Fluid Systems - ROPRO, Dow - ROSA, USF Acumem - RO System Designer). However, because of the assumptions made in obtaining the results the predictions made should be a starting point rather an end point of the design process.

5.8.4   Gas Separation

The final transport model considered is that for gas separation of a two component ideal mixture. The starting point for most models is that the component fluxes through a membrane are proportional to their respective partial pressure differences, viz.,

$$J_j = A_j \left(p_j^{(f)} - p_j^{(p)} \right),$$

(5.8.19)

where A_j is the permeability of component j, p_j is the partial pressure for component j, and the superscripts (f) and (p) indicate whether it refers to the feed or permeate respectively. This equation is based on the solution-diffusion model and Henry's law for partitioning between the gas and the polymer. One further assumption is required to calculate the permeate concentrations. This is that the rate at which components are removed from the permeate side is proportional to their partial pressures. It follows that the selectivity is given by (for $\alpha < 1$)

$$\alpha_{1/2} \equiv \frac{c_1^{(p)}/c_2^{(p)}}{c_1^{(f)}/c_2^{(f)}} = \left(\frac{y_f}{1-y_f}\right)\left(\left(\frac{1-\alpha}{2y_f}\right)\left(\frac{\alpha}{1-\alpha}+y_f+\phi+\sqrt{\left(\frac{\alpha}{1-\alpha}+y_f+\phi\right)^2-\frac{4\phi y_f}{1-\alpha}}\right)-1\right),\qquad (5.8.20)$$

where y_f is the fraction of component 2 in the feed,

$$\phi \equiv \frac{P^{(p)}}{P^{(f)}} \qquad (5.8.21)$$

is the pressure ratio, and

$$\alpha \equiv \frac{A_1}{A_2} \qquad (5.8.22)$$

is known as the "ideal" separation factor. In the limit of zero pressure ratio it follows from equation 5.8.20 that the selectivity equals the ideal separation factor, viz.

$$\lim_{\phi\to0}\alpha_{1/2} = \alpha, \qquad and \qquad \lim_{\phi\to0}y_p = \frac{y_f}{\alpha\left(1-y_f\right)+y_f} \qquad if \quad \alpha < 1. \qquad (5.8.23)$$

For economic reasons the pressure ratio is rarely chosen to be close to zero. The smaller the pressure ratio the larger is the recompression costs (the power requirement varies as the log of the pressure ratio). Consequently, the separation factor used will invariably be less than the ideal separation factor. For example in the case of creating oxygen enriched air the practical pressure ratios used are typically 0.2 to 0.4. As can be seen from figure 5.8-6 a material with a separation factor of 2 barely reaches 30 %.

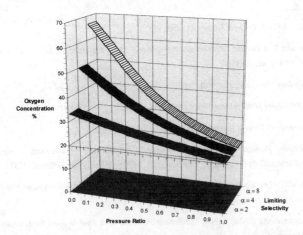

Figure 5.8-6 *Variation of fraction of oxygen in permeate versus pressure ratios for membranes with selectivities of 2,4, and 8 from air (21% oxygen).*

In practice even this is not achieved due to the other losses that occur when one-scales up the problem to work on a module scale (see Hwang and Kammermeyer & [18] for details).

Polydimethylsiloxane which has a separation factor of 2.2 has been used on the small scale for producing enriched air for medical uses. However, for combustion applications a higher separation factor is required, ideally something greater than 8. Unfortunately, any enhanced selectivity is nearly always accompanied by a loss of permeability, and hence a significant increase in cost.

This analysis, like the two previous ones, indicate the full separation capability of a membrane is never achieved, and that the actual separation is dependent on both this limiting value and the pressure ratio at which it is operated.

5.9   References

1   C Gostoli, S Bandini, Di Francesa, and G Zardi *"Analysis of a Reverse Osmosis Process for Concentrating Solutions of High Osmotic Pressure: The Low Retention Method"* Tran I Chem E 74 Part C Jun 1996, 101-109

2   H W Blanch and D S Clark, *"Biochemical Engineering"* Marcel Dekker, 1996

3   J M Coulson and J F Richardson *"Chemical Engineering"* Vol III, Pergamon Press, 1971

4   E L Cussler *"Diffusion, Mass transfer in Fluid Systems"*, Cambridge University Press,1984

5   G Segre and A Silberberg *"Behaviour of macroscopic rigid spheres in Poiseuille flow"* J Fluid Mechanics 14 (1962) 115-131, 136-157

6   P Meares *"Fundamental Mechanisms of Transport of Small Molecules in Solid Polymers"* in 4th BOC Priestely Conference "Membranes in Gas Separation and Enrichment" Sept 1986, Leeds. Publ Royal Society of Chemistry.

7   H B Hopfenberg, and H I Frisch Polymer Letters 7 (1969) 405

8   D W Van Krevelen and P J Hoftzer *"Properties of Polymers: Their Estimation and Correlation with Chemical Structure"* Publ. Elsevier (1976)

9   F Hellfrich *"Ion Exchange"* Publ. McGraw-Hill, 1962

10 S G Schultz *"Basic Principles of Membrane Transport"* Publ. Cambridge University Press 1980

11 R J Hunter *"Zeta Potential in Colloid Science"* Publ. Academic Press, 1981

12 P M Bungay *"Transport Principles - Porous Membranes"* in *"Synthetic Membranes: Science and Engineering and Applications"* ed P M Bungay, H K Lonsdale, M N de Pinho, Publ: D Reidel, 1983

13 Brenner, H, and Gaydos *"The Constrained Brownian Movement of Spherical Particles in Cylindrical Pores of Comparable Radius"* J Colloid and Interface Science 58 (1977) 312-356

14 M Nystrom, M Lindstrom, and E Matthiasson *"Streaming Potential as a Tool in the Characterisation of Ultrafiltration Membranes"* Colloids and Surfaces, 36 (1989) 297-312

15 A E Childress and M Elimelech *"Effects of Natural Organic Matter and Surfactants on the Surface Characteristics of low Pressure Reverse Osmosis and Nanofiltration Membranes"* 717-725 in Proceedings of Membrane Technology, Feb 1997 New Orleans

16 H K Lonsdale, U Merten and R L Riley, *"Transport Properties of Cellulose Acetate Osmotic Membranes"* J Appl Poly Sci 9 (1965) 1341-1362

17 T K Sherwood, P L T Brian, and R E Fisher, *"Desalination by Reverse Osmosis"* In Eng Chem Fund 6 (1967) 2-12

18 S-T Hwang & K Kammermeyer *"Membranes in Separations"* (Techniques of Chemistry; v 7) Publ. John Wiley, 1975

Chapter 6

POLARISATION

Contents

6.1   Introduction

Polarisation is a phenomenon which is of major significance in cross-flow membranes. As the productivity of a membrane increases, an inhomogenous layer develops on the input side of the membrane. Eventually, this polarisation layer becomes rate limiting. Thus, to understand the design and performance of membranes is very much about understanding polarisation.

In batch filtration a cake of rejected material develops on the membrane surface, and cross-flow causes the cake layer to be continuously moved along the membrane device. A steady state is reached, where the deposition rate due to filtration and that which is washed away balances. Under these circumstances the flux of water through the membrane approaches a limit, rather than continuing to fall as in direct filtration. Hence, the process can be used in a continuous method of filtration. The key performance question is how much energy will be gained from the thinner cake layer, and the consequential increase in permeability, versus the energy required to remove the particles with a cross-flow.

As one descends down the particle size range the sharp divide between a deposit and solution disappears and is replaced by a concentration profile. This phenomenon is known as ***concentration polarisation***. A characteristic of treating colloids is that at a sufficiently high concentration the material at the surface can undergo a phase change to form a cohesive solid. The cake that forms is often referred to as ***gel polarisation***, on account of the high water content and mechanical properties of the layer. These phenomena add an additional degree of complexity to the transport models developed in chapter 5, and indeed in many cases dominate the process. Understanding polarisation is therefore a key aspect to understanding why membrane / elements /systems do not reach their full potential. Good element design minimises the worst excesses of polarisation, but they cannot remove it. By operating elements in certain ways performance can be improved, but at some increased operating cost. Polarisation needs to be taken into account for all constituents in the feed. For example, in reverse osmosis polarisation needs to be considered for both the small ionic species and the large colloidal materials that are present in all feeds. This material gives rise to a fouling layer which can have an additional impact on membrane performance. Thus, polarisation can impact on membrane fouling and accelerate the need for cleaning. The problem of fouling lies not only in the operation of the membrane but also in the feed. Even natural waters present a wide range of different potential fouling factors, and the potential foulants from water provide a useful example The management of these is a critical part of process design as we push the technology to its practical limits.

Concentration polarisation forms quickly and can be seen as a reversible process. In contrast, gel polarisation is frequently associated with slow irreversible changes, and this contributes to more permanent fouling.

6.2   Concentration Polarisation

6.2.1   Introduction

The introduction of a cross-flow over a membrane surface removes depositing material from one location and passes it on to another. If this is all that happens then there would not be much benefit, since the accumulation of material would increase rapidly as one progresses down the membrane channel with a consequential reduction in productivity. The resolution of this is that some of the rejected material does not accumulate at the membrane surface, but diffuses back into the bulk flow via Brownian motion, driven by the concentration gradient created between the surface and the bulk flow. The concentration enhancement produced at the membrane surface is called the **polarisation**, M. The effect is called **concentration polarisation** and is an important concept in membrane processes (see figure 6.2-1).

Table 6.2.1 Typical conditions operating in membrane filtration devices operating in spiral, plate and frame, hollow fibre elements

Property	Units	Conditions		
		1	2	3
Technology		RO	UF	MF
Cross-flow velocity, u	mm/s	200	1,000	2,000
Hydraulic Diameter, d_h	mm	1	1	1.5
Flux, J	μm/s	83	220	420
Channel length, L	mm	1,000	1,000	1,000
Axial Reynolds Number, Re		200	1,000	3,000
Wall Reynolds Number,		0.08	0.22	0.42

Because of its significance it has been extensively studied both theoretically and experimentally [1] to [6]. The knowledge gained has meant that the worst excesses are usually avoided by good element/system design, and correct choice of operating conditions. Ultimately though, concentration polarisation limits the operability of membrane processes. These phenomena occur to different extents in different membrane processes, and vary with application and membrane type. Only some of the more salient points and conclusions will be addressed here. For simplicity, unless otherwise stated, the results refer to a plate and frame geometry, i.e. flow passing between two membranes spaced a distance $2h$ apart.

6.2.2 Concentration Polarisation in Laminar Flow

The balance between the convection processes bringing the solution to the membrane surface and the back diffusion processes removing rejected material leads to the concentration polarisation layer. The simple picture of polarisation portrayed in figure 4.2.1 is that adjacent to a membrane surface there is a concentration polarisation layer of thickness, δ_c.

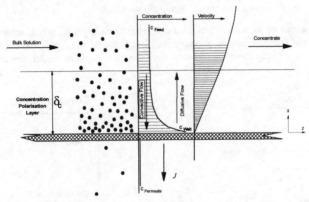

Figure 6.2-1 shows concentration of solute close to membrane surface increases in its proximity and there is a balance between the material convected to the surface and that diffusing away.

Within this layer the balance of a convective flow to the surface and the back diffusion gives an exponential concentration profile

$$c(z) = c_f \exp\left\{\tfrac{J}{D}(\delta_c - x)\right\},$$ (6.2.1)

where x is the distance away from the membrane surface, and J is the flux of water passing through the membrane. The level of concentration enhancement at the membrane surface is known as polarisation, M,

$$M \equiv \tfrac{c_w}{c_f} = \exp\left\{\tfrac{J}{D}\delta_c\right\}.$$ (6.2.2)

This indicates that the polarisation increases with increasing permeate flux rate, and increases with reducing diffusivity. Since larger particles diffuse more slowly than smaller ones polarisation is expected to be more significant in microfiltration processes than in reverse osmosis. In chemical engineering a key concept is that of the mass transfer coefficient. In this problem the "mass transfer" coefficient is defined as the ratio of the diffusivity to the thickness of the boundary layer:

$$k \equiv \tfrac{D}{\delta_c} = J/\ln\left(\tfrac{c_w}{c_s}\right) = J/\ln\left(M\right) \quad \Rightarrow \quad M = \exp\left\{J/k\right\}.$$ (6.2.3)

Thus, in this model the polarisation is the exponential of the ratio of the water flux rate to the mass transfer coefficient. The better the mass transfer coefficient the higher the flux that can be operated at equivalent polarisation. By carrying out experiments in which the polarisation is measured as a function of the flux values the thickness of the polarisation layer can be estimated if the diffusivity is known.

This polarisation layer concept raises a number of questions

- *How thick is the polarisation layer?*
- *How does it vary along the channel?*
- *How does it vary with membrane process?*

The picture conveyed by figure 6.2-1 of a mass transfer boundary layer which separates the bulk fluid from the membrane surface is appealing but is insufficient to answer these questions. The starting point to answering these questions requires the more complex model illustrated in figure 6.2-2.

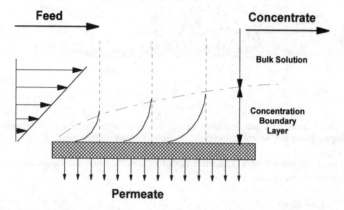

Figure *6.2-2 shows the development of a concentration polarisation layer. The key characteristic is that the concentration profile has a self-similarity varying in magnitude and thickness but maintaining its form.*

As material is conveyed along the surface the accumulated material at the surface must increase, since in addition to the matter being convected to the surface from the bulk there is the material being tangentially pushed along. To accommodate this increase the boundary layer increases in thickness as it evolves along the pipe. Consequently, the magnitude of the polarisation increases as one progresses along the membrane.

In the most general case not only is the mass distribution evolving down the channel, but the velocity distribution is also evolving. Eventually, the velocity profile will approach the familiar parabolic profile (in laminar flow case)[1]. Concentration polarisation has many similarities with the momentum transfer problem. In principle the development of both the momentum transfer and mass transfer boundary layers should be considered together. It can be shown that the relationship between these

---

[1] Since mass is being constantly removed at the walls there is in the more general case a modification to the parabolic profile which has to be taken into account.

two boundary layer phenomena is governed by the Schmidt number, which is the ratio of the kinetic coefficients for momentum and mass transfer

$$Sc \equiv \frac{v}{D}. \qquad (6.2.4)$$

Typical values for Schmidt numbers are given in table 6.2.2, where for the particles the diffusion coefficient, D, has been estimated from the Stokes-Einstein equation. This table indicates that in liquids Schmidt numbers are very large, and hence it is expected that the relevant boundary layer thickness will be quite different in these problems.

Table 6.2.2 shows values of Schmidt number for various materials.

Case	Schmidt Number
Oxygen in air	1
Na ion in water	560
Sucrose in water	2,200
10 nm particle in water	23,300
100 nm particle in water	233,000

In the early stages of development the momentum and mass transfer boundary layers are related by

$$\delta_c \approx \delta_v \, [Sc]^{-1/3}. \qquad (6.2.5)$$

Thus, with the exception of gases, the mass transfer boundary layer is very much thinner than that of the momentum transfer boundary layer. It follows that for liquids the momentum transfer reaches a fully developed range much more quickly than that for mass transfer. More formally the length scale for the velocity profile to become fully developed in a channel is given by

$$L_v \sim \frac{u_o h^2}{30 v} = \frac{1}{120} Re\,h, \qquad (6.2.6)$$

where h is half the channel spacing. Typical values indicate that, in laminar flow, the velocity boundary layer is established within a centimetre or so of the entrance. Thus, the velocity profile is usually well established before mass transfer even begins (most membrane devices have several centimetres of inactive surface associated with glue lines or end pots). When the analysis is carried out for concentration polarisation layer a similar looking formula is obtained.

$$L_c \sim \frac{u_o h^2}{20 D} = \frac{1}{80} Sc\; Re\; h. \qquad (6.2.7)$$

The ratio of these two length scales is approximately the Schmidt number, which indicates for liquids that the distance for the concentration profile to become established will vary from meters to kilometres as one considers ions to particulates. It follows that with the exception of RO devices the concentration profile rarely extends across the full channel width in membrane devices, and only boundary layer considerations need to be analysed. The various stages in the development of the velocity and concentration profile are shown schematically in figure 6.2-3

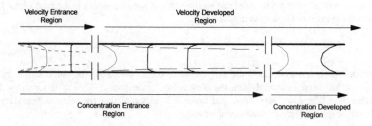

Figure 6.2.3 showing schematic of the various stages in the development of the velocity distribution and concentration distribution along a hollow-fibre membrane.

The equations describing the developing region have been analysed, and these indicate that in the early stages the thickness of the boundary layer is approximated by

$$\delta_c \cong \left[\frac{20 D h z}{\bar{u}} \right]^{\frac{1}{3}},\qquad (6.2.8)$$

where z is the distance along the channel, and u is the average tangential velocity. This equation shows that the thickness of the boundary layer

- *increases with decreasing cross-flow velocity*
- *decreases with decreasing space between membrane surfaces*
- *increases as the third power of the distance along the channel*
- *is smaller for larger molecules*

The latter two points are illustrated in figure 6.2-4.

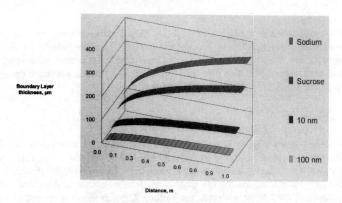

Figure 6.2-4 Shows how the boundary layer thickness increases with length along feed spacer for sodium, sucrose, and particulates. This has been calculated for conditions 1 given in table 6.2.1.

Practically, the more important question is what happens to the polarisation. Detailed analysis of the equations shows that the polarisation is given by a formula

$$M \equiv \frac{c_w}{c_f} \cong 1 + f(\xi), \qquad (6.2.9)$$

where $f(\xi)$ is a montonically increasing function of

$$\xi \equiv \frac{J^3 h \, z}{3 \, u_o D^2} . \qquad (6.2.10)$$

The various parameters in equation 6.2.10 can be grouped in to design factors, operating factors, and physical properties of the rejected species. In particular it shows that for a 10 % increase in flux one would have to make a 30 % increase in cross-flow velocity to maintain the same level of polarisation.

In this asymptotic region no concentration boundary layer exists, and the value calculated for the developing region overestimates the degree of polarisation[2]. In this asymptotic region the key dimensionless group that governs the mass transfer is the Peclet number, the ratio of the convective flux to the diffusive flux in that region.

$$Pe \equiv \frac{J h}{D}, \qquad \left(Pe \cong \frac{\text{Convective Flux}}{\text{Diffusive Flux}} \approx \frac{c_w J}{D \left(c_w - c_f \right) / h} \sim \frac{J h}{D} \right). \qquad (6.2.11)$$

Using the various formula for the boundary layer thickness, polarisation etc one can make predictions for various cases. Such calculations have been done for reverse osmosis, and the results summarised in table 6.2.3 .

Table 6.2.3 Shows the theoretical length scales for and a spiral wound element operating under typical conditions (200 psi) and ignoring influence of spacer.

Case	L_v, μm	L_c, m	$\bar{\delta}_c$, μm	M_{exit}	Pe	M_{lim}
Na	4	0.3	262	1.76	1.16	1.73
Sucrose	4	1.2	166	4.25	4.6	6.67
10 nm particle	4	12.8	76	184	48.	17.29
100 nm particle	4	128.4	35	11600	482.	161.87

The table shows that for small molecules such as sodium, and sucrose the boundary layer is fully developed in standard devices[3], while for larger molecules the boundary layer is in the developing region and is more compressed, with the polarisation factor significantly larger.

While this analysis provides some quantitative predictions it is apparent in the complexity of real devices that an analysis will not be so amenable. Nevertheless, it provides a useful limiting case and through it one can understand some of the design issues. In particular, equation 6.2.11 shows that to maintain the same state of polarisation the cross-flow velocity has to be increased at the third power of the permeate flux, i.e. if we were to double the permeate flux we would need to increase the cross-flow by a factor of eight! Thus, in going from RO to MF it will be inevitable that polarisation

---

[2]   In the asymptotic region the polarisation definition has to be modified since no boundary layer exists. Instead in the aysmptotic region polarisation is defined as the ratio of the wall concentration to the velocity averaged concentration across the channel.

[3]   These calculations ignore the effect of the spacer. Measurements indicate that the spacer produces a substantial benefit which is discussed in section 6.2.4

will become more significant since the diffusivities will decrease by several orders of magnitude. This degree of change cannot be accommodated by an increase in cross-flow velocities. One further caveat to this is that for large particulates shear at the membrane surface enhances the back diffusion.

6.2.3 Sherwood Number

The above analysis shows how different factors play a role in determining mass transfer in membrane devices. In particular, the analysis emphasises that the usefulness of various dimensionless groups can be used to provide simple scaling equations, and the importance of different operating domains. In order to progress mass transfer problems when analytical solutions are not available chemical engineers have used a particular dimensionless group to characterise it called the Sherwood number. This number is defined in terms of the mass transfer coefficient.

$$Sh \equiv \frac{kd_h}{D}. \tag{6.2.12}$$

In the developing region a boundary layer exists and the mass transfer coefficient is defined in terms of the thickness of the boundary layer (see equation 6.2.3). Thus,

$$Sh = \frac{d_h}{\delta_c}, \tag{6.2.13}$$

i.e. in the concentration developing region the Sherwood number is equal to the ratio of the characteristic channel width to the boundary layer thickness. Thus, on entry the Sherwood number is large, and steadily falls to an asymptotic value. In practice mass transfer measurements are made over a fixed length of membrane which means that the measurement relates to the average value. Thus when considering a device it is more useful to consider the average Sherwood number. In the laminar flow case considered in the previous section it can be shown that the Sherwood number can be expressed in terms of a number of other dimensionless groups viz.

$$\overline{Sh} = 1.988 \, [Sc]^{\frac{1}{3}} \, [Re]^{\frac{1}{3}} \, [Ge]^{\frac{1}{3}}. \tag{6.2.14}$$

This equation shows that the Sherwood number is related to the

- *physical properties of the liquid and solute (Sc),*
- *the hydrodynamic characteristics (Re),*
- *the geometric features($Ge \equiv d_h/L$).*

Such a relationship could have been argued on dimensional grounds, and similar relationships are found for other cases, but with different exponents. For problems involving more complex characteristics the measured Sherwood must be related to the various dimensionless groups viz.

$$Sh = \alpha \, [Re]^a [Sc]^b [Ge]^c \ldots \ldots \tag{6.2.15}$$

where additional terms may be added which reflect for instance geometric features (e.g. features of spacers). Once values of the exponents have been obtained from experiment these expressions can be used to extrapolate to other conditions. This is most commonly done by observing how rejection data changes with cross-flow conditions.

6.2.4   Spacers, Turbulence Promotion, and Turbulence

In spirals, and plate and frame devices spacers are used to keep neighbouring membrane surfaces apart. The effect of the spacers is to increase pressure losses, and to provide nooks and crannies for deposits to build up. However, they also have a significant impact on mass transfer. In a study by Schock and Miquel [4] they found that their data on a channel filled with a spacer could be fitted by an expression of the form

$$Sh = 0.0065 * Re^{0.875} * Sc^{0.25}$$

(6.2.17)

Typical values calculated from this expression yield Sherwood numbers five times higher than that for an empty channel i.e. the spacer improves the mass transfer. Two features of Schock and Miquel's equation is that it is independent of length, and the much higher Reynolds number dependence that was derived in the concentration developing regime. The physical picture that is proposed to explain these features is that as the feed passes over each spacer it partly "refreshes" the concentration profile. The practical consequence of the spacer is that for RO spiral wound elements the polarisation is much less than that for an open channel. Typically polarisation is expected to be of the order of only 10-15 %. The turbulent promotional effects of spacers should be recognised as quite distinct for devices which are operated at Reynolds numbers where turbulence is dominant, i.e. as normally operated tubular membranes. The mass transfer benefits that derive from the spacers have to be weighed against an increased pressure drop, and lost surface area.

In tubular membranes the flow conditions are such that the flow is invariably turbulent. The flow in such devices follows a chaotic pattern, and while there is no constancy in the velocity, a time average value exists. The constant fluctuations in velocity mean that diffusivities are enhanced. While close to the membrane surface the velocity falls to zero, turbulence still encroaches into this region. The concept of a concentration polarisation layer disappears and mass transfer is homogeneous along the length of the pipe, i.e. polarisation becomes independent of its progression down the membrane. Various analyses have been carried out using different models of turbulence close to the wall boundary (surface renewal, Prandtl model). The expression which is most widely used is [3]

$$Sh = 0.024 \, [Sc]^{0.25} \, [Re]^{0.875}.$$

(6.2.20)

The importance of polarisation and factors that influence it have been the subject of numerous studies. A whole range of methods have been devised to improve or enhance it. For example, in tubular membranes, which have a low surface area per unit volume, static mixers have been used to deliberately disrupt the boundary layer and thereby improve mixing. The use of different types of turbulence promoters an other methods of modifying the polarisation layer is an active area of research, and new devices are continually being developed.

Equations like the above are often derived by analogy to the heat transfer equations. However, this only holds when the transport processes follow a similar form. It is clear for large colloidal particles (i.e. greater than 100 nm) shear on the particle considerably enhances the transfer back into the bulk (see section 3.5). In these circumstances the Sherwood number is expected to be much greater than that predicted by the above formulae.

6.2.5 Impact on Rejection

In reverse osmosis and nanofiltration concentration polarisation can have a significant impact on the rejection properties. This property is determined by the concentration at the membrane surface, which due to polarisation is higher than that in the bulk. Consequently if the membrane rejection properties are independent of concentration[4] the observed rejection, R_o, will be lower than the measured one R_m, since polarisation has increased the concentration driving force. Now the concentration polarisation is simply related to the ratio of the observed transmittance to the intrinsic transmittance viz.

$$M \equiv \frac{c_w}{c_f} = \frac{c_p/c_f}{c_p/c_w} = \frac{1-R_o}{1-R_m}. \qquad (6.2.21)$$

This relationship can be used to obtain some of the exponents in any empirical dimensionless group correlation. For instance, by measuring the rejection at various feed flow rates one can determine the exponent for cross-flow velocity. Similarly, making measurements at a range of pressures allows one to determine the flux exponent.

One caveat that must be considered in doing this is to note that the theory given in section 6.2.2 and 6.2.3 is related to ideal membranes. Extensions have been developed for non-ideal membranes. One of the simplest approaches is to extend the exponential model described in section 6.2.2 to allow for some solute transmittance. Solving the mass balance equation gives

$$\left(\frac{1-R_o}{R_o}\right) = \left(\frac{1-R_m}{R_m}\right) \exp\left\{\frac{J}{D}\delta_c\right\}. \qquad (6.2.22)$$

Concentration polarisation is a major effect in other membrane processes. Most notably in electrodialysis. In these cases increasing the potential would be expected to increase the current. However, if the membrane is driven too hard then these will be insufficient current carriers in the polarisation layer near the membrane surface. As a result the impedance of the system would rise rapidly.

6.3   Gel Polarisation/Flux Limitation

When some colloidal materials are concentrated they become viscous and gelatinous. The simple convection and back diffusion of matter no longer applies, and further concentration of colloidal material stops. Instead, material accumulates and grows from the membranes into the bulk flow. This fouling layer adds an extra hydraulic resistance to the permeate flow. This process continues until the material arriving, at the fouling layer, through convection is balanced by the back-diffusion into the bulk, i.e. the permeate flux falls owing to the additional resistance of the deposited material. In the concentration layer above the fouling layer the solute distributes itself as before in an exponential manner:

$$c_g = c_f \exp\left\{\frac{J}{D}\delta_c\right\}, \qquad (6.3.1)$$

where c_g is the limiting concentration. Re-arranging shows that the limiting permeate flux is given by

[4]   This is a reasonable approximation for most high rejecting membranes. Some membranes show a decrease in rejection with concentration (eg cellulosic), and is particualrly noticeable for nanofilters.

$$J = \frac{D}{\delta_c} \ln\left(\frac{c_g}{c_f}\right) = k \ln\left(\frac{c_g}{c_f}\right). \tag{6.3.2}$$

The fouling layer is called **gel polarisation**, even though it may not strictly be a gel.

Equation 6.3.2 implies that the permeate flux is pressure independent and is determined by the concentration of the feed, and the factors that determines the thickness of the concentration polarisation layer. This formula implies that increasing the feed concentration will reduce the permeate flux, and increasing the cross-flow velocity will increase the permeate flux. Experimental work on a wide range of proteinacious materials has shown this result (see figure 6.3-1). One of the most important prediction to note of equation 6.3.2 is that at high pressure the **flux is independent of the membrane**, and instead a characteristic of the molecules and hydrodynamic conditions in the feed channel.

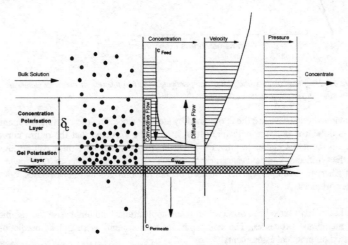

Figure *6.3-1 Schematic showing what happens to the concentration, velocity, and pressure close to the membrane surface when a gel layer forms.*

The model predicts that as the applied pressure increases the productivity will increase until the concentration at the membrane surface reaches the gel concentration. At this point further increases in pressure see no further gain in flux. If the feed concentration is increased then the onset of gelation will be seen at lower pressures.

Blatt [6] provided an early demonstration of how concentration and cross-flow velocity alter the productivity-pressure relationship in the way predicted by these models (see figure 6.3-2). His data showed that the flux changes from a linear dependent function of pressure to a pressure independent function as the pressure increases. The threshold for this transition varying with concentration and cross-flow velocity.

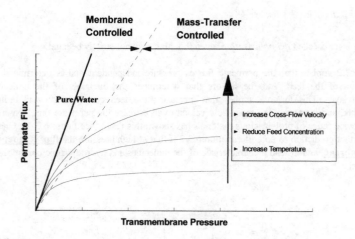

***Figure** 6.3-2   shows a schematic plot of the permeate flux as a function of the trans-membrane pressure, and the effect of various processing factors on flux limitation.*

The curves mapped out above are those obtained by steadily increasing the transmembrane pressure. If once this is done, the pressure is lowered then it might be expected that the return curve would superimpose on the initial curve. In practice, however, the permeate flux rates will be lower. This hysteresis effect reflects kinetic issues associated with consolidated phases (see page 76 and following of Cheryan, 1986,[13]). This situation occurs not only for inorganic phases, but also for oils, and other organics.

Curves like that of Blatt have been observed in many applications, and emphasise the significance of gelation, in membrane processes. The complexity of real systems is such that predictions must largely be based on practical assessments.

Gelation is not the only mechanism that can cause the permeate flux to become pressure independent. As solutes concentrate the solution viscosity can increase and can take on a non-Newtonian character. This means that the shear conditions at the membrane surface are significantly altered and lead to less transfer of material back into the bulk phase.

Another consequence of increasing concentrations at the membrane surface will be an increase in osmotic pressure. While this will obviously be the case for small molecules it can also be significant for much larger molecules on account of the high concentrations being generated (circa 50 %) and a non-linear increase in the osmotic pressure curve with concentration.

The conclusion from this section is that there is a practical limit to the productivity that can be achieved in any given system. The purpose of pilot testing is to identify this limit, since it provides the basis of design and scale-up of such filtration systems.

6.4 Polarisation in a Dialysis Process

An example of a membrane process which is totally dominated by polarisation is the transfer of a volatile from one liquid stream to another across a membrane with an air gap. This is specific type of dialysis that utilises microporous membranes such as polytetrafluoroethylene, or polypropylene. The hydrophobicity of the polymers prevent the liquid bridging the gap, but allows volatiles to diffuse across. The mass transfer rates for the extraction is limited by polarisation in the liquid layers, since gaseous diffusion is a lot faster than aqueous diffusion. The process for a plate and frame device of length L, configured for counter-current, a mass balance on an element gives the pair of equations[5]

$$+d_1 v_1 \frac{\partial c_1}{\partial z} = -k(c_1 - c_2) \quad , \qquad (6.4.1a)$$

$$-d_2 v_2 \frac{\partial c_2}{\partial z} = +k(c_1 - c_2) \quad , \qquad (6.4.1b)$$

where d is the space between membranes, v is the velocity, and the subscripts 1 and 2 refer to the two streams. The mass transfer coefficient k is the made up of three terms, viz.

$$\frac{1}{k} = \frac{1}{k_1(z)} + \frac{1}{k_{mem}} + \frac{1}{k_2(L-z)} \quad . \qquad (6.4.2)$$

Because diffusion of a gas in a gas is very rapid compared to the diffusion of gases in liquid it can be concluded that mass transfer is dominated by the diffusion in the liquid, i.e., the membrane exerts a negligible resistance. The mass transfer coefficient is maximum at the entrances and least at the centre. However, as can be seen from figure 6.4 the variation along the length is small. A useful simplification is to take an average mass transfer coefficient.

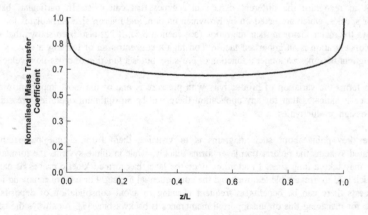

Figure 6.4 Variation of normalised mass transfer coefficient along length of dialysis unit.

---

[5]   From the analysis in section 6.2 it is clear that the mass transfer coefficient will vary with position in the membrane.

With this assumption the solution to these equations is readily obtained and shows that there is a uniform concentration increase in one channel and a uniform decrease in the other channel. The amount of volatile that is recovered (transferred) is given by

$$Y = \left[\frac{1-e^{-\alpha(1-\theta)}}{1-\theta e^{-\alpha(1-\theta)}} \right], \qquad (6.4.3)$$

where

$$\alpha \equiv \frac{k}{d_1 v_1}, \qquad and \qquad \theta \equiv \frac{d_1 v_1}{d_2 v_2}. \qquad (6.4.4)$$

Inverting this equation gives

$$\alpha = -\frac{1}{1-\theta} \ln\left[\frac{1-Y}{1-\theta Y} \right]. \qquad (6.4.5)$$

This equation can be used to calculate the mass transfer coefficient from measured recoveries. By making measurements at various feed flow-rates the dependency on velocity can be quantified. Equation 6.4.3 also provide a useful equation for design purposes.

6.5   Summary

This section has shown how in filtration devices cross-flow has a major impact on what happens above the membrane surface, and how this impacts on both rejection and throughput of the membrane. Although there are a large number of variables their effect can be characterised through a small set of dimensionless groups, which provide methods of extrapolation [11]. However, it is important to recognise the different dynamical regimes that can exist. In particular, between molecular species, which are acted on by Brownian motion, and micron size particulates for which shear plays the major factor in their migration (see section 5.5). It has also been shown that in MF and UF concentration is an important factor. Too high a concentration or too low a cross-velocity and throughput becomes no longer a function of pressure, but is a function of cross-flow velocity.

In design terms the variation of productivity with pressure is one of the most important practical issues since it indicates that for any application there will be an optimum operating pressure and working design productivities.

The other key point about such diagrams is to visualise them from a time perspective. In concentrated systems the polarisation layer forms quickly, while in dilute systems the formation of the layer may take a long time. In addition, once the layer has formed kinetic processes can take place which lead to irreversible features, and the consequential fouling. For ease of removing foulant this suggests there can be benefits in frequent cleaning to avoid consolidation of deposits. One technique for managing this on unsupported membranes is backwashing [7]. As always the benefits of back-washing have to be weighed against reduced efficiencies.

A key conclusion is that membranes have to be operated around a flux basis. If one exceeds this flux one can get a short term benefit but with a long term penalty. For the production of potable water by reverse osmosis, and similar applications, good estimates of the flux limits have been obtained by experience from various feeds, and these can be a guide to new sites. However, experience shows that sites vary. Thus the more critical the performance, and the more original the application, the greater the need for on-site pilot trials [7].

The importance of polarisation is not limited to filtration processes. For example, polarisation is a critical aspect of electrodialytic processes, where polarisation provides a limitation to the flux densities that can be worked with. The full recognition of the significance of polarisation has driven a large amount of research and developments to overcome the limitations it imposes. Developers have experimented with different spacers, including static mixers [8], dynamical effects (i.e. pulsetile flow) [9] [10] , moving the membrane surface, reticulation of membrane surfaces. The key question is not can the limitations in traditional membrane devices be reduced but can this be done cost effectively?

The most important consequence of polarisation and fouling is that membrane designs need to be based around an operating flux, rather than an operating pressure. It is one of the key objectives of pilot testing to determine what this flux is for all the variations in feed that it may encounter.

6.6   References

1   T K Sherwood, P L T Brian, R E Fisher, L Dresner *"Salt Concentration at Phase Boundaries in Desalination by Reverse Osmosis"* I&EC Fundamentals 4(2) 1965,113-118

2   W N Gill , D Zeh, C Tien, *"Boundary Layer Effects in Reverse Osmosis Desalination"* I&EC Fundamentals 5(3), (1966),367-370

3   W N Gill, L J Derzansky, and M R Doshi *"Convective Diffusion in Laminar and Turbulent Hyperfiltration (Reverse Osmosis) System"* in Surface and Colloid Science Vol4 ed E Matijevic, Wiley, New York 1971

4   G Schock, and A Miguel *"Mass Transfer and Pressure Loss in Spiral Wound Modules"* Desalination 64 (1987) 339-352

5   M C Porter *"Concentration Polarisation with Membrane Ultrafiltration"* I &EC Production Reserach and Development 11(3), 1972, 234-248

6   W F Blatt, A Dravid, A S Michaels, and L Nelson *"Solute Polarisation and Cake Formation in membrane Ultrafiltration: Causes, Consequence, and Control Techniques"* in "Membranes Science and Technology" ed J E Flinn, Plenum Press, NY, 1970

7   T J van Gassel and S Ripperger *"Crossflow Microfiltration in the Process Industry"* J Membrane Sci 26 (1985) 373-387

8   J Hiddick, D Kloosterboor and S Brudin *"Evaluation of Static Mixers as Convective Promoters in the Ultrafiltration of Dairy Liquids"* Desalination 35 (1980) 149-167

9   H Bauser, H Chmiel, N Stroth, and E Walitza *"Control of Concentration Polarisation and Fouling of Membranes in Medical, Food, and Biotechnical Applications"* J Membrane Sci 27 (1986) 192-202

10   T J Kennedy, R L Merson, and B J McCoy *"Improving Permeation Flux by Pulsed Reverse Osmosis"* Chem Eng Sci 29 (1974) 1927-1931

11 V Gekasm B Hallstrom *"Mass Transfer in the Membrane Concentration Layer Under Turbulent Conditions. I Critrical Literature Review And Existing Sherwood Correlations to Membrane Operation"* J Membrane Sci 28 (1987) 153-170

12 L C Racz, J Groot Wassink, and R Klaassen *"Mass Transfer, Fluid Flow and Membrane Properties in Flat And Corrugated Plate Hyperfiltration Modules"* Desalination 60 , 1986, 213-222

13 M Cheryan *"Ultrafiltration Handbook"*, Publ. Technomic Publishing Co, 1986

14 J G Wijams, S Nako, and C A Smolders *"Flux Limitation in Ultrafiltration: Osmotic Pressure Model and Gel Layer Model"* 20 (1984) 115-124

15 G M van den Berg and CA Smolders *"Flux Decline in Membrane Processes"* Filtration & Separation March/April 1988

Chapter 7

FOULING & CLEANING

Contents

7.1   Introduction

Fouling presents the most difficult challenge to the design and operation of membrane plants. Fouling results in loss of productivity and loss of quality. The success of membrane technology is critically dependent on how this single issue is dealt with. At one end of the spectrum are the methods of *prevention* while at the other there are the methods of *cure*. While total prevention is desirable it is not always achievable technically. The issue is not to stop fouling, but how to achieve the best compromise between performance loss and additional cost; a balance which varies with technology and application. In microfiltration a productivity loss of 90 % is quite acceptable, while in reverse osmosis a fall of 40 % is unacceptable. The design of systems to minimise fouling and to manage the changes that result is one of the most difficult aspects of membrane system design, largely due to the variable nature, the complexity of the foulants, and their poor characterisation.

The time-scale over which performance is lost due to fouling varies from a fraction of a second to many months. What can be said is that the harder a membrane process is run the faster will be the decline. Not only is the decline faster but the cumulative production is less i.e. fouling is not simply the cumulative deposition of material. The performance of membranes is recovered through physical and chemical cleaning. Again, an economic balance has to be achieved between performance and the costs of cleaning / element replacement. Fouling is more than the accumulation of particulate debris at the membrane surface (see figure 7.1-2). Fouling occurs on and inside membranes, by deposition, reaction, precipitation, and or microbiological processes.

The art of good membrane/element/system design is to liberate the potential of membranes and maximise the cumulative output, by minimising the effect of the various factors that influence the loss making processes. To do this requires an understanding of the processes which limit this realisation. These issues are complex and varied. Some simple models however exist and can provide us with some useful insights which can help guide design and operation.

The starting point for understanding fouling in filtration systems is conventional filtration. At its simplest level the fouling layer is seen as a stead build up of debris on the membrane surface to form a cake. The flux through the membrane and cake is related to the pressure drop by Darcy's equation

$$J \equiv \frac{1}{S}\frac{dV}{dt} = \frac{K}{\eta}\frac{\Delta P}{L} = \frac{1}{R}\frac{\Delta P}{\eta} = \frac{1}{R_m + R_c}\frac{\Delta P}{\eta} \qquad (7.1.1)$$

where K is the permeability (see Annex D), R is the hydraulic resistance of the membrane, and S is the surface area. The latter is made up of the membrane resistance, R_m, and the cake resistance, R_c (these two are additive since flow is in series). Operating at constant pressure, the flow drops with time due to the development of the cake. If the cake resistance increases uniformly with thickness (i.e. no compression) equation 7.1.1 can be solved to give a formula for the volume of material treated as a function of time:

$$V(t) = \frac{SR_m}{r}\left(\frac{2\,t/\tau}{1+\sqrt{1+2\,t/\tau}}\right) \qquad (7.1.2)$$

where r is the cake resistance per unit volume, and

$$\tau \equiv \frac{\eta R_m^2}{r\,\Delta P} \qquad (7.1.3)$$

is a characteristic time-scale. For times small compared to this characteristic time the membrane dominates behaviour, while for long times the cake is the key factor. It follows that using a less permeable membrane or lower operating pressure extends the time over which the membrane is the dominating factor, as expected. It follows from equation 7.1.2 that

$$V(t) \approx \frac{SR_m}{r}(\tfrac{t}{\tau}) = \frac{S}{R_m}\frac{\Delta P}{\eta}\,t \quad , \qquad\qquad t \ll \tau, \text{ membrane dominant} \qquad (7.1.4a)$$

$$V(t) \approx \frac{SR_m}{r}\sqrt{\frac{2t}{\tau}} = S\sqrt{\frac{2\Delta P}{\eta r}t}, \qquad\qquad t \gg \tau, \text{ cake dominant} \qquad (7.1.4b)$$

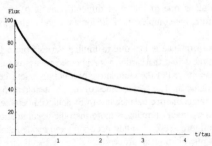

Figure 7.1-1 Shows how the flux (normalised to initial value) varies with time for conventional filtration with a material producing an incompressible cake.

Equation 7.1.2 provides a quantitative prediction of how the instantaneous flux falls with time (see figure 7.1-1). In particular, it predicts that as the cake dominates the flux will become inversely proportional to the square root of the filtration time.

In reality, there are a number of complications when treating colloidal materials. Firstly, the membrane resistance also changes with time (most significantly in an initial period) due to absorption of materials within its interstices. The second issue is that once colloids are concentrated they can undergo irreversible changes to form a cohesive solid, which is more difficult to remove. As this layer consolidates, its filtration characteristics will usually get worse.

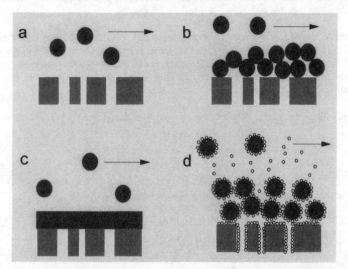

Figure 7.1-2 Schematic diagram showing some simple mechanisms of how particles affect performance, a) ideal - no effect, (b) concentration of material at interface (polarisation), (c) material at interface transforms, and forms a secondary membrane (gel polarisation), (d) adsorptive fouling - small solutes can line surface of pores reducing flow, but can prevent particles from coalescing.

Concentration polarisation forms quickly and is a reversible process. In contrast gel polarisation is associated with slow irreversible changes, and this contributes to more permanent fouling. Evidence of these irreversible changes show up in continuous cross-flow processes as a flux decline, or in batch processes as an increase in the batch time. Fouling with its eventual impact on performance is a feature of nearly every membrane process. While good system design and appropriate choice of membrane can reduce the worst excesses, three other aspects need to be considered

- *Pre-treatment of the feed*
- *Physical cleaning*
- *Chemical cleaning*

7.2   Foulants and Fouling

7.2.1   Introduction

Fouling becomes significant when either productivity or rejection is lost. The most common causes are a combination of deposition and adsorption of material in the membrane interstices, adsorption of material on the surface, and in the most general sense, phase changes in colloidal material immediately above the membrane brought about by the concentrating action of the membrane[1]. In addition to these processes there are secondary processes such as biological growth, which can help bind surface residues together and make them difficult to remove.

In general, the decline of performance can be assigned to physical, chemical, or biological processes (see figure 7.2-1). By assessing the feed water from these three aspects it is possible to pre-treat the feed so that the problems can be minimised. However, a cure for one problem can create a problem of another sort e.g. chlorine can be added to contain biological growth problems but can create a corrosion problem, and might chemically attack the membrane .

Different applications give rise to different foulants. A typical classification is to break down foulants into

- *Gelatinous Foulants - compressible/incompressible*
- *Precipitates*
- *Biological*

Usually one of these factors will dominate the design problem.

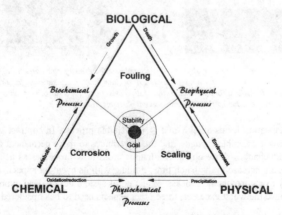

Figure 7.2-1. Diagram of the relationship between fouling materials (adapted from Walton [1])

A variety of formulas have been used to parameterise the loss of flux with time. A widely used form is the power law expression which gives a scaling exponent

---

[1]   Sometimes the effect is not caused by the membrane but by some other process. For example chlorination will precipitate heavy metals such as manganese. If insufficient time is given to precipitate such metals they will disipate themselves on the membrane.

$$J(t) = J(1) \; t^{-n}$$

As already discussed in the introduction, conventional filtration would predict an exponent of $n = \frac{1}{2}$. Measurements on many different systems have given a wide variety of values ranging from 0.001 to 0.5. Under some conditions the flux increases with time. This effect is usually associated with chemical attack, or with gradual wetting of the smaller pores.

7.2.2 Plugging and Coating

Two processes that can significantly affect the permeability of microporous membranes are coating, and plugging. In "coating", materials present in the feed prefer the environment of the surface. This can occur on both the external and internal surface of the membrane. The consequence of this coating is that the pores are reduced in size, and hence the permeability of the membrane is reduced. By considering flow through cylindrical pipes it can be seen that the reduction of flow varies with the fourth power of the pore size. Thus a reduction of 10 % in pore size will produce a 34 % reduction in permeability (see figure 7.2-2).

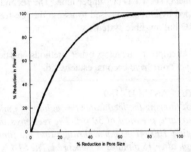

Figure 7.2-2 Effect of reduction in pore size on reduction in permeate flow rate

Another mechanism of loss of permeability is plugging of pores. This is particularly noticeable in "clean" systems where the initial permeability is observed to fall slowly to a lower value. In essence the time to come to a new steady state flux is a reflection of the number of particles in the feed, and the number of pores in the surface of the membrane. The probability that a hole is plugged will depend on the likelihood of a particle lodging in the appropriate pore. Since a perfect match between the particulate and pore is unlikely then the consequence of plugging is a reduced flow-rate. Hence it is expected that the flow-rate follows an expression of the form

$$Q = Q_i \exp\left[-\tfrac{t}{\tau}\right] + Q_f\left(1 - \exp\left[-\tfrac{t}{\tau}\right]\right)$$

where τ is a time constant. This time constant is expected to be inversely proportional to the concentration of particles. Operating the membrane at higher pressures will mean that in any given time the membrane sees more particles and thus faster loss of flux is expected. This characteristic is most acute in asymmetric membranes.

7.2.3 Particulate and Colloidal Fouling

During operation, particulate and colloidal material deposits itself on membrane surfaces. In natural waters this material usually consists of a cocktail of materials such as iron, manganese, silicates, silica, humics, and cellular debris. In process applications, contaminants such as pigments, cellular debris, rust, etc. will find their way to the membrane surface. This unwanted debris occurs in all membrane applications, and if not addressed will produce severe loss in performance. Membrane system designers take various steps to deal with the problem

- *pre-treatment of feeds (Filtration, dispersion)*
- *optimal selection of design parameters (e.g. high cross-flow, and low flux)*
- *chemical cleaning*

The problems are most widely seen in reverse osmosis membranes. In the development of reverse osmosis membranes a method was sought to characterise the fouling potential of water. Only two methods are widely used to assess water suitability for RO applications. The first is turbidity. Most manufacturers of membrane elements set a 1 NTU[2] upper limit. The second measurement is the Silt Density Index. This was developed by DuPont devised to assess the pre-treatment requirements for water to be processed by their RO hollow-fibre systems.

There have been a number of attempts to develop other methods, particularly to incorporate a cross-flow characteristic, but as yet none have become established.

Silt Density Index (SDI) - ASTM D 4189
The SDI test consists of filtering the feed water through a 0.45 micron filter of 1350 mm^2 surface area, at a pressure of 30 psi. The feed pressure is set at 30 psi. The time for the first 100 mL to come through and the time required for 100 mL to come through after 15 minutes is measured. The SDI is
$$SDI_T = 100*(1-t_1/t_2)/T$$
where

t_1 - *is time required to filter the first 500 mL of feed solutioun*
t_2 - *is time required to filter 500 mL of feed solution after*
T   - *time of continuous filtration*

Typically a time of 15 minutes is used, and it follows that the SDI must lie between 0 and 6.6. In practice, the filtration rate can become so slow that it is impossible to measure the flow rate at 15 minutes. In these circumstances (and also if SDI>5) the time period, T, for the test is reduced. Typically a 5 minute period might be used (SDI$_5$ ranges from 0 to 20) and in more extreme cases a 1 minute time period (SDI$_5$ ranges from 0 to 100). The test conditions chosen for the SDI are such that for the applications to which it applies it is the cake resistance rather than the membrane that is dominant.

While there are clear shortcomings with the SDI, its long use by the membrane industry as a preliminary guide for the quality of an RO feed water. While SDI is a useful characterisation of fouling potential it is not sufficient. The general guide is that waters with SDIs less than 3 are considered to be suitable for most RO applications, while values greater than 5 are considered to be unsuitable without pre-treatment.

---

[2]   NTU stands for nephelometric turbidity unit (see appendix B)

By measuring the flow continually over a 15 minute period it is possible to calculate how SDI will vary with the time period. The results of such a study are shown in Figure 7.2-3 for three types of water. As would be expected the SDI value falls with time, as would be predicted by theory given in the introduction. Interestingly the SDI figures obtained from a laboratory high purity water supply system gave SDI figures which were worse than the tap water, indicating that the unit, has so often happens, was not properly cared for. The water with the lowest SDI was a cooling water blowdown which had been treated by cross-flow microfiltration (the SDI of the cooling water blowdown was in excess of 40) and gave an SDI of less than 1. For low SDI measurements special precautions need to avoid contamination issues e.g. from particles attached to the glass walls.

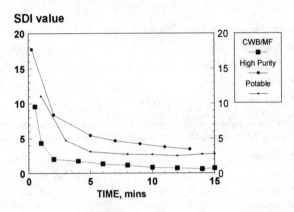

Figure *7.2-3 SDI$_T$ values calculated from data collected over 15 minute period for tap water, high purity water supply, and a cooling water blowdown water that had been treated with cross-flow microfiltration.*

Another useful thing to do when carrying out an SDI test is to examine the accumulated debris on the membrane under a microscope or even better under an SEM. A whole range of particulates can occur which derive from the source water (e.g. silts from surface waters), the treatment (e.g. oxidation producing precipitates, coagulation) and delivery (e.g. Fe corrosion, bacteria). Even "clean" water can readily contain 100,000 particles/L > 0.2 microns, and while in mass terms this does not amount to very much it will become significant if unremoved. Typical municipal waters rarely contain more than a few ppm of solids. While this might not appear to be much when one considers how much fluid each square metre of membrane treats it will in a few days give rise to a layer several microns thick. Fortunately, most of this material does not stay on the membrane surface but gets washed out. However, the problem is made more difficult in elements which contain spacers. In these devices there are dead spots around where the spacers contact the membrane surface. Deposits can accumulate in and grow from these regions. An additional complication occurs if cellular material is allowed to breed in these regions since they can help bind the particles together making them more difficult to remove. More often than not, it is not the particulate matter that creates the problem but the biological and organic material that builds up in these deposits which make it difficult to remove physically and sometimes chemically. Despite several shortcomings SDI remains a popular method for quickly evaluating feeds, and setting pre-treatment requirements.

7.2.4   Precipitates

In reverse osmosis the precipitation of inorganic salts poses an additional problem. This occurs because by removing the water one is concentrating the salts and thereby creating a supersaturated solution, i.e. it is thermodynamically more favourable for the system to precipitate. The most common precipitates to be generated when treating potable waters are

- *calcium carbonate,*
- *calcium sulphate,*
- *silica.*

Other phases that sometimes can form are

- *barium sulphate*
- *strontium sulphate*
- *calcium fluoride*
- *calcium and magnesium silicates*

There is a good theoretical understanding of precipitation, based on water quality and solution thermodyanmics. The key factors which effect the solubility of these compounds is

- *Temperature*
- *pH*
- *Alkalinity*
- *Hardness*
- *Ionic Strength*

However, for accurate predictions an analysis of the key constituents (see Table 7.2.1) is required.

Table *7.2.1 Measurements parameters required for feed water analysis for an RO application*

	Cations	Anions	Other
Feed Temperature (min,max,avg) C	Na^+	Alkalinity- M	Fe
Feed Conductivity uS.cm^{-1}	K^+	SO_4^{2-}	Mn
pH	Ca^+	NO_3^-	Si
TOC (mg/L)	Mg^+	PO_4^{3-}	
Turbidity (NTU)	Sr^+	F^-	Cl_2
	Ba^+		O_2

The solubility of ionic compounds is expressed through the solubility product, K_{sp}. Thermodynamic equilibrium is expressed by relating the solubility product to the activities of the solution species. For example, in the case of barium sulphate equilibrium with a solution occurs when the solubility product equals the product of the activities of barium and sulphate viz.

$$K_{sp} = \{Ba\}\{SO_4\}$$

$$(7.3.1)$$

The solubility product is a thermodynamic property of the solid, and thus is independent of the solution composition. However, the solution compoistion does effect the activity coefficient. For

7 - 8

insoluble materials like barium sulphate a general increase in the solution composition (ionic strength) can increase its solubility substantially. These factors are best calculated with computer programmes.

Table 7.2.2 *Solubility Product Data at 25 C (from Stumm and Morgan[2])*

	Compound	pKsp	ΔH kcal/mol
Celesite	Strontium Sulphate	4.36	-1.71
Gypsum	Calcium Sulphate Dihydrate	4.58	-0.109
Calcite	Calcium Carbonate	8.48	-2.297
Barite	Barium Sulphate	9.97	6.35
Fluorite	Calcium Difluoride	10.6	4.69

Concentration polarisation exacerbates any problem associated with precipitation, since the extra enhancement close to the membrane surface increases the potential for precipitation. Typically, polarisation is expected to produce a 15 to 20 % concentration increase at the surface than at the bulk in the worst spots in the system. Thus, precipitation is expected to occur at the surface in the last element in the system, and propagate back down the system from there. While for particulate the reverse is expected. In the initial phase, growth is limited by nucleation and available crystal surface area. Once established though, the system becomes limited by supply. In material terms it is the difference between the actual concentration and the saturated value that provides a measure of the amount of material that will be deposited. In this respect more soluble compounds such as calcium carbonate can provide a more sudden transformation from unfouled to fouled.

7.3   Fouling Remedies

7.3.1 Introduction

The successful application of membranes requires good management of fouling. Indeed one of the main goals of pilot work is to identify and quantify fouling issues, and evaluate various approaches that can be used to minimise it. Fouling is managed at the process definition stage by

- *design of pre-treatment*
- *membrane selection*
- *system design*
- *specification of operational conditions*
- *defining cleaning procedures*

Good system design minimises fouling problems but rarely eliminates it. The first line of defence is the design of pre-treatment. Invariably a balance between the additional capital cost of pre-treatment versus operating costs of cleaning (lost time, chemical costs, disposal charges). For an established plant the options are less and concern physical and chemical procedures (see table 7.3.1). Physical methods provide the main line of defence. Chemical methods are usually the last line of defence, and carry the additional burden of supply and disposal of chemicals. Nevertheless if chemical cleaning is required then it is important to design the pumps, storage vessels, valves etc. into the process

Problems on fouling can be addressed at three levels:-

- *design of pre-treatment*
- *design of membrane system*
- *operationally*

In order to design a suitable pre-treatment for a filtration processes an assessment of the potential foulants is required. In the case of reverse osmosis a water analysis can provide some indication of the potential solids that might precipitate. Most manufacturers provide computer programs which can calculate the extent of the problem. These programmes are needed so that all the various factors that play a role can be taken into account, e.g. ionic strength, pH, composition changes. These programmes though, are restricted to the common precipitates. Assessing colloidal fouling is far more difficult. In the case of reverse osmosis the SDI method is widely used guide to whether additional pre-treatment is required. In ultrafiltration and microfiltration the data largely comes from pilot trialling.

Table 7.3.1 Summary of methods used to deal with fouling.

METHODS OF REDUCING FOULING		
	Physical	**Chemical**
Pretreatment	• *Prefiltration* • *Precipitation* • *Coagulation* • *Flocculation* • *Carbon sorption*	• *Ion-exchange* • *Dispersant* • *Disinfectants eg chlorine* • *Antiscalants* • *pH adjustment*
Design	• *Flow regime (laminar/turbulent)* • *Element Design* • *Inserts, roughness, reticulation* • *Pulsetile* • *Moving surfaces* • *Combined field*	• *Membrane materials* • *Surface modification*
Operation	• *Limit production rate* • *Maintain high cross-flow* • *Periodic flushing* • *Mechanically clean*	• *Cleaning frequency* • *Cleaning chemicals*

Through theoretical and practical studies element manufacturers and systems manufacturers have derived the optimum balance between hydraulic losses getting the water to and from the membrane, and fouling. In the case of reverse osmosis these design parameters are built into the computer models.

Now the analysis given in section 2 shows that for both concentration polarisation and gel polarisation the key factor is the ratio of the flux through the membrane to the mass transfer coefficient. This formula can be examined in terms of design factors for the simple case of the developing regime in laminar flow and in the absence of a spacer. This analysis shows that

$$J/k \;\propto\; [d_h \, V]^{1/3}$$

where V is the enclosed volume. Thus polarisation becomes smaller if the enclosed volume can be reduced and or the separation distance can be reduced. Hence the interest in making small compact devices. This benefit though has to be equated against any increase in hydraulic losses incurred by pushing the liquid through a reduced volume.

Similarly, one can look at the problem in terms of operational parameters the flux and cross-flow rate;

$$J/k \;\propto\; J/Q_F^{1/3}$$

Thus, if a membrane is driven harder by increasing the pressure then rejection will fall. The equation shows that increasing cross-flow can reduce the effects of polarisation on rejection. When analysed from the total element there is clearly an optimum since increasing cross-flow means increasing pressure drop and thus the back end of the element will operate at lower pressure which will produce a rejection and flux penalty.

The formula can also be used to examine in terms of the application parameters:

$$J/k \propto D^{-2/3} \, v^{-1/3}$$

Thus polarisation is larger for more slowly diffusing materials (larger) and for more viscous materials as expected. A consequence of the fact that viscosity decreases with temperature is that increasing the temperature reduces polarisation.

From the analysis given earlier on polarisation it is apparent that for processes like reverse osmosis the operational conditions will be set to minimise concentration polarisation factors, while pre-treatment will be designed to remove the bulk of colloidal materials which could lead to a gel layer. Since pre-treatment is not perfect, operational procedures can be used to remove these particulates before they make their presence felt. This includes flushing, rinsing etc. In the case of ultrafiltration and microfiltration the issue in many process applications is gel polarisation

One more insidious fouling problem derives from life forms. The major issue is bacteria. Clearly, the growth is controlled by the amount of nutrients that are convected along with the water. One of the problems is not the bacteria, but the sugary sticky molecules they exude.

7.3.2 Pre-treatment Processes

The better the pretreatment, the better the operation of the membrane process. For the high purity waters that are required for the electronics industry there can be up to 20 different process stages of which membranes might provide one or more steps. Perhaps, more typically, there maybe half-dozen stages in a typical large scale process application (see figure 7.3-1). As ever there are a variety of different ways to achieve the effect

- *Solids removal*
- *Hardness reduction*
- *Chlorine removal*

- *pH adjustment*
- *Pathogen removal*
- *Pyrogen removal*
- *Organics removal*

In addition to these removal processes, there are processes which add chemicals in order to inhibit precipitation, bacterial growth, etc. For example, the use of polyacrylics to sequestering calcium and hence, to extend the recoveries to which RO can be used. Invariably, designs have to be customised to meet the specific users situation eg the cost of discharging waste streams, the feed water chemistry.

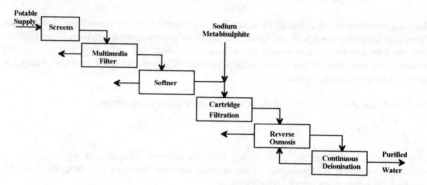

Figure *7.3-1 Schematic of operations used to create pharamaceutical grade water from a potable water supply.*

Filtration

Invariably, before any membrane plant there is some sort of pre-filtration or screen to remove particulates that would block up the narrow spaces between membranes. The solution depends on size. On the small scale, extensive use is made of cartridge filters. Frequently, there is a staged reduction in size (e.g. 5 microns followed by a 1 micron device). For larger systems multimedia filters are the primary methods of removing particulates. For microfiltration systems 50 micron screens are all that is usually required. The recent developments of large low cost microfiltration provide an opportunity to use the technology as a pre-treatment to reverse osmosis. While more expensive than the multi-media filters the quality of the feed is substantial.

Coagulation

For large scale applications colloidal waters can be treated with alum or ferric salts. This increases the size of colloidal matter, which is then removed in a clarifier or depth filters of some type.

Antiscalant

In reverse osmosis salts of hard ions (e.g. calcium carbonate, calcium sulphate, barium sulphate) can readily precipitate. One method is to add a chemical which sequesters the hard ions. One of the early chemicals used for this was hexametaphosphate. This compound binds hard ions. (e.g. Ca, Sr, Ba) and heavy metal ions, (e.g. Fe, Mn) and makes them unavailable for precipitation. This group of compounds has now largely been replaced by polyacrylic type molecules (e.g. Flocon) which have better performance.

Recently, companies have marketed products to combat silica precipitation. These usually involve chemicals designed to either inhibit nucleation, or growth, and frequently contain a dispersant to encourage separation.

Softener

In reverse osmosis the presence of hardness (i.e. calcium, barium, strontrium, magnesium) can lead to precipitation within the elements as the salts are concentrated. One solution is to use a water softener. This is certainly quite a convenient method on small to medium size operations. The major problem is to insure that the system operates well and avoids breakthrough on the ion-exchange beds. The important feature here is that with time the capacity of the beds fall Another problem is that if a municipal supply is being used, then water quality can change dramatically as the municipality switches sources, and this can lead to vastly different loads. Thus the performance of the beds should be regularly checked.

Lime softening is popular in the water industry. This involves adding calcium hydroxide which leads to the precipitation of calcium carbonate, which is then removed from the water. A small amount of acid is then required to remove the threat of calcium carbonate precipitation within the reverse osmosis unit.

pH adjustment

pH adjustment is done either because of the pH stability of the membrane, (e.g. cellulosic membranes can operate between pH 4.5 and 7.5), or to prevent precipitation of calcium carbonate. The potentiality to precipitate is determined by calculating the Langelier index (essentially the supersaturation) from the water analysis. Since carbonic acid is a weak acid, calcium carbonate will increase its solubility with increasing acidity. Thus, the addition of sulphuric acid provides a simple and cost effective method of dealing with the precipitation of calcium carbonate, particularly on large scale operations. For some waters, however, sulphuric acid addition can increase the risk of calcium sulphate precipitation. In these cases hydrochloric acid can be considered. If silica is present and the pH is low and there is no calcium carbonate problem (i.e. saturation pH is very high) then addition of alkali can improve its solubility.

Carbon Sorption

Activated carbon is a popular method of getting rid of organic carbon prior to small devices and systems. For membranes which are sensitive to chlorine (e.g. FT-30) the carbon also serves to remove chlorine.

Precipitation

A common problem in a number of waters is manganese and other heavy metal ions. One approach to this is oxidation, often with air, which leads to precipitation. These precipitates can then be harvested by growing on a suitable substrate. Another approach is to use chlorine, and then remove it later in the process.

Sanitisation

The addition of low-levels of chlorine in the form of hypochlorous acid is widely used as a disinfectant to inhibit the development of bacteria on the membrane surface. The extent to which it can be used is dependent on the compatibility of the membrane to chlorine, operating conditions, and warranties. One alternative to continuous chlorine dosing is intermittent chlorine dosing. In its simplest form chlorine is dosed continuously and sodium metabisulphite is dosed periodically downstream in excess of that required for chlorine.

UV

In the last 10 years ultra-voilet radiation has been seen as a convenient method of destroying bacteria. To use it effectively requires feeds with low solids. Unlike chlorine there is no residual disinfection, and colonies of bacteria can often be found post UV treatment feeding on the bacterial residues created by the UV radiation. However, it is seen as a way of killing pathogens without the need for chemicals. Hence the interest in its use for small domestic systems, and for tertiary off-take discharges from waste water treatment works.

7.3.3 Design

For many applications the manufacturers of membrane elements have optimised their designs to perform well for targeted applications. The key decision for the user or system designer is not the details of the design but what element type to use.Once decided most manufacturers provide detailed guidelines to equipment manufactures as to how these should best be configured. In the case of RO most manufacturers supply design programmes to assist system designers. For ultrafiltration applications the problem is more involved due to fouling.

The most widely used devices are those that simply pass the feed over the membranes surface. In the case of tubular membranes the velocities and size scale is such that the flow can be turbulent. In turbulent flows the scouring action from the turbulent eddies provides additional scouring, which can be important to mitigate problems with high fouling feeds.

In addition to the conventional approaches a number of devices have been developed which exploit other principles, to enhance the mass transfer coefficient. The question with these technologies is not whether they will work, but will the processing benefit be sufficient to pay for the extra cost.

Moving Surfaces

As well as moving the fluid a number of inventors have considered moving the membrane surface tangentially to the direction of flow. One of the simplest ways of doing this is to combine membranes with centrifugal technology. Such an approach has been developed by Alfa-Laval.

Another approach is the use of counter rotating cylinders. When fluid is passed between the two cylinders the rotation of the inner surface which contains a membrane sets up a stable set of vortices enhancing the shear at the membrane surface (see figure 7.3-2).

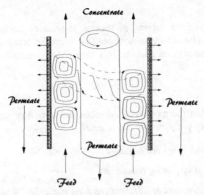

Figure 7.3-2 *Schematic of the "biodruckfilter" by Sulzer Corp.*

A recent developments is the use of torsional oscillations by New Logic. The basic idea is the introduction of torsional oscillations to tubular membranes. In a similar vein Pall have designed a flat sheet stack of stainless steel sheets on which is mounted PTFE membrane. Torsional vibration occur around the axis of the stack (Pallsep VMF). At the present time these systems seem geared to small scale applications in which the added value in achieving higher solids concentration out-weigh the higher capital cost.

Researchers report that the use of moving surfaces allows substantially higher fluxes to be achieved, and applications can achive a higher level of dewatering. These benefits have to be weighed against the higher capital costs of such devices.

Pulsetile

A number of researchers have looked at the effect of applying a periodic pressure fluctuation in addition to the fixed one. The response to the fluctuations is a velocity profile that is much flatter than the familiar parabolic profile of Poiseuille flow. These fluctuations in velocity enhance the back-diffusion process, and thereby allow higher flux conditions to be used. The penalty of course is the extra energy required to transport the fluid through the device.

Another adaptation of the technique is to use a reticulated membrane surface(e.g. dimpled surface). As pulses pass over the surfaces they create vortices which in turn help to scour the membrane surface, and disturb sediment that has built up in quiescent parts of the device.

Reticulation

During the 70's Prof Bellhouse patented a novel way of exploiting pulsetile flow to enhance mass transfer in the oxygenation and dialysis of blood. One of the features of the device was that the membrane was reticulated. As the flow fluctuates, vortices generate in the membrane dimples, enhancing the mass transfer [3].

Combined Field

An approach that has shown various degrees of success in research laboratories is to combine electric with pressure fields. Electrodes are so arranged that when the field is turned on they attract solids away from the membrane surface into the bulk flow. Most commonly the electric field is pulsed.

Summary

The effects of fouling depend on the membrane materials and how they are packaged. For new applications it is essential to carry out pilot work to establish

- *Decide on the membrane area required*
- *Define the range of operating conditions*
- *Establish the effectiveness of different cleaning methods*

7.3.4 Operational

All manufacturers provide guidance as to the operating conditions in which their products can be used (see table 7.3-2). These guidelines come from past experience of how different water types give rise to different fouling problems, and what tolerance can be set for fouling before cleaning is required. Manufacturers usually recommend a flux limit for a particualr application, and as the membrane fouls one can increase the pressure to maintain constant production output. This can continue until a pressure limit is reached.

Table 7.3.2 *Operating limits recommended by Fluid Systems for design of RO systems*

Feed Water	Flux Limit		Fouling Allowance
	GFD[1]	L/m²/hr	
RO permeate	51	87	0
Well water - deep	44	75	15
Well water - shallow	38	65	20
Surface water - lake	35	60	20
Surface water - river	31	53	25
Sea water - beachwell	35	60	15
Wastewater	26	44	30
Sea water - surface	27	46	25

[1]   GFD = US Gallons.ft⁻².day⁻¹

Implicit in these operating conditions is an assumption of the pre-treatment that has been applied. By improving the pre-treatment, e.g. by using a microfiltration pre-treatment instead of a multi-media filter, it is possible to operate at higher production levels.

Cleaning is an intrinsic part of the definition of a membrane process because of the inherent problem of fouling that continually challenges them. Left unchecked, fouling will lead to loss in process efficiency and effectiveness which if left untreated will lead to an irreparable loss in performance. Cleaning can allow one to maintain a higher productivity (see figure 7.3-4). However, frequent cleaning brings a cost penalty of lost operating time, and chemical costs; work time will also make the process unattractive. Thus, a balance has to be struck between cleaning procedures, loss of performance, operating conditions, and system design. This balance varies with application and customer requirements. For example, in small applications, which need to be operated with a low level of technical supervision will be designed with good pre-treatment, very safe operating conditions (e.g. low pressure), and low recoveries. Such an approach brings a large cost penalty which, if applied to many larger applications, could make the process unattractive. Chemical remedies might have a lower capital cost, but this has to be weighed against a higher operating cost. In practice, the process designer must look to design around the minimal performance anticipated from the membrane in use.

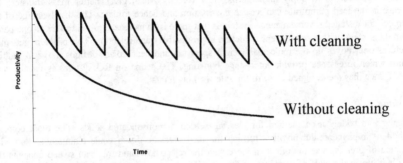

Figure 7.3-3 Effect of cleaning on productivity.

The first consideration in the design process should be an assessment of the potential foulants. Data for this comes from

- *Past experience on similar applications*
- *Analysis of feed*
- *Pilot trials*

From such an assessment it is possible to provide a process to minimise the fouling problem. The options open are

- *selection of operating conditions*
- *pre-treatment*
- *cleaning procedures and frequency.*

There is no universal cleaning procedure since options will vary with application, membrane choice, and customer requirements (e.g. disposal costs, discharge consents). An essential element in a new application is both an analysis of the water, and an audit of the chemicals. This list should include not only those that are used on a routine basis, but those that arise irregularly through cleaning equipment etc. There is also a need for a general appreciation that membrane processes can be sensitive to the unconsidered use of a chemical introduced upstream for cleaning equipment etc.

There are a variety of options available to clean membranes. In practice a combination of these will be used which depends on system design and costs.

Operating Conditions

One of the key decisions in designing a plant will be to select the design flux. This is invariably based on past experience and or pilot trial results. Since no two feeds are ever the same there is always some element of uncertainty. If too ambitious a flux is set, then the point at which cleaning is required will soon be reached. If fouling does occur too rapidly one option would be to down-rate the plant productivity. Alternatively, one could try to improve the performance of the pre-treatment and or the cleaning process.

Another approach is to apply higher cross-flow velocity which gives better mass transfer. The increase in the feed pump rate can reduce polarisation and hence fouling. The disadvantage of this approach for a staged reverse osmosis system is that there is an increased pressure loss as one passes down the system, with a consequent reduction in productivity and rejection. This problem gets more severe as one moves up the molecular weight spectrum and hence plant designs move to multiple pumps which in essence provide inter-stage boosting. The scope on RO systems is usually fairly limited since they are designed close to the safe working limits

Flushing

Flushing is a widely used method for clearing settled or consolidated solids. The most common method of implementation involves turning the feed pressure down while maintaining flow. The dropping of pressure has the effect of increasing the reject flow, and thus can sweep deposits that have built up in the ends of elements. With centrifugal pumps, lowering the pressure results in an increased flow-rate. However, caution must be exercised to avoid applying a flow-rate that exceeds the manufacturers limits. This is particularly important for spiral elements where a consequence of exceeding the maximum flow-rate can be to burst the wrapping or telescope the element. Manual flushing can be used on those systems which can be disassembled, but the cost both in time and manpower is usually too costly to carry out on a routine basis except on small volumes, or research applications.

Backflushing

Backflushing is another technique that is quite widely used where consolidated solids have formed. This technique is limited to unsupported membranes like hollow-fibres.

For hollow-fibres, in which fouling has created a greater pressure loss down the fibre relative to that through the membrane wall, a variant of the backflushing method can be used. In this case water can flow through the membrane wall and then along the outside of the fibre and back through the membrane.

Backwashing is a critical part of the large membrane systems now being put in for surface water treatment. Memcor have successfully used air in their microfiltration technology, while USF use water in their microfiltration product. Typically, these systems are backwashed every 15 minutes. The water used for the backwashing is product water and consumes up to 10 % of production. Usually, the backwash water is further treated so that water recovery figures are as high as 99 %. Another variant reported is to use a 2 phase mixture like water and carbon dioxide.

For supported membranes the back-flow can lead to delamination of the membrane from its backing cloth and this, ultimately, will lead to the membrane surface becoming damaged. Also spacers can cut into the membrane surface due to the spacers cutting into the membrane.

Mechanical Scouring

If the device is amenable to it then mechanical scouring of the surface can be carried out, but care has to be taken not to damage the membrane surface. In the case of tubular membranes PCI have developed an automatic method which involves passing a soft rubber ball along the membrane tube at periodic intervals. As it moves it disturbs deposits which get discharged with the reject.

Sterilisation/Sanitisation

Sanitisation is a process that reduces the number of microbial contaminants to safe levels as judged by public health standards. Sterilisation is a physical or chemical cleaning process that destroys all forms of life, and can be regarded as an extreme form of sanitisation. The most common methods for achieving this are heat, and or an oxidant. This is one of the areas that ceramic membranes excel over polymeric membranes. In addition to killing the micro-organisms it is essential to remove the residues that can accumulate in membrane systems since they provide breeding areas for bacteria. Thus, applications in the food and pharmaceutical industries usually employ a combination of sanitisation and rigorous physical cleaning. The cleaing requirements in these industries as favoured the simpler designs of elements with little or no dead space. The chemical sanitsers that have been used

- *Sodium hypochlorite*   *<50 mg/L*
- *Hydrogen peroxide*   *<200 mg/L*
- *Ethylene oxide*   *Used in haemodialysis*
- *Formaldehyde*   *<5000 mg/L*
- *Gluteraldhyde*   *<5000 mg/L   Recent replacement for formaldehyde*

Chemical Cleaning

While physical methods remedy the loss of performance they usually do not give a full recovery. Also the speed at which fouling occurs usually becomes quicker after each successive clean. Ultimately the only remedy left is chemical clean. The use of chemicals is less desirable but in some areas like disinfection they are seen as providing the guarantor on other factors like cleaninless. To meet these needs a number of manufacturers (see table 7.3.3) have developed special formulations for the membrane industry. Some membrane manufacturers provide their own formulations, but more frequently these are simply rebadged chemicals from the main suppliers.

Table 7.3.3 Suppliers of chemicals for pretreatment or cleaning of membrane elements

• FMC	• Argo Scieintific	• Pacific Aquatech	• Grace-Dearborn-Betz
• Houseman	• Heinkel	• King Lee	• BF Goodrich

A number of generic cleaners also exist (see table 7.3.4). These chemicals can sometimes be used together or sequentially:

- *acids/bases*
- *oxidants*
- *enzymes*

- *chelatants/sequesterants*
- *detergents*

Pilot trials are an essential element for new applications. Indeed, along with identifying appropriate operating conditions to avoid significant fouling, their purpose is to identify the key fouling processes, and appropriate treatments. Pilot trials however do not mimic all the conditions of a full scale plant, and thus there is still a level of uncertainty which must be guided by experience. Prior to pilot trials it is sometimes useful to carry out simple laboratory tests. These can provide an immediate indication of a chemical compatibility issue. However, since the treated volumes are usually only small the effect of a low level contaminant might not show up in such tests, thus, wherever possible, pilot testing is ultimately required.

Table 7.3.4 *A selection of cleaners and sanitisers for membranes. It is essential to consult with manufacturers as to the compatability of products with membrane and other components before use*

Foulants	Cleaner/Sanitisers	
	Generic	**Manufactured**
Mineral scale (carbonate, & sulphate)	• *HCl, pH3* • *2% citric acid*	• *Floclean[1] MC3*
Iron and manganese	• *1 % sulphuric acid* • *2 % sodium metabisulphite* • *0.5 % phosphoric acid* • *0.2 % sulphamic acid* • *0.5 % sodium metaphosphates*	
Silica	• *0.1 % NaOH, 0.1 % Na EDTA, pH 12, 30 C*	
Organics, Silts, Biological materials	• *1% orthophosphate, 0.1 % EDTA, 0.1 % SDS , pH9* • *1 % NaOH*	• *Floclean MC11*
Polymers, latex products		• *MICRO[2] cleaner*
Fats, Oils,and Greases		• *1 % Teepol[3] GD53* • *1 % ULTRA-SWIFT[4]*
Microbial Growth	• *0.1 % sodium metabisulphite* • *200 ppm NaOCl, pH 8-10* • *200 ppm peracetic acid, pH7*	

[1]   Floclean is a trademark of FMC (UK) Ltd

[2]   MICRO is a trademark of International Products Corp., NJ, USA

[3]   Teepol is a trademark of Shell Chemicals

[4]   ULTRA-SWIFT is a trademark of Hygene Laboratories Ltd. UK

Summary

On new applications the user needs to carry-out pilot trials to maximise the overall performance from a membrane system. The aim of these tests is to

- *Assess which membrane type and packaging is appropriate.*
- *Define any pre- or post-treatments requirements.*
- *Decide on the methods of cleaning to be employed and the frequency of use.*
- *Decide which cleaning chemicals are to be used.*
- *Define the cleaning protocol(s) to be used.*

7.4   Membrane Failure Mechanisms

Ocaasionally membrane processes fail. The causes of failure lie with the membrane materials, and the application.

Thermal Failures

All materials have a temperature above which mechanical failure develops. The temperature limits of cellulosic membranes is typically set at 35 C. Interfacial membranes have a higher temperature limit. Typically, this is quoted as 45 C. This limit, however, stems not from the membrane, but from the various materials that make up the membrane element, which highlights the fact that the temperature limit is set by the weakest link. By carefully selecting materials Desalination Systems have been able to make RO membrane elements that will operate in water at 80 C (Duratherm range). This is most vividly illustrated for ceramic membranes. The membrane materials can withstand temperatures of several hundred degrees, but the glues and seals usually employed result in much lower operating limits.

Biological Failure

Membrane materials derived from a biological sources are vunerable to depolymersisation by the enzymes generated from bacteria. Cellulosics are particularly vunerable to this sort of degradation. The extent of the problem depends on the feed. If the feed is a rich source of nutrients frequent sanitisation will be required. This is usually done with sodium hypochlorite either continuously or periodically dosing it at low levels.

Chemical Degradation and Failure

The two most important chemical factors that create failure are attacks by acid/base, and oxidation/reduction. Cellulosic membranes which are created by esterifying the hydroxyl groups on cellulose have a limited pH stability range of 3 - 8. Outside this range they will rapidly hydrolyse back to their native form and with it loose their properties. The pH stability of interfacials is vastly superior to cellulosics. The upper pH limit of 12 for interfacials derives not from the active layer of the membrane but from the backing cloth, which is made of polyester.

Glater and Zachariah [4] have investigated the effect of halogens on the DuPont polyamide membrane. They concluded that halogen attack occurred on the aromatic rings by electrophilic substitution. The effect of this is to change the nature of the hydrogen bonding forces within the polymer.

The oxidative resistance of polymers varies widely. The resistance of a membrane is frequently quoted as a total exposure tolerance which is the integrated amount of chlorine seen by the membrane

7 - 21

Chlorine Exposure (ppm-hrs) = [Exposure Time (hrs)] * [Chlorine Conc (ppm)]

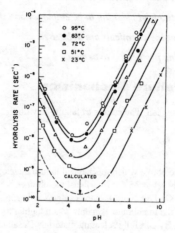

***Figure** 7.4-1 Hydrolysis rate of cellulose acetate versus pH of feed. Reprinted by permission of John Wiley & Sons, Inc from reference [5].*

However, this oversimplifies the chemistry of chlorine in water. Under normal conditions most of the available chlorine is present as hypochlorous acid or hyochlorite ion (see figure 7.4-2), the former of which has the greater biocidal effect. At low concentrations it is also important to differentiate between total and available chlorine. This difference is largely made up of chloramines which are formed by the reaction of ammonia with chlorine. The chloramines are much weaker oxidants than hypochlorous acid. The failure mechanism varies with membrane type. and trace transition metal ions such as Fe can catalyse the transfer process. The chemistry is also encouraged by higher temperatures. This makes it hard to quote an absolute tolerance.

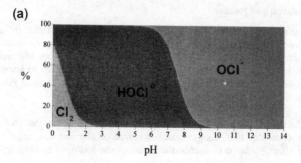

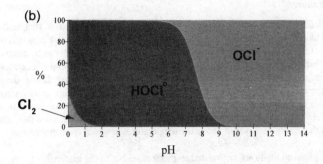

Figure 7.4-2 Distribution diagram for 1 ppm of chlorine in the presence of (a) 50 ppm chloride, and (b) 5 ppm of chloride. Under normal pH conditions in natural water the predominant solution species is hypochlorous acid. The amount of chlorine present is actually very small, but is significantly affected by the chloride level, with high chloride levels significantly increasing the amount of chlorine present.

Chemical Compatibility

Chemical attack is not the only reason why a membrane might fail. If a molecule has a high affinity between it and a polymer there will be uptake of the component by the solvent. the polymers of the membrane and materials of construction and the feed application. The consequence of this is that the materials may swell. This may be sufficiently large to dissolve them or to create physical damage. While compatibility tables for polymers and "solvents" provide an indication of the individual they do not tell us what may happen for a mixture. It is quite common for two non-solvents to act as a solvent when mixed together.

Table 7.4 Compatibility of various membrane materials.

	Membrane Material Type[1]			
	TFC	CA	PS	PTFE
Alcohols	✓	✗	✓	✓
Ketones	✗	✗	✗	✓
Ethers	✗	✗	✗	✓
Esters	✗	✗	✗	✓
Acids & Bases	✓	✗	✓	✓
Aliphatic hydrocarbons	✓	✓	✓	✓
Aromatic hydrocarbons	✗	✓	✗	✓
Halogenated hydrocarbons	✗	✗	✗	✓

[1]   TFC= thin film composite, CA= cellulose acetate, PS= polysulphone, PTFE=polytetrafluoroethylene

As ever, a balance has to be reached between the chemical compatibility required and other properties. For example, although PTFE has exceptional chemical resistance, it is relatively weak mechanically, and is very hydrophobic.

Mechanical Failures

A feature noticed early on with cellulosic reverse osmosis membranes is that under pressure permeability is slowly lost. This loss in performance appears to be associated with creep giving rise to compaction of the microporous sub-layer [6]. The extent of the problem can be altered in various ways, such as thermal or solvent annealing the polmer, using a polymer blend of acetates etc. These procedures manipulate the level of crystallinity and water content, and hence the mechanical properties of the membranes. However, increasing the level of crystallinity or reducing the water content produces in itself a reduction in permeability. Thus, as ever, a balance has to be maintained. The more modern polysulphone membranes do not suffer the same problem since the glass transition temperature of these polymers is well above the temperatures at which the membrane are operated.

Some of the mechanical limits are not related to membranes but to the materials of construction. For example in reverse osmosis three different types of permeate spacers are used depending on the operating pressure. These structure differ in the degree of openness. For high pressure applications the tight simplex structure is used since it provides the required mechanical support for the membrane. The more dense structure of the permeate spacer means that there will be greater pressure drop in the permeate channel. If the more open structure is used, then the pressures are such that the membrane will be pushed into the permeate spacer, resulting in restriction of flow, increase in pressure drop, and sometimes, mechanical damage to the membrane.

Another mechanical limit in spiral elements is the differential pressure that arises between the inside of the element and the outside, due to the hydraulic losses which occurs when water flows through the element. Ultimately, the generated pressure difference has to be taken up by the outer wrap. The pressure generated is related to the maximum flow through the element, and the feed spacer. Typically, in RO a 12 psi limit is set for each element, and 50 psi for a single stage of 6 elements.

For ultrafiltration and microfiltration backwashing is commonly employed to clean membranes. Such a procedure cannot be used for flat sheet or spiral reverse osmosis membranes since the backing cloth can delaminate from the membrane with a subsequent failure in the membrane. For this reason systems which use such membranes have an operational limit of 0.35 bar (5 psi) back-pressure. This has particular consequences for design when membranes fail.

In hollow-fibre systems fluctuations in pressure can result in the fibres breaking producing a slight flux increase, but a substantial loss in performance. In these circumstances the flux rejection trdae-off is described by the equation

$$Flux*Rejection = constant.$$

Summary

In order to minimise failures

- *Work within membrane limits and supplier guidelines.*
- *Avoid unecessary shocks, both thermal and mechanical.*
- *Do not let a system vegetate when process is stopped.*
- *Keep membranes moist*

> • *Make sure that no new chemicals are used in any of the preceding processes without reviewing its potential impact on membranes.*

7.5 Membrane Storage

A features of all process operations is that there are unscheduled failures when the membrane unit will have to be turned off. Leaving a unit with a warm organically rich feed is inviting problems. The key guidance here is to design the system so that the feed side of the membrane can be flushed with water, preferably permeate quality water of low TOC, which more often than not is the permeate water.

Whether in batch or continuous application, there frequently arises a scheduled period when the system might be left unsed for a period of time. Leaving the membrane in contact with the feed can be disastrous and might leave one with little choice but to replace elements. The problems that arise are

> • *"Consolidation" of foulant*
> • *Bacterial growth*
> • *Fungal growth*

The minimum requirement is usually to flush the feed thoroughly from the elements, usually using permeate water. This could then be then followed by use of a disinfectant. A common treatment is the use of a solution of sodium metabisulphite (typically 1.5-2 %) which is an oxygen scavenger. If the membrane is to be left a very long time an osmotic inhibitor (concentrated solution that will osmotically destroy bacteria, and other life-forms, (e.g. 20 % propylene glycol) might be required. A common combination in reverse osmosis would be a mixture of sodium metabisulphite and an osmotic inhibitor. It is essential before any such materials are used that these are validated for chemical compatibility with suppliers. Historically, membranes have been stored with formaldehyde (0.2 %) and latterly with gluteraldehyde. In recent years concerns have been raised about the toxicity of these compounds. Thus, there has been a reduced tendency to use such materials, though they are particularly good at destroying fungal material.

> • *Degassing*
> • *Oxygen scavenger*
> • *Disinfection*
> • *Osmotic inhibitor*
> • *Fungicide*

Some companies supply special formulations which address a number of issues simultaneously.

In summary, if membranes are to be left for any significant time then the user should

> • *Follow the guidelines provided by suppliers.*
> • *Replace the feed with a clean water with low organic content - permeate water might suffice.*
> • *If the membrane is to be left for several days consider the use of a disinfectant.*

• *If the membrane is to be left for longer periods consider the use of a bacteriostat and fungicide.*

7.6   Leachables

When first used the components of membrane elements invariably shed organic, and inorganic residue related to their origins e.g. monomers, solvents, stabilisers, catalysts, wetting agents, preservatives, adhesives. These residues which come-out with the permeate are of particular importance in a number of industries

- *potable water*
- *dialysis water*
- *high purity water*
- *food processing*

There is no common standard to which membrane elements are tested. Each industry, and each country tend to have their own legislation. For example membranes to be used in the production of potable water in the UK must satisfy the requirements of regulation 25 which is adminstered by the Drinking Water Inspectorate (DWI). As such this is not a standard but a procedure which allows the DWI to assess each product. In the US the local State EPA provides the permits to use a technology. Results obtained by the supplier with the National Sanitary Foundation (NSF), and independent organisation, can facilitate approval by the State EPA.

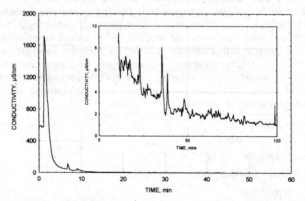

***Figure** 7.6-1 Plot of permeate conductivity from an interfacial membrane element pressurised with high purity water. The initial conductivity spike rapidly falls as salts are purged from the element. Magnification of the low conductivity area (see inset graph) shows however that the decay is not uniform nor exponential. The fluctuations are attributed to small areas of the membrane wetting out.*

The components, which are simply in the interstices, are quickly flushed from the elements (see figure 7.6-1). However, in spiral elements rinsing is slightly harder, since the transport regime on the permeate side is more akin to a stirred tank reactor than plug-flow. In addition small areas of membrane take time to wet-out (see inset graph in figure 7.6-1). However, the major problem comes from components dissolved in the polymers. The release of these materials is governed by diffusion with the polymer. This can lead to the release of organics at low levels for much longer times. For some membranes it can take more than a day for organic levels in the permeate stream to fall below 0.2 ppm. Another problem is that some parts of the membrane are not fully wetted out. As a result small bursts of organics can be released from these areas sometime after the initial purge of material.

For critical applications the user should

- *Establish the quality standard required in the critical stream*

- *Agree the period and methods required for conditioning/rinsing the elements with the supplier*

- *In the absence of supplier information carry out an evaluation*

7.7   References

1   N G R Walton *"RO Pre-treatment - Injecting a Little Chemical Control and Management"* pages 281- 301 in Volume 3,Desalination and Water Re-Use, Proceedings of the 12th International Symposium, Malta

2   W Stumm & J J Morgan *"Aquatic Chemistry"* Publ. J Wiley, 3rd Ed 1996

3   J W Starmand and B J Bellhouse *"Mass transfer in a pulsating turbulent flow with deposition onto furrowed walls"* In J Heat Mass Transfer 27 (1985) 1405-1408

4   J Glater, and M R Zachariah *"A Mechanistic Study of Halogen Interaction with Polyamide Reverse Osmosis Membranes"* in ACS Symposium Series 281 "Reverse Osmosis and Ultrafiltration" ed S Sourirajan, and T Matsura 1985

5   K D Voss, F O Buris Jr, and R L Riley "   " J Appl Poly Sci 10 (1966) 825

6   W M King, D L Hoernschemeyer, and C W Saltonstall Jr, *"Cellulose Acetate Blend Membranes"*

7   B J Rudie, T A Torgrimson and D D Spatz *"Reverse-Osmosis and Ultrafiltration Membrane Compaction and Fouling Studies Using Ultrafiltration Pretreatment"* in ACS Symposium Series 281 "Reverse Osmosis and Ultrafiltration" ed S Sourirajan, and T Matsura 1985

8   G M van den Berg and CA Smolders *"Flux Decline in Membrane Processes"* Filtration & Separation March/April 1988

9   Samon, D C *"Hyperfiltration Membranes, Their Stability and Life"* in "Synthetic Membrane Processes" ed G Belfort, Publ Academic Press, 1984

10   J C Shippers and J Verdouw, " " Desalination 32 (1980) 137-148

11   J Murkes and C-G Carlsson *"Crossflow Filtration - Theory and Practice"* John Wiley 1988

12   D Comstock *"Membrane Fouling: Reduction of Colloidal Fouling in Membranes Systems of Organic Antifoulant Formulations"* Industrial Water Treatment Jul/Aug (1991) 39-42

13   T J Kennedy, R L Merson, and B J McCoy *"Improving Permeation Flux by Pulsed Reverse Osmosis"* Chem Eng Sci 29 (1974) 1927-1931

Chapter 8

FEASIBILITY, SCALE-UP & DESIGN

Contents

8.1   System Basics

8.1.1   Introduction

System design requires knowledge of the, membrane, the challenge, and the requirements of the application. The first decision to make is how should the process be operated, batch, continuous, or cyclically. The choice depends on a number of factors of which

- *scale,*
- *variability of feed,*
- *demand cycle,*

are the more important. Even once the mode of operation has been decided there is a considerable choice in the elements to be used, the format of choice, and how they should be configured. These different ways give rise to differences in

- *performance*
- *membrane area*
- *residence time*
- *complexity*

Ideally, for optimum performance independent control of pressure and cross-flow conditions is desirable throughout the system. The discrete nature of the units, and the practical implications of such a solution preclude this. The art of the system design is to reach the best compromise, and this varies with application, and technology format. In the following section details of the various types of systems will be outlined along with their advantages and problems.

8.1.2 Batch Processes

Batch processing is preferable when either demand or feed is variable. The simplest type of system is the *open loop recycle* system where the reject from the membrane is returned to the feed tank (figure 8.1-1a). This process continues until the required volume reduction is achieved. A practical problem with such a design is that the flow-rate and pressure are determined by the pump specification. One way of achieving better control is to introduce a second pump and a closed recycle loop (*closed loop recycle*), as shown in figure 8.1-1b. This system allows independent control over the cross-flow and the pressure.

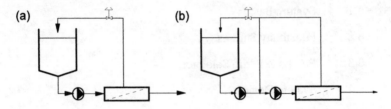

Figure 8.1-1 Schematic of batch processes, with (a) there is an open loop recycle, and with (b) there is a closed loop recycle

8.1.3 Continuous Processes

The simplest continuous system design is the *single pass*. The two extremes of this design are the all in series, and all in parallel. These designs illustrate some basic problems with designing continuous systems. If the membrane area is placed all in parallel then there will be a low cross-flow velocity, while if it is placed in series the cross-flow will fall through the system so the last elements will be operated at low cross-flows. The maintenance of cross-flow is a key aspect of the membrane process, particularly in the latter stages of treatment. Thus, a key aspect of continuous systems is how to maintain reasonable cross-flow throughout the system.

For large systems the feed is generally passed over several stages of elements (*multi-stage*). In the case of reverse osmosis plants a single high pressure feeder pump is used at the head of the plant, and the banks are tapered in order to maintain the average cross-flow. Each bank may contain up to 6 elements in series. The general design guide line is that the outflow from one stage should not fall below 50 % of the maximum inlet flow. This is done to ensure that polarisation effects in the latter stages of the system are not signficantly different to that entering the system. Also it reduces the quanitity of solids that would otherwise accumulate in the latter part of the system. To maintain the appropriate levels of flow the system the cascade is tapered (see figure 8.1.2).

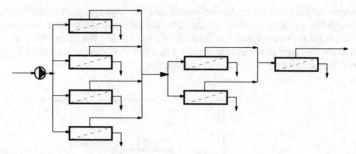

Figure *8.1-2 Schematic continuous process showing a 3-stage tapered design, popular in reverse osmosis plants.*

The flow guideline provide a simple way of estimating the number of stages required given a target recovery (see table8.1.1). However, these values are only a guide. In particular they are based on the assumption that the productivity of each element is the same. In reverse osmosis system there is often a substantial pressure loss across each stage (up to 3.5 bar). This means that the latter stages will tend to operate at lower production rates than might be expected and hence higher cross-flows are achieved.

Table *8.1.1 Number of stages required for target recovery range for reverse osmosis process.*

Recovery Target	Number of Stages Required
0 - 50 %	1
50 - 75 %	2
75 - 87.5 %	3
87.5 - 94 %	4

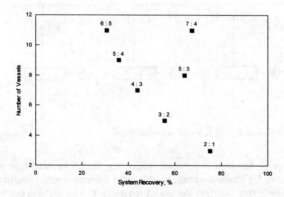

Figure *8.1-3 Graph shown how optimal degree of tapering alters depending on required system recovery.*

These continuous designs however are confined to relatively large systems. For small systems there maybe insufficient elements to allow staging. This can be overcome by introducing a recycle in which reject is fed back to the feed (see figure 8.1.4). Recycle like this saves adding an extra stage e.g. if there are only enough elements to construct an a two stage design, but a three stage recovery is required, then introducing recycle allows on to do this.

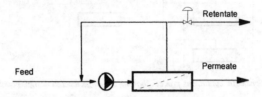

Figure *8.1-4 Schematic of a continuous system utilising a recycle to maintain cross-flow conditions (feed and bleed process)*

Whilst recycle allows one to maintain better cross-flow conditions within the elements, there is both a quality and energy penalty.

Mass balance equations for the system are readily established. The recycle allows on to operate the system recovery, Y, at a reasonable value while operating the stage recovery at a more modest value, Y_s. A simple relationship can be established between the system and stage recoveries.

As one moves from RO to MF the feed pressure requirements fall, and the pressure losses in passing through the elements become more significant. In addition to the head loss in many applications the rheological properties of the feed are changing with concentration leading to high pressure losses. This challenge is met by providing independent pumping at each stage. In essence there is a single pump to provide the required pressure head, and a circulating pump to control the cross-flow conditions.

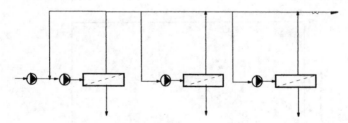

Figure *8.1-5 Schematic of a cascade feed and bleed design*

In some applications the key issue is permeate quality, e.g. landfill leachate, high purity water for electronics, high purity for pharmaceuticals, potable water from sea water. Unfortunately, elements are not always available that can meet the quality requirement. This problem can be overcome by using a *multi-pass* system in which the permeate from one bank of membranes is passed through a second bank. This can be done by back-pressurising the first stage (see figure 8.1.6) or introducing an inter-stage pump to boost the pressure. In this design the reject from the second pass is returned

8 - 4

to the feed of the first stage. The quality of the permeate from the first pass is such that this secondary membrane system can frequently operate at much higher recoveries and fluxes than the primary separation stage.

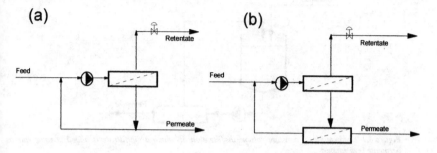

(a) **(b)**

Figure 8.1-6 Schematic of a permeate recycle system (a) with direct recycle (b) with additional membrane stage on permeate line.

The size of some systems might be too small to have a second pass as in figure 8.1-6 (b). However, the effect can sill be achieved, though less efficiently, by feeding some of the permeate from the first pass directly back to the feed to the first pass (figure 8.1-6 (a)). The effect of this is to reduce the feed concentration to the element, and thereby improve the system rejection. While twin pass systems can achieve higher quality they do at the expense of higher energy and more membrane area for a given permeate flow. For a given rejection target it can be shown that for every halving in transmittance a doubling in membrane area is required.

> **Sea Water Desalination**
> *In sea water desalination the high osmotic pressures of the feed, and its low cost, mean that sea water reverse osmosis plant need operate only at low recoveries (circ 30 %). The high hydraulic pressure required (800 psi) means that there is a large amount of energy in the reject stream. Various devices are available to transfer up to 90 % of the energy back to the feed, thereby significantly improving the energy requirements. To produce potable water from sea water in a single pass requires a membrane with rejection of greater than 99.4 %. The lack of such membranes meant that the early designs were invariably had a second RO stage on the permeate.*

8.1.4   Diafiltration

In some applications the issue is not to concentrate a component but to remove a low molecular weight component. If water is removed with the low molecular weight components then the concentration effects can lead to unnecessary precipitation and or fouling problems. This is a problem that occurs in a wide range of areas, biotechnology, pharmaceuticals, fine chemicals, and food industries. Membranes can be used to remove low molecular weight components and yet not significantly effect the concentration of higher molecular weight components in a process called *diafiltration* in which water is added to make up for material lost in the permeate. The simplest conceptual version of this is *constant volume diafiltration* in which the diafiltrate is added to maintain a constant head in tank (see figure 8.1-7).

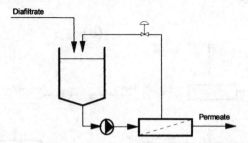

Figure 8.1-7 Schematic of constant volume diafiltration, in which a dialfiltrate is added at same rate as permeate is produced.

Colloidal systems can become unstable if the ionic strength of the solution is changed. To avoid this destabilisation the diafiltrate is sometimes a solution of an equivalent ionic strength to that of the feed.

A mass balance predicts that the concentration in the permeate will fall with time according to the expression

$$C(t) = C(0) \ \exp[-N(t)] \qquad (8.1.1)$$

for solutes which are totally rejected by the membrane, where $N(t)$ is the turnover number;

$$N(t) \equiv \frac{\text{Volume of Liquid Added}}{\text{Volume of Feed}} \qquad (8.1.2)$$

An alternative version of this process is *sequential dilution diafiltration*. In this batch process the feed is diluted prior to treatment. Diluting the feed provides a benefit of easier operation of the membrane plant, less fouling, lower pressure drops. In many cases diafiltration is combined with concentration in a three step process

- dilution
- diafiltration
- concentration

For large applications *continuous diafiltration* process can be devised, in which the diafiltrate is the permeate from next stage (see figure 8.1-8).

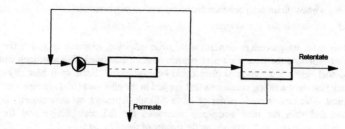

Figure 8.1-8. *A schematic of 2-stage continuous diafiltration process*

A detailed theoretical analysis of diafiltration can be found in a paper by Beaton and Klinowski [1].

Table 8.1.8 *List of diafiltration applications (from [1])*

Process Type	Application	Comment
Fractionation	Milk	Removal of lactose/ash prior to cheese making
	Whey	Reduction of of lactose/ash to produce wheyt concentrates
	Water Soluble Polymers	Separation of low moelcular weight fractions in polyelectrolytes, polysaccharides, proteins
Purification	Enzyme	Reduction of sugar and mineral content
	Protein	Purification of proteins from various sources eg fish, blood,egg
	Antibiotic	Removal of high moelcular weight components such as proteins, pyrogens, and suspended solids
Clarification	Fruit Juice	Separation of colloidal materials in fruit juices
	Beverage	Separation of haze components/precursors from wins and spirits
	Fermenation Broths	Separation of cells/cell debris with purification of fermenation derived products eg amino acids, antibiotics
Demineralisation	Gelatin	Reduction of the ash content

8.2   Feasibility Studies

8.2.1   Introduction

A feasibility study covers both the technical and the economical aspects of a process. This chapter is concerned only with the technicality of a process. Economic feasibility of a particular process is dealt with in the next chapter.

It has been generally accepted that the design of a commercial membrane plant cannot be accomplished by a purely theoretical approach alone. Technical aspects which should be examined during the initial phase of a project include,

- *a characterisation of the feed stream,*

- *a quantification of the membrane separation performance*
- *an outline process design*

A satisfactory scale up procedure requires a stepwise empirical approach in which the effect of various factors affecting the separation performance must be quantified. This approach would result in time and cost before a commercial plant could be constructed. Scale up is basically concerned with ensuring that the operating conditions that applied in the pilot scale trial are reproduced in the full scale plant. This task is made easier or more difficult depending on how closely, in terms of configuration and size, the pilot equipment resembles the full scale design and the level of confidence in the scale up result depends on the quality of the pilot data.

The analysis given here is restricted to filtration processes.

8.2.2 Membrane Aspects

A membrane processing plant cannot be considered as a collection of individual process units; the design must be taken as a whole, and full consideration must be given to

- *the choice of membrane,*
- *the module configuration,*
- *the nature of the material to be processed,*
- *the process duty specification of the plant,*
- *the most suitable plant configuration,*
- *the nature of process upstream and downstream of the membrane plant,*
- *environmental impact,*
- *safety issues,*
- *constraints on ancillary equipment.*

In the following sections discussion is given to aspects which are considered to be most significant in the design and scale up of a membrane plant.

Membrane selection should be carried out during the feasibility studies. Ideally, the most suitable module type and size to be used on the full scale plant should have been identified along with the optimum operating conditions and performance data. This is particularly important where investment in the final plant is very large and where errors would be extremely costly. The modular design of membrane elements considerably facilitates the scale up proces. However, one should not be over-ambitious with the scaling factors since any error in the data would also be multiplied by the same factor. As a general guide scale-ups of greater than 10 times should not be undertaken unless one is confident with the quality of the data, and are prepared to accept a possible large margin of error.

8.2.3 Feed Characterisation, and Membrane Selection

The starting-point of any scale-up is an assessment of the feed. From this an outline flow-sheet can usually be devised and assessment can be made as to what further information is required from laboratory/field testing. Field testing is particularly important in assessing the fouling potential. Such testing also provides data on the pre-treatment requirements and cleaning effectiveness. An obvious requirement that is sometimes overlooked is a quantitative statement of the separation requirement. This should be stated before any separation process is specified. The composition of the stream should be determined if it is not already known. The condition under which the separation is to be carried out must be stated. Feed parameters and their significant in the feasibility study are indicated in Table 8.2.

Table 8.2 Feed parameters and their significance

Parameters	Significance
Particle size or molecular weight of material to be retained by membrane	Choice of membrane pore size
Settleable solids (size & quantity)	Means of pre-treatment
Shear sensitivity	Module type and pump type
Range of product viscosity	Module type and pump type
Temperature range	Membrane type; heating / cooling requirement; material of construction
pH range	Membrane type; material of construction
Organic solvent content	Membrane type; material of construction

Membrane Materials

The main membrane materials used in reverse osmosis are cellulose acetate and polyamide, though a great variety of other polymeric materials can be used. The cellulose-based membranes offer very much lower chemical and thermal performance than the polyamide types and often have lower water fluxes. Membranes most commonly employed are of a 'thin-film composite' type. The polyamide membranes are normally supported on a layered structure, in which the selective polyamide layer is physically supported on a material such as polysulphone. There is a greater choice of materials for use as UF membranes some of which have been designed to withstand the wider variety of feeds. Polysulphone is by far the most popular membrane material having good chemical resistance, high temperature performance and great tensile strength.

Membrane Properties

Fundamental data on membrane properties provides a useful guide in the initial stages of membrane selection. However, prediction of membrane performance in actual operation, from this information, is not generally possible. Indeed, in many cases, factors such as the composition of the feed stream and operating conditions have a greater influence on performance than the membrane itself. Nevertheless, membrane selection is still important since some membranes will foul more severely than others leading to lower fluxes or reduced service lifetimes. Fundamental membrane data therefore need to be supplemented by application-specific performance data in order to assist in selecting membrane types and to enable the required membrane area to be estimated.

Membrane Selection

The main objective when selecting a membrane is to carry out a performance comparison in order to determine the most cost effective option. The main factors that influence the choice are:

- *chemical and thermal compatibility*
- *cleaning regimes*
- *separation performance*

8 - 9

- *fouling behaviour*
- *module characteristic*
- *unit cost*

These factors are inter-related to a large extent, for instance pore size and chemical compatibility both affect the rate of membrane fouling. In practice the selection procedure may well prove relatively straight-forward since membrane manufacturers can usually advise on likely system equipment. In non-standard situations (such as treatment of a mixed effluent) or where the feed contains unknown foulants, testing to determine operational characteristics is normally advised. Fundamental data is provided for the purposes of membrane characterisation, such membrane performance data and specifications are often quite independent of separation applications. Similarly the same information is used by membrane manufacturers to define the specification and performance of a membrane in technical literature distributed to membrane users. The following standards may be used for defining membrane performance and for comparing the separation performance across a range of membrane materials:

- *Clean Water flux*
- *Salt rejection / MWCO*
- *Nominal pore size (particle challenge / bacterial challenge test / bubble point)*
- *Pore size distribution*
- *Asymmetry and/or Dirt Holding Capacity*
- *Chemical compatibility (pH range, chemical resistance)*
- *Burst pressure / collapse pressure*
- *Working temperature and pressure*
- *Wettability / Contact angle*
- *Non-specific binding*
- *Capability for withstanding steam*

Care should be taken when comparing the membrane specifications from a range of manufacturers. It is important that the user is satisfied that data can be directly compared from different sources. If the data is not directly comparable the consequences of making an erroneous selection decision must be appreciated.

The relationship between the physical properties of membranes and the performance in practical applications is extremely complex. It is therefore more usual to describe the properties of membranes in terms of simple operating parameters which are more directly related to actual separation performance.

8.2.4   Experimental trials

There is no unique scale-up procedure for cross-flow filters, since in some cases alternative designs and alternative membranes can meet the same application requirement. Within a customers specification there is however an optimal solution. To come close to this solution requires data or experience.

Without previous experience or test data, scaling up information based on data sheets can only be achieved with excessive over-design or excessive risk of under-performance. Testing at various intermediate stages provides a simple method of reducing risks. Only by correctly assessing the risk factors and carrying out appropriate testing can the level of risk be assessed. SO doing allows allows an appropriate design to be drawn up that is more likely to meet the customers needs.

A major weakness of laboratory trials is that they do not give a full picture of the fouling issues. This is because for laboratory trials the feed will be often reconstituted or over "worked". Since foulants are usually well rejected and accumulate on the membrane the fouling challenge is not assessed. It is for this particular reason that pilot trials are of such great value.

One of the most important aspects of any testing work is to define the required pre-treatment, and cleaning methods. Appropriate sizing of the screens, or pre-filters is an essential element in all system designs and can usually only be assessed by carrying out pilot studies. In some plants the pre-treatment can make up 50 % of the total capital cost. However, a good pre-treatment can make substantial savings in the sizing of the membrane plant, and operational lifetime between cleaning or element replacement.

The fouling factor is the major unknown in most systems designs. Fouling factors for reverse osmosis membranes typically encountered are 30 %, while for ultrafiltration processes values of 80 % are likely, and for MF values encountered can easily exceed 90 %. The extent to which a membrane fouls depends both on the operating conditions and the feed.

An additional complicating factor concerns the effect of cross-flow mode on the rejection properties. At the surface of the membrane there is a balance between the rejection of material and the back-diffusion into the bulk. This phenomenon is known as concentration polarisation, and is discussed in detail in Chapter 6. The consequence of concentration polarisation is that as it increases, the concentration is at the interface increases, and the rejection falls.

8.2.5   *Estimation of membrane area requirement*

Batch operation

The batch size, V, membrane area requirement, A, and the batch cycle time, t are related by the following relationship:

$$A_2 = A_1(V_2/V_1)(t_1/t_2). \tag{8.2.1}$$

where suffix 1 denotes the pilot plant and 2 denotes the full scale plant. The above relationship only hold true if the starting concentration and the desired final concentration in the full scale plant are the same as in the pilot trial.

Alternatively, the membrane area requirement may be estimated from a knowledge of the relationship between the flux and the solid concentration. Such an equation may be estimated from a flux versus concentration plot (as shown below). Where a plot is not linear, the flux relationship may be approximated by two or more lines. The basic routine below provides a useful tool for an estimation of the membrane area requirement. By substituting in the appropriate flux equation and the system parameters into the program code, the program would generate a series of retentate volume and concentration values versus time. These may be plotted as shownin figure 8.2-2, and from which the membrane area requirement may be estimated.

```
10 REM ROUTINE FOR ESTIMATION OF MEMBRANE AREA REQUIREMENT
20 REM FOR SIMPLE BATCH OPERATIONS
30 GOTO 50
40 J= 60-6.89*LOG(C): RETURN: REM FLUX EQUATION
50 C0=2.5: C1=40: V0=20000
60 A=30: H=0.005
70 T=0: N=0: V=V0: C=C0
80 LPRINT "MEMBRANE AREA (M2)=", A
90 LPRINT "TIME (H); RETENTATE VOLUME (L); RETENTATE CONC. (%)"
100 T=T+1: N=N+1
110 GOSUB 40: V=V-J*A*H
120 C=V0*C0/V
130 IF N=300 THEN LPRINT T, V, C: N=0
140 IF C>C1 THEN GOTO 160
150 GOTO 100
160 LPRINT T, V, C
```

Figure 8.2-1 Program for calculating membrane are required for a simple batch process, written in a simple form of Basic.

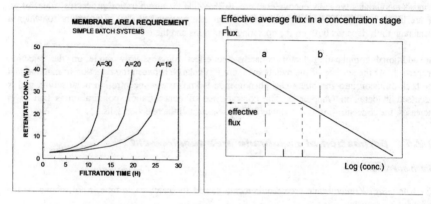

Figure 8.2-2 Left graph shows impact of membrane area of filtration time on retentate concentration for a batch process. Right graph shows that effective flux stage falls at each concentration stage.

Continuous operation

For a continuous operation the concentration of solid within any given stage is constant. The membrane area in a separation stage is given by the following equations:

$$C_a Q_a = C_b Q_b, \qquad Q_p = Q_a - Q_b, \qquad A = Q_p / J_e \qquad (8.2.2)$$

where C_a is the inlet concentration; C_b is the outlet concentration; Q_a is the feed flow rate; Q_b is the concentrate flow-rate; Q_p is the permeate flow rate; A is the membrane area required and J_e is the effective flux for the separation stage. The effective flux is may be estimated by an empirical expression as follows:

$$J_e = J_b + (J_a - J_b)/3 \qquad (8.2.3)$$

where J_a and J_b are the flux values at inlet and outlet concentration respectively.

8.2.6   Fouling Allowance

It is the responsibility of the process design engineer to ensure that the membrane plant is fully capable of delivering it process duty throughout its service life. Given that the membrane has a limited life, say replacement every three years, a projected cash flow is predicted from this scenario. However, a membrane performance deterioration through fouling may severely affect the financial forecast. It is therefore very important that sufficient allowance is made for long term fouling in sizing the membrane area for an application. In many industrial applications a membrane replacement frequency of once per year is generally adopted. For such applications, a design margin of 15-20% is usually sufficient. It is not unusual, however, for membrane to last up 6 years in applications with a non-aggressive environment. Larger margins should be allowed for applications that require longer replacement intervals.

Clearly, it is impractical to wait for up to 6 years for the long term fouling data from a pilot study. Theoretical modelling provides a useful tool for the prediction of long term membrane performance. A number of fouling models are available. The simplest model to use which provides a good estimate of fouling is described below. This model was originally devised for activated sludge systems [3], but it has been found equally useful in many other applications.

$$J = \Delta P / (R_m + R_c + R_a) \qquad (8.2.4)$$

where R_m is the membrane resistance
R_c is the gelling layer resistance
R_a is the additional gelling layer resistance due to consolidation with time

J is the flux and ΔP is the trans-membrane pressure. R_a is treated as an empirical function that has to obtained from data. A widely used expression for fitting this term is:

$$R_a = R_c \, A \, t / (B + t), \qquad (8.2.5)$$

where A and B are the function constants for a particular type of sludge and hydrodynamic condition. The parameters A and B may be determined by fitting the R_a function to the pilot data.

8.2.7   Module arrangement

Module size

The total number of modules required for a plant is can be determined esily given the total membrane area requirement. However, the choice of the module size may be influenced by operational considerations such plant availability, ease of maintenance (see **operational considerations**). While the use of the largest modules may cut the engineering cost, it makes it less flexible for a cascade design. Also, if a module fails in operation, a large portion of the permeate flow must be isolated or returned to the feed source for re-processing. A choice must be exercised between the different number of modules available taking into consideration any operational risks and preferences.

Series or Parallel flow

The pumping costs for a membrane process, both in terms of the capital investment and energy consumption, are considerable. Modules should be arranged in such a way so as to minimise the costs of pumping (see *Optimum Recirculation Rate* below). However, any module arrangement may be influenced by the following factors:

- *Recirculation rate*
- *Total pressure drop*
- *Product viscosity*
- *Maximum allowable working pressure for the module*

The total pressure drop and the maximum allowable working pressure for the modules determines the number of modules that can be used in series. The pressure drop, of course, is dependent on the feed viscosity, temperature and the recirculation rate. Note that, since pressure drop is dependent on viscosity and temperature, consideration must be to the pressure changes over expected working concentration and temperature ranges.

In desalination applications with spiral wound RO modules, it is usual to have as many as 12 elements in series even though the pressure drop per element may be as much as 2 bar, this is because RO modules have very high working pressure (typically 40 bar gauge). Similarly, systems with tubular UF membrane commonly employs over 18 tubes in series. In the latter case the tubes themselves are limited to a maximum working pressure of under 7 barg, but pressure drop per tube length is quite small (typically about 0.1 bar at the optimum cross flow velocity). Hollow fibre modules, however, have very high pressure drop, but low burst pressure, therefore they are invariably employed in a parallel configuration.

Cascade flow

In a cascade flow arrangement is widely used in RO applications where the recovery rate per pass is high. This design is used to overcome a significant reduction in the cross flow velocity that would otherwise occur in the later stages, because of the reduction in the volumetric flow-rate. In order to maintain a high cross flow velocity, the number of modules is reduced in proportion to the reduction in the volumetric flow-rate.

8.3   Pumps & Pump Selection

8.3.1   General

When selecting pumps for any service, it is necessary to know the liquid to be handled, total dynamic head, suction and discharge heads and in most cases, the temperature, viscosity, vapour pressure, and specific gravity. The task of pump selection is frequently complicated by the presence of solid particles and corrosion characteristics demand special materials. Solid particles accelerate erosion, have a tendency to agglomerate, or may need delicate handling to avoid degradation.

Range of Operation: Because of the large variety of pump types available and the number of factors which affect the selection for any specific application, the designer must first eliminate all but those reasonable possibilities. Since the range of operation is always an important consideration, the pump coverage chart below should serve as a useful guide.

Pump Materials: In the chemical industry the selection of pump materials is dictated by considerations of corrosion, erosion, personnel safety, and liquid contamination. The experience of pump manufacturers is often valuable in selecting materials.

Presence of Solids: Hydraulic performance and use of the most durable materials may not always be sufficient to produce the most satisfactory pump selection. When solid particles are present there are other considerations of equal importance. All internal passages must have adequate dimensions. Pockets and dead spots where solids could accumulate must be avoided. If the solids are abrasive, close internal clearances between stationary and moving parts are undesirable. Means should be provided for flushing with a clean liquid before shut down.

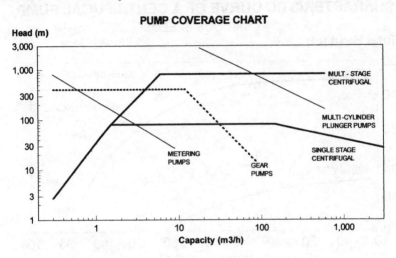

Figure 8.3-1 Pump coverage chart.

8.3.2 Centrifugal pumps

The centrifugal pump is the type most widely used in different membrane processes. They are available through a vast range of sizes; in capacities from 0.5 m³/h up to 27,050 m³/h for discharge heads (pressures) from a few m up to 100 barg. The size and type best suited to a particular application can be determined only by an engineering study of the problem. The primary advantages of a centrifugal pump are simplicity, low cost, uniform (non-pulsating) flow, small floor space, low maintenance expense, quiet operation, and adaptability to use with motor or turbine drive.

A centrifugal pump, in its simplest form, consists of an impeller rotating inside a casing. The impeller consists of a number of blades, either open or shrouded, mounted on a shaft that projects outside the casing. The pump shaft may be horizontal or vertical, to suit the work to be done. Closed or shrouded impellers are most efficient. Open or semi- open type impellers are used for viscous liquids or liquid containing solid materials. Casings are of three general types The most common type of casings take the form of a volute which has an increasing cross sectional area as the outlet is approached. The volute converts the liquid momentum imparted to it by the impeller into pressure, with comparatively little losses.

Centrifugal pump characteristics: The chart below shows a typical characteristic curve of a centrifugal pump. It is important to note that at any fixed speed the pump will operate along this curve and at no other points. On pumps with variable-speed drivers, it is possible to change the characteristic curve. It is important to remember that the head produced will be the same for any clean liquid of the same viscosity. The pressure rise, will vary in proportion to the specific gravity. Viscosity less than 50 centipoise do not affect the head materially.

CHARACTERISTIC CURVE OF A CENTRIFUGAL PUMP
AT VARIOUS SPEEDS

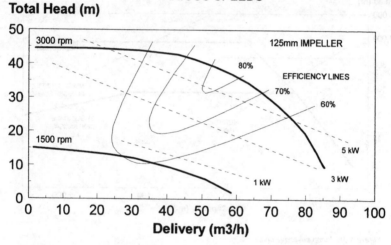

Figure 8.3-2 Characteristic performance curve of a centrifugal pump at various speeds.

Sealing of Centrifugal Pumps: Current practice demands that packing boxes be designed to accommodate both packing and mechanical seals. With either type of seal, one   important consideration is that the liquid at the sealing surfaces must be free of solid particles. Consequently, it is necessary to provide a secondary compatible liquid to flush the seal or packing whenever the process liquid is not absolutely clean. The use of packing necessarily results in a continuous escape of a small amount of liquid past the seal, as its operation depends upon the maintenance of wetted surfaces to minimise the generation of heat. If the liquid is toxic or corrosive, quench glands or catch pans are usually employed. Mechanical seals cannot be adjusted without shutting down the pump, whereas packing can be tightened while operating. On the other hand, mechanical seals pass only infinitesimal amount of material, usually in the form of vapour. In general, the more effective performance of mechanical seals when properly applied is gaining them increased acceptance.

8.3.3 Pump Suction Performance

Cavitation occurs when the static pressure within the pump falls below the vapour pressure of the liquid. Some of the liquid then vaporises giving rise to the formation of vapour bubbles. These bubbles are entrained in the liquid and transported until they reach an area where the static pressure is greater than the vapour pressure. At this point the vapour bubbles implode since they can no longer exist as vapour. Each implosion causes a severe pressure wave and can cause mechanical damage to the pump materials. Furthermore, the hydraulic performance of the pump is reduced by the onset of cavitation. Cavitation is therefore undesirable and should be avoided.

In connection with cavitation the concept of Net Positive Suction Head (NPSH) is encountered. NPSH has two forms; one is a system form and the other a pump form. The system form is designated NPSHa, Net Positive Suction Head Available. NPSHa is calculated from the suction system configuration and is the suction pressure minus the vapour pressure at the NPSH datum. NPSHa informs the pump manufacturer how much head is available for losses in the pump suction. The pump form is designated NPSHr, Net Positive Suction Head Required. NPSHr is a characteristic of the pump and specifies how much head is necessary for losses in the pump suction. NPSHa and NPSHr are difference between two pressures or heads; there is no requirement to specify gauge or absolute. Both forms of NPSH are variables, not constants. Both are a function of flow. In a fixed suction system configuration, NPSHa will increase as flow is reduced because friction losses and the velocity head requirement are reduced. Conversely, NPSHa will reduce as flow increases because the same losses increase. NPSHr for a pump is predicted during the design phase and confirmed by testing. The permissible suction lift of a pump is controlled by:

- *liquid vapour pressure*
- *liquid density*
- *the pressure on the liquid surface*
- *velocity in the pipe*
- *friction losses in the pipe*
- *NPSHr of the pump.*

8.3.4 Optimum Recirculation Rate

There is an optimum recirculation rate for each process. A higher recirculation rates will result in greater productivity (flux), therefore less membrane area required and lower capital cost. Membrane life may also be extended due to greater flux stability. However, larger flow means larger pumps and increased energy costs. Figure. 5.3.2 shows the effect of the pumping rate (recirculation rate per tube) on the operating costs for a plant with 12 mm ID tubular membranes [2]. Increasing the cross flow velocity through a module decreases the membrane replacement costs due to less area requirement and longer life membrane life. Indeed, all costs associated with the size of the plant will go down, but the energy costs go up. The optimum design point will depend on the current cost of power and membranes along with their productivity. For a multistage feed and bleed system, the optimum cross flow rates can be different for each stage. The later stages operate at higher concentration than the front stages. Often, the first stage is not fully gel polarised (due to the low concentration) and high cross flow velocities are not as critical as high operating pressures (until full gel polarisation is reached).

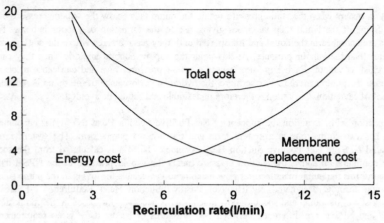

Figure *8.3-4 Effect of pumping rate (recirculation rate per tube) on operating costs*

8.3.5 *Pressure Drop Calculation*

Head losses in straight pipes

The pressure drop for water flow in straight pipes may be estimated from charts such the one shown here for stainless steel of various diameters. For other process liquids the pressure drop may be estimated from the following expression:

$$\Delta P = 0.032 f Q^2 \, L/D^5 \qquad\qquad (8.3.1)$$

where
ΔP = pressure drop (bar)
f = Fanning friction factor = $16/R_e$ for $R_e < 2000$ (i.e. laminar flow)
R_e = Reynolds number
Q = flow rate (m³/s)
L = pipe length or equivalent flow path length (m)
D = pipe diameter or equivalent diameter (m)

The fanning friction factor for turbulent flow can be determined from the graph 8.3.5.

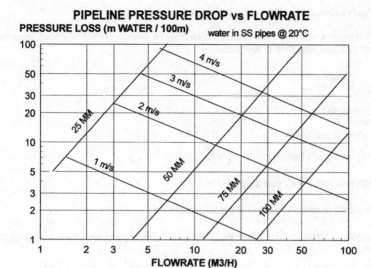

Figure 8.3-4 Pipeline pressure drop as a function of flow-rate.

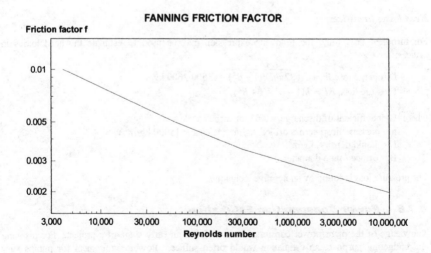

Figure 8.3-5 Fanning friction factor as a function of Reynolds number.

Head losses in bends & sudden changes in pipe diameter

For turbulent flow only, the following expression may be used to estimate the head losses in bends (90°) and sudden contraction in pipe diameter:

$$\Delta P = KU^2/g \qquad (8.3\text{-}2)$$

where ΔP = pressure loss (m)
  $g = 10 \text{ m/s}^2$
  $U = \text{m/s}$
  $K = \text{loss coefficient}$

For bends with radius = pipe diameter $K = 0.4$

For bends with radius > 4x pipe diameter $K = 0.2$

For U-bend the pressure loss is 2x the pressure loss in a right-angled bend.

For a sudden contraction K is a function of the ratio of the pipe cross sectional areas (A2/A1) as shown below:

A2/A1	0	0.2	0.4	0.6	0.8
K	0.5	0.45	0.36	0.21	0.07

For sudden expansions:

$$\Delta P = (V_1 - V_2)^2/g = V_1(1 - A_1/A_2)/g \qquad (8.3.3)$$

Head losses in orifices:

For turbulent flow only, the following expression may be used to estimate the head losses in orifices:

  Discharge rate, $W = CA\sqrt{2ghD/(1 - B^4)}$ = 3500/3600 kg/s
  Pressure loss, $\Delta P = h(1 - B^4)/(1 + B^4)$,

where C= coefficient of discharge = 0.63 for turbulent flow
  h = pressure drop across orifice, kg/m^2 (1.5 bar = 15000 kg/m^2)
  D = liquid density, kg/m^3
  B = orifice / pipe diameter

The pressure loss is found by an iterative technique.

8.3.6 Power Consumption Estimation

Knowledge of the plant power consumption is useful at an early stage of a project. For planning and budgeting purpose, an estimation would often suffice. Power requirement for pumps vary considerably depending on the pump types, duties and suction condition.

Power Requirement (W)= Delivery Rate (m³/s) * Differential Pressure (N/m²) - for an ideal pump.

For actual power requirement divide the figure by the pump efficiency (15-70% for centrifugal and 50-85% for positive displacement).

The total power requirement also depends on the type of motors being employed. Typical efficiencies for induction motors are: 80-90% for motors up to 20 kW and 90-97% for motors in the range 20 - 200 kW.

For centrifugal pumps power consumption will depend upon the curve shapes and the intersection points. Pump efficiency depends upon flow and head or pressure, driver efficiency depends upon the load. Manual change gearboxes are relatively simple so the mechanical efficiency should be about 95 to 97%. The basic efficiency of semi-automatic gearboxes will be slightly lower because of the greater complexity. Also the fluid coupling/fluid flywheel may lose up to 4% when not locked up. Manual gear changing is very efficient at transmitting power but poor at changing gear. Semi-automatic gearboxes are good at changing gear at the expense of reduced transmission efficiency. The type and size of the pump package, limitations on storage volumes and cost, considered with the operational requirements will show which gearboxes are suitable.

8.4   System Design

8.4.1   System Configurations

Open loop: In this design (see figure 8.1-1a) the feed is drawn from the batch tank and is then passed across the membrane surface by the recirculation pump, which provides both the cross-flow velocity and the feed pressure.   The permeate is collected and the retentate is recycled via a pressure control valve to the batch tank. In this mode of operation the concentration of solute increases with time, very slowly at first then rapidly toward the end of the process.   The high recirculation rate ensures that the solute concentration in the module is not significantly higher than that in the tank in order to achieve the highest possible productivity. This process has the advantage that it is the simplest design.   However, the disadvantage is that it uses only a single pump. If a centrifugal pump is chosen then the operating pressure and the hydrodynamics within the membrane module cannot be varied independently and thus it may not be possible to operate simultaneously with optimum trans-membrane pressure and crossflow velocity. However, if a positive displacement pump is selected then it becomes possible to vary the pressure and crossflow independently, but the delivery rate of such pumps may be too low to maintain the designed crossflow velocities.

Closed loop: This form is an alternative design of the batch process (see figure 8.1-1b). In this arrangement the feed is pressurised by pump P1 and is fed into the closed loop. The feed rate is designed make up any liquid loss as permeate is withdrawn from the system plus a sufficient flow in order to maintain a low differential in concentration between the loop and the batch tank. Recirculation across the membrane surface is provided by pump P2. This system has the advantage over the total open loop system that the feed pressure and the hydrodynamics can be varied independently, and   also means that the pressure control valve can be left out of the main recirculation loop. Since only a small portion of the flow (the feed rate) has to be pressurised and de-pressurised as it enters and leaves the recirculation loop, total energy consumption is minimised. The disadvantage, however, is that the concentration within the loop increases more rapidly than in the above design which leads to a lower mean flux. The differential in concentration between the loop and the batch tank is governed by the ratio between the feed rate and the permeation rate. This relationship is illustrated by figure 8.4.1. For practical purposes the ratio between the feed rate

8 - 21

and the permeation should be greater than 5. In this and the open loop design above the concentration increases with time; slowly at first than rapidly toward the end of a batch run. Problems with pumping and heat generation may be experienced during the end of the batch.

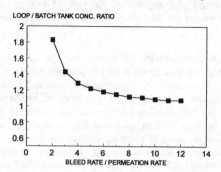

Figure 8.4-1. Feed permeate ratio effect on loop tank concentration differential

Multi-stage system: For very large plants and for continuous operation the cascade design as illustrated by figure 8.1-5 provides an elegant solution. In this system concentrate from one stage becomes the feed to the next stage. Steady state operation is possible since the concentration in each stage remains practically constant with respect to time given a constant feed composition. Thus, only the last stage experiences the final product concentration. In operation, once a feed-and-bleed-bleed system has reached steady-state operation, the product concentration within the system is equal to the final, desired product concentration. Concentrate is drawn from the loop at a rate such that the ratio of feed flow-rate to bleed rate is equal to the designed concentration factor. As the last stage is also the smallest stage with the least volume of material to processed, since most of the water has been removed in the front stages, only a small pump would be required with advantage of a minimum heat input. The mean system flux increases as the number of stages increases and approaches that of a batch process. However, there is an optimum number of stages, usually about 3-7 depending on the process scale because of a trade-off between reduced capital cost of membrane modules and the increased cost of the ancillary equipment such as pumps and fittings.

Single pass system: The above designs are sometimes referred to as multi-pass designs because the same feed material is recirculated through the same set of modules many times until a certain portion of the water has been removed. Muti-pass design is a key feature in cross flow operations which is the most common method for concentration polarisation control in the separation of feeds with high solid contents. For feeds which have a very low solids content or where the solutes have high diffusivities a very high rate of recovery per pass may be operated. In such situations a single pass system may be more appropriate than a recirculated or multi-pass system. A single pass design is the same as the open loop design shown above but the concentrate is not returned to the tank. In a single pass design the concentration of the feed stream increases gradually along the length of the membrane module achieving the desired value at the end of the module. Since there is no recycle, this design has a minimum residence time. However, because of the need to maintain a minimum crossflow in order to limit the effect of concentration polarisation, such a process must either have a very limited overall concentration factor or must be used with a membrane cascade with a large number of modules in series.

8.4.2   *Types of Operations & Process Selection*

General considerations: The selection of a process flowsheet is influenced by the nature of the feed streams to be processed. If the constituents of the stream are very susceptible to denaturation it is better to use a feed-and-bleed process in order to minimise the product's exposure time to the operating conditions. If, however, the process stream is relatively stable, minimising the membrane area is likely to be a more important consideration. In which case a closed-loop batch operation may be better option choice. It is not sufficient, however, to consider the membrane separation process in isolation, since it would generally be part of a larger system. For instance if upstream unit operations are batchwise, it is likely to be better to adopt a batch process for the membrane unit also. Similarly, a continuous upstream process is likely to be more beneficial when coupled to a continuous feed-and-bleed membrane system.

In effluent processing applications, the main factor deciding the nature of the process flowsheet will be one of economics. Unlike other uses of membrane processing, such as pharmaceutical applications, there are few constraints on the way the process stream can be handled. The most important consideration is that the capital and operating costs should be minimised.

Batch: This is the simplest type of operation with minimum control requirement. As the name implies, the feed material is processed in batches, usually of a fixed volume. In sample preparation for laboratory uses batches as small as 100 ml are routinely processed. At the other extreme, for example, fermentation product processing often involves batch volumes of up to 25,000 L. Although any of the system configurations described above may be deployed in a batch operation, in practice cost consideration and volume constraint dictate the choices of systems. For very small batches the ideal systems would be the open loop type. Using Hollow fibre modules or thin cassettes and peristaltic pumps for recirculation, the system hold up volume may be reduced to under 10 ml. For very large batches, where hold up volume is not a significant consideration, energy consumption and heat generation may be more considerable. The closed loop system design provides the most energy efficient process, although capital cost is slightly higher than the open loop system for the same duty.

In operation the initial feed volume is pumped continuously through the membrane modules at the process conditions. Changes are caused mainly by the gradual increase in the concentration of retained substances in the recirculating liquid. This occurs when no concentrate is removed. Increasing concentration alters the rheological properties of the liquid; the pressure drop within the module increases and the circulating flow rate is reduced where the centrifugal pump speed is not automatically adjusted. Filtration is interrupted when the desired concentration has been reached and the concentrated liquid is removed.

Semi-batch: There are a number of possible variations to the operation of a batch process. The batch tank may be re-filled a number of times before the end of a batch so that the batch volume is effectively several times larger than the actual batch tank. Instead of adding the feed step-wise, it may also be introduced to the tank continuously. In the latter case the total batch volume must be monitored by a flow totaliser. Semi-batch operation is an ideal choice when the availability of the feed is intermittent or when it is not possible to have a large batch tank because of space limitations. It is also most suitable when the feed is very dilute so that a large concentration factor is required.

In most wastewater applications the solid concentration varies widely from hour to hour and from day to day. Balancing tanks are often used to reduce the variations. Additionally, semi-batch operations provide the most practical solution to a highly variable process condition. Since in the operation of a semi-batch, e.g. where the batch tank is continuously top up until the final desired concentration has been reached, the membrane has to process at the high concentration for a longer period, flux efficiency is worse than in the case of a simple batch as the chart below illustrates. In order to improve the overall efficiency it would be better to carry out a top up only during the early part of the batch. For example, to achieve a 20 fold concentration of a feed volume one may concentrate the volume 5 fold in a top up mode, then completing the job by a simple batch concentration. This would produce an overall process efficiency of 80% compare to 20% had the task be completed with a top up mode of operation throughout.

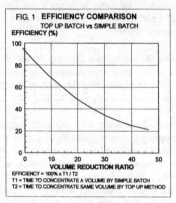

Figure 8.4-2 Efficiency comparison between top up batch versus simple batch.

Continuous Processes: Continuous operations may be carried out with either the 'feed-and-bleed' system design or single pass design. The main advantage of continuous operations over batch operations is the greatly reduced residence times of the feed within the systems. This is of considerable importance in applications involving sensitive products since the extent of product denaturation will increase with increasing time of exposure to the processing conditions. Unlike batch processes where concentration increases with time, in continuous processes the solute concentration only vary along the feed flow path which make a steady state operation possible. Continuous operations also lend themselves conveniently into process automation. For very large scale operations, continuous operations or cyclic operations (see below) offer the most practical process solutions. Feed-and-bleed systems are the most widely adopted design in the dairy and food processing industries. On the other hand Single pass designs are often used in desalination or pyrogen removal applications. For sea water desalination the recovery per module is typically 3%, therefore for a 30% water recovery more than 10 modules in series may be used in a cascade design. Such designs have fewer control; they are relatively cheap to built and simple to operate; they clean easily and can maintain sterility for a long period which also find them favour in ultra-pure water applications.

Cyclic Operations: Cyclic operations are traditionally the domain of filter presses and sand filters with periodic backwashes. More recently they have found important applications in surface and bore hole water treatment. These operations are akin to batch operations but with an important difference. In a cyclic operation there is no batch tank. In operation the feed from a reservoir is passed to the membrane module and as the water permeates across the membrane, the solids are retained inside the module. The solids are purged from the membrane module at a regular time interval or when there is a fixed change in either the flowrate or pressure. Since the membrane operates in a dead-end fashion these processes are only suitable for water with very low suspended solids or solids which form cakes with very high permeability such as ferric or alum flocs.

8.4.3   Pre-treatment

Pre-treatment are treatment steps which are required before membrane separation to enhance or maintain the membrane performance over an acceptable duration of operation. The types of pre-treatment required to reduce the potential for membrane fouling will depend on both the process stream and the membrane type. The following are examples of some common possibilities:

Physical screening or pre-filtration is of particular importance in reverse osmosis or when thin-channel membrane equipment is used. In a mixed effluent stream containing suspended solids, this technique is generally recommended.

In both water and wastewater applications, the use of coagulants or precipitation processes may help to reduce fouling in subsequent membrane operations. Where colloidal materials or natural organic matters (NOM) are present in the feed stream, they may be removed by such methods. Indeed, colours due to humic substances in surface waters which are normally too small to be retained by microporous membranes are easily filtered out by coagulation with ferric sulphate.

For RO applications the feed water pH is usually adjusted to a pH of between 4 and 6 for the following reasons:

- *It is required /usually as a condition of warranty to minimise the rate of hydrolysis of the cellulose acetate membranes. Cellulose acetate hydrolysis reduces the useful life of the membrane by increasing the flux but reduces the rejection of the membrane.*

- *The flux and rejection of some composite membranes are a function of pH and the optimum pH is between 4 and 6.*

- *Many natural waters are saturated in calcium carbonate which is highly rejected by the membrane. Consequently, it is concentrated in the retentate during the process and will precipitate on the membrane decreasing the flux and rejection. Lowering the pH of the feed water to between 4 and 6 converts some of the carbonate or bicarbonate ions to carbon dioxide and this prevents carbonate precipitation.*

Sterilisation of the process stream by chlorination, ozonation or UV treatment is a technique by which biological attack of membranes may be minimised. Chlorination of RO feed water is used to alleviate this problem. However, care must be taken to ensure that the membranes are comparable with this process, for example activated carbon must be used to remove chlorine for polyamide membranes which cannot tolerate chlorine.

8.4.4   Heat Generation & Temperature Profile

Temperature affects the choice of material of construction. Membrane performance is also affected by temperature, therefore, it is useful to be able to predict the rate of temperature rise during filtration:

For a feed and bleed system (continuous operation), at steady state:

$$MC_p \Delta T = Po - U \qquad (8.4.1)$$

Where M = Mass throughput rate
C_p = Specific heat capacity of the retentate
ΔT = Temperature difference between the outlet and inlet
Po = Power supplied to the recirculation pump
U = Total heat loss through vaporisation, convection & radiation

At steady state the process temperature attains a constant value.

For batch operations, unless cooling is applied, the process temperature would rise until the rate of heat loss through radiation and water evaporation equate to the power supplied to the pumps. Accurate prediction of the temperature profile is difficult without precise data, but a good approximation may be obtained by adding the following lines to the previous basic routine:

```
71 K=15: REM INITIAL TEMP
72 P=2.1: REM MAX. PUMP POWER
101 P1=P*(1-0.65*(C-C0)/(C1-C0))
102 K=K+P1*3600*H/V/4.18
103 IF K> 55 THEN LPRINT " COOLING REQUIRED"
104 IF K> 55 THEN K=55
105 F= 0.5+0.02*K: REM TEMP CORRECTION FOR FLUX
110 GOSUB 40: V=V-J*A*H*F
130 IF N=300 THEN LPRINT T, V, C, "TEMP=", K: N=0
```

Figure 8.4-4 Modification of programme in figure 8.2-1 to estimate the extent of heating, written in Basic.

Line 102 assumes that the recirculation pump is a centrifugal pump operating at maximum efficiency at the initial concentration C0 and power consumption drops inversely with increasing feed concentration (see Centrifugal Pumps Section).

8.4.5   Process control & Instrumentation

Pressure and flowrate are the two most common process variables often requiring control in membrane processes. Concentration, level and temperature control are also sometimes required. Time control is often associated with cleaning or back flushing operations.

In most systems the pressure and flowrate are inter-dependent so that only one variable may be controlled. In such a situation pressure control is often adopted because it is the most straight forward of the two variables to measure and control. For clean feeds spring-loaded diaphragm pressure regulating valves provide the simplest and lowest cost option. Feeds containing solids often cause blockage and prevent diaphragm operation. For such troublesome feeds it is best to use a feed back control loop with a pressure transducer and an actuated ball or butterfly valve.

Concentration control is probably the most complex and difficult target to achieve, yet often the aim of a process is to produce a product of a consistent quality. The main problem here is that concentration is a quantity which cannot be determined directly, but must rely on a secondary property such as conductivity, turbidity, colour intensity, density, viscosity, etc. for its determination. Even so, suitable transducers for the secondary properties are often not available for all the products of interest. Under certain circumstances an associated property may be used instead of the secondary property, for instance, pressure transducers have been used (instead of viscosity transducers) successfully to control the solid concentration of paints and lattices.

In batch operations level control is widely used for volume determination. Various types of level sensing devices are available. The simplest and cheapest types are the float switches which combine a reed switch encapsulated in a ball or cone (the float). The float is anchored to the side of the tank or may be suspended from the end of a cable. The main disadvantages of the float switches are their wide dead band, susceptibility to liquid turbulence and low tolerance to solid deposits. Conductivity probes as level detector fare better than floated switches in these regards, but they are particularly prone to fouling by film forming liquid such oils and paints. For those highly fouling feeds, non contact level detectors are preferred. In this class of detectors pressure pads are the most common. The pressure pads are essentially pressure transducers which measure the head of the liquid and because the sensing element is isolated from the liquid by a diaphragm there is no danger of fouling by any contaminants from the feeds. For underground sumps or in situation where a pressure pad cannot be fitted easily, ultrasonic level sensors are available. These devices equate level by sending out an ultrasonic wave and measures the time taken for it to be bounced back by the liquid surface.

In many applications involving biological materials temperature control (cooling) is often required to prevent product damage rather than to regulate any system performance. In these cases a suitable heat exchanger is installed in series with the membrane module, usually downstream of the membrane modules. The product temperature is maintained by regulating the flow of the coolant through the heat exchanger. For coolant temperature under 5°C water with over 20% ethylene glycol is used, otherwise water is used. The heat load must generally be removed by a refrigeration system, however, for products which can processed at temperature over 30°C air cool chillers often provide a more economic alternative.

Nowadays, any discussion on process control would not be complete without the mention of PLC (programmable logic controllers). Their low cost and versatility make their presence in any control scheme inevitable. Almost without exception, all membrane plants are now built with a PLC in their control panel. Quite apart from replacing a large number of timers, relays and wiring which would be necessary in a traditional control panel, the PLC enables the management of any process a simple matter. Many tasks or operations may be easily automated. Furthermore, any changes in the operations can be quickly and simply accommodated by a software change.

8.4.6 Materials of construction

The materials of construction depend on the desired applications. Pressure rating, thermal expansion and concentration, temperature stability, and mechanical strength are some of the factors to be considered to ensure that the materials of construction are compatible with the chemicals which are used. Some components of a membrane plant may be sensitive to metallic corrosion, e.g. water and wastewater duties. The most effective way to minimise corrosion is to limit the use of ferrous metals. Engineering plastics and plastic composites should be used in such cases. Polypropylene covers a wide range of the requirements. In many cases u-PVC can be used. Some applications, e.g. in the food, beverage and pharmaceutical industries and in biotechnology a steam sterilisation and hot cleaning steps are required. For such applications grade 316 stainless steel is used as the material of choice.

8.4.7   Protective System & Relief Policies

Any system design must provide a safe operation. This means that suitable equipment must be included in the design to afford the necessary protection for any personnel in the plant vicinity, the environment and the plant itself against any identifiable risks. The likelihood and the seriousness of any risks must be considered before a solution may be designed. In any case, any safety solution must be consistent with the company's health and safety policy.

The relevant risks associated with the operation of membrane plants include extremes of pressure and temperature, loss of containment and membrane rupture. The significance of each of the risks depend on the applications and they must be dealt with according their particular circumstances.

Pressure risks are most prominent where positive displacement pumps or compressed air are used in the process. Where positive displacement pumps are employed high pressure trips as well as pressure relief valves are recommended as two separate lines of defence. Where compressed air is used, the air supply line should be fitted with a pressure relief valve which is isolated from the process by a non-return valve. Any enclosed vessels should be fitted with a vent filter to prevent accidental vacuum generation or air pressurisation.

For equipment operating outdoor, frost protection must be considered. Where the process temperature may exceed the operating limits of any of the materials of construction, temperature trips must be included in the protection system.

Loss of containment can range from a small pump seal leak to an overflow situation from the batch tank due to a level control failure. In many cases a small leak is usually regarded as acceptable, except for instance when dealing with materials with bio-hazards or radio-activity. On the other hand a major tank overflow could pose a serious pollution threat. In general, hazardous materials are only processed in sealed facilities specially designed to contain the agents and enable their de-contamination. For all other industrial processing, bunding or interceptor tanks are the usual practice.

Membrane rupture or seal failure are small possibilities. In many cases, small leaks of the feed into the permeate stream do not cause a significant deterioration in the permeate quality and such leaks may go undetected. The significance of a feed leakage may depend on whether the permeate or the permeate is the desired product stream. Generally, if the permeate stream must meet a set quality criteria then the ratio of the leakage rate to the total permeation rate becomes a critical parameter. For instance, for a sterilisation application requiring a log reduction factor of 6, a pin hole in the membrane which produces a leakage rate only 1/1,000,000th the main flow rate would cause a quality failure. On the other hand in an application requiring only 99% solid removal from a stream, the system could tolerate up to 1% leakage. Pin holes in membranes, broken hollow fibres and failed seals may be detected by the formation of air bubbles through water after the normal process is suspended. On line detection of any membrane failure must rely on the detection of the presence of a contaminant in the permeate. Techniques such as UV absorbance, turbidity and conductivity measurements are available, but they are only able to detect fairly gross failure which are adequate for most purposes.

8.5 Operational Considerations

8.5.1 Introduction

A key consideration in the design of a membrane system is the way it will be operated. This should take into account all the local constraints and practices. Before a design philosophy may be conceived a full set of operational requirements must be drawn up. Such list should include the following main items:

1. *Throughput rate*

2. *Feed composition and variability*

3. *Feed flowrate and variability and size of balancing tank if available*

4. *Product composition requirement including any acceptable quality level, e.g. consent limits*

5. *Will the plant be manned or not. Any shift pattern and weekend working should be stated*

6. *Consequences of plant failure or shut down*

7. *Area classification*

8. *Whether operation is indoor or outdoor*

9. *Availability of services, including power, compressed air, water, sewer, floor space*

10. *Plant location*

8.5.2 Design philosophy

The design philosophy sets out both the design objective and how it may be achieved. For a membrane system, it is not enough to simply specify a batch or continuous membrane process to treat a certain feed in order to produce a certain stream. Wider considerations are to be given to the manner in which this is achieved; process reliability; the economics of the process and not least the health, safety and environmental issues. While the different system characteristics, the investment analysis and safety matters have been discussed at some length elsewhere in this publication, the following sections are concerned with accommodating the operational requirements of the process in the design.

8.5.3 Plant Control

As a minimum a membrane plant should facilitate the following operations:

1. *Plant start up*

2. *Batch or continuous filtration*

3. *Normal and emergency shut down*

4. *Cleaning*

5. *Draining and venting of membrane modules, pipework and tanks*

During start up any air in the recirculation path must be purged out. This is particularly important for systems which incorporate a closed loop design. Although any trapped air will eventually dissolved and emerge in the permeate, the process could take up to an hour and during this period severe cavitation or pressure fluctuations may occur as the air bubbles circulate between regions of high and low pressure. An air bleed valve or vent valve (which may be actuated) should be fitted at the highest point where the air may collect in each loop. Air purging and system priming may be accomplished with permeate or clean water. For membrane modules which can tolerate a very low reverse pressure (i.e. pressure from the permeate side) consideration should also be given to venting the permeate manifold where there is a possible existence of air pockets which may be pressurised by the permeate during operation.

Any sudden surge in the system pressure is likely to reduce the longevity of the membrane. For this reason it is desirable to reduce initial flowrate from the pumps at the start up. The larger the pumps the greater the need for this provision. For centrifugal pumps the flow rate can be reduced by simply throttling the delivery of the pumps. For positive displacement pumps a reduction in speed or a bypass valve would be required.

8.5.4   Batch Cycle Time

The cycle time for a batch process is primarily determined by the ratio of the membrane area to the volume of the batch tank. It is one of the most important parameters in the operation of the batch process and must be carefully designed to achieve optimum results. For a manned operation, the cycle time may be designed to fit in with the shift work patterns, but this should not be the only or the over-riding criteria. Consideration must be given to the system hold up volume and the volume of product that remains in the equipment after a normal product removal procedure. For a simple batch operation the following relationship must be observed:

$$(Batch\,volume) > (Hold-up\,volume) * (conc.factor) \qquad (8.5.1)$$

Ideally the batch size should be at least twice the product of the hold up volume times the concentration factor in order to ensure that the residual volume of product that remains in the equipment after a product removal does not significantly affect the initial composition of the feed in a subsequent batch. This would also reduce the possibility of air entrainment due to turbulence in the batch tank at low liquid level.

For operation in the so-called top up batch mode, the best flux performance is achieved by topping up batch tank continuously or doing so frequently in very small steps. The batch tank should also be as large as practicable.

8.5.5   Maintenance

A well-designed membrane plant will be easy to maintained and when maintenance is required it will allow any repair procedures to be carried out conveniently. The following are equipment items that normally require a scheduled maintenance programme:

- Pre-filters / strainers
- Pumps (seals, gear box, couplings, check valves, diaphragms, impellers, etc.)
- Membrane (element replacement)

- *Actuated valves*
- *on-line sensors (e.g. turbidimeters, conductivity meters)*

All such maintenance can normally be carried out by the process operators, except for pump maintenance which would require skilled fitters.

The pre-filters should be sized so that the frequency of service is acceptable to the process operator. If a change of pre-filter is expected during the process then a valving arrangement should be provided to enable a set of standby pre-filters to come into service which the fouled ones are being replaced. Ideally, differential pressure gauges should be installed on the pre-filters to indicate their condition.

One of the criteria for pump selection should be ease of maintenance. Pumps may occasionally require a complete strip down, therefore, this should be born in mind when the plant layout is designed. Pumps which allow a back pull-out leaving the motors and pipework undisturbed would be an ideal choice. Pump seals can often be a source of aggravation if an unsuitable selection is made for the application. For clean water applications, lip seals are generally acceptable. For dirty water applications packed glands are the favourite. Packed glands are designed to weep slightly in order to keep the sealing surface cool; they also allow the seal to be loosened or tightened even when the shaft is rotating. Furthermore, they can be quickly and easily replaced with the pump in situ thereby keeping any down time to a bare minimum. Note that traditionally isolation valves are installed both upstream and downstream of a pump to facilitate pump maintenance, if this is implemented then a suitable pressure relief or trip should be provided in case the valves are closed accidentally. Until recently, the generally held view was that pressure relief was not required if system pressure rating was higher than the pump's shut off head, however, it has been shown that certain centrifugal pumps could implode when the liquid in the closed space vaporises.

Ideally if a membrane module fails in service, the operator should be able to detect the failure, isolating the failed module to allow the plant to continue operation without an interruption. In practice, the costs of providing isolation valves for all individual modules would be too prohibitive. The general approach is to group a small number of module into sections. All the modules in a section will share common feed and permeate manifolds. Each section would allow separate sampling and isolation from the rest of the system if required. For a very large scale application it is quite usual for each section to be completely self-contained with its own pumps and control circuitry. In principle, the overall system reliability is enhanced, since at any point in time any number of sections could be isolated from the main operation and be undergoing cleaning or maintenance or be put on standby.

8.5.6 Cleaning

The main aim of a cleaning procedure is to restore the flux and membrane selectivity. However, cleaning also help prevent any possible consolidation of the fouling layer which may lead to an irreversible loss of performance. The decontamination and sanitising action of the cleaner is apparently essential for the control of the microbial population in food related applications. Indeed, in many dairy and food processing applications, membrane equipment is cleaned and sanitised on a daily basis. In other applications the frequency of cleaning depends on the rate of flux decline and the performance margin built into the design of the process. A weekly cleaning regime is common, but many processes have operated successfully with cleaning only 2-4 times a year.

The cleaning procedure is usually performed under manual control. Where the procedure is applied frequently it may be automated. The cleaning programmed can be carried out after a certain time or number of batches or it may be triggered by a fall in the throughput rate.

Before a chemical clean is introduced the plant should ideally be decontaminated by rinsing out once or twice with the permeate or clean water. This enables the cleaning chemicals to work much more effectively and also reduces the quantity of reagents required. After the chemical clean it is usual to rinse the plant out again once or twice with the permeate or clean water to avoid product contamination. In some wastewater applications such as oily wastes or latex washwater it may be possible to combine the spent cleaning chemicals and the rinse water with the normal feed so that a secondary waste stream is not generated. In other cases the spent cleaning chemicals and the rinse water must be collected and disposed by an alternative method.

Many cleaning formulations have been developed in the past 15 years for membrane cleaning and are available commercially. However, with a large number of different membranes and a countless number of different applications it must be up to each user to evaluate the effectiveness and suitability of a specific cleaning formulation for his particular application. The following table offers a guide which may be used as an aid in the selection of the appropriate cleaning agents.

8.6   References

1. N C Beaton and P R Klinkowski *"Industrial Ultrafiltration Design and Application of Diafiltration Process"* J Separ Proc Technol 4 (2) (1983) 1-10

2. Le, M.S. and Howell, J.A., *"Ultrafiltration"*, in Comprehensive Biotechnology, Cooney, C.L and Humphrey, A.E, (eds), Permagon Press, N.Y. 1982.

3. A R Cooper (ed), *"Ultrafiltration Membranes and Applications"*, Plenum Press, N.Y. 1980, pp 631-658.

Chapter 9

MEMBRANE PROCESS ECONOMICS

Contents

9.1   General investment evaluation

9.1.1   Factors Affecting An Investment Decision

A decision to invest in a membrane facility carries with it the burden of continuing depreciation, taxes, insurance, maintenance, etc. and also reduces the Company's capital for other purposes. Such a decision, therefore, must be made with great care as its success affects the company's profitability. In the final analysis, an investment decision may include intuition (strategic decision) as well as quantified parameters, but most factors can be quantified to enable a final assessment to be made without assistance. There are three basic steps in quantifying these decisions. First, one must decide which factors will affect future profitability should the proposal be accepted. Next, forecasts must be made for each of the factors, for both the proposed case and base case. Lastly, all the forecasts must be combined in a consistent and understandable format so that the overall profitability and return on investment on for each case can be compared. Risk and other factors must also be considered.

In general, every investment decision requires forecasts for at least two cases, one or more assuming an investment is made and one assuming no investment is made. All significant factors affecting the company cash flows must be considered for each investment option. Factors which affect the profitability of an investment may be as follows:

1. *Capital cost*
2. *Working capital*

3. *Construction period*

4. *Start up costs*

5. *Sales or savings forecast*

6. *Economic life*

7. *Depreciation life*

8. *Depreciation method*

9. *Minimum acceptable rate of return*

10. *Tax rate*

11. *Inflation rate*

12. *General business condition*

13. *Risks*

The significance of each factor is different for each investment option. It is important to note the significance of each factor and concentrate on areas of greatest concern. Further discussion of the more relevant factors to a membrane investment is given in the following sections.

9.1.2 Cash Flows

If an investment is to be made on a completely new site and a group of balanced facilities is installed (process equipment, auxiliaries, offices, etc.) which are designed for next year's operation only and no more, the amount of fixed investment to be used in the economic analysis is clearly defined. However, most new investments are made on existing sites and are integrated with existing facilities which are already in use. Existing equipment, buildings and personnel may be shared by the new installation. However, it is not appropriate to add a proportion of the value of the existing investment to the new investment so that the new project is seen to be paying for its share of the available facilities. Instead, the value of the existing investment should be reflected in the cash flows it can generate. The new investment should be justified by creating an increase in cash flows. Conversely, if the use of existing facilities with the new investment forces one to forego other profitable opportunities for using them, then it should be reflected by reducing the differential estimated cash flow. Thus, the value of alternative uses of existing equipment should be considered in determining the cash flow streams in the investment and no investment cases. No value for the existing investment should be included in the profitability forecast estimates (see profitability measures).

The above discussion is not applicable to a considerable number of non-productive facilities needed on many manufacturing sites such as administrative buildings, car parks, roads, security services, air compressor, steam boilers, etc. which are generally used throughout the site but which cannot be justified on the ground that it will produce a profitable product for sale. Some accounting policy must be agreed and followed consistently as to how profit earned by the productive facilities will be used to cover all the non-productive investments. There are a number of ways for doing this. One method is to charge each of the productive facilities an overhead cost or a transfer price high enough to show an acceptable rate of return on the non-productive investments. Another method of handling non-productive facilities is to require a high enough minimum acceptable rate of return on the productive investments which are profit-justified to produce a satisfactory profit level on the total investment for the company. The choice of methods for handling non-productive facilities is up to the company; the main requirement is that a consistent system is used which takes account of these facilities in the overall profit scheme. Again, no value for the existing investment should be included in the profitability forecast estimate.

If an expansion to an existing plant is made which will results in an addition to the overhead staff, only the costs which really change should be allocated to the expanded plant in justifying the new investment, regardless of the cost accounting procedures. The accountant will allocate a fair share of overhead to the expanded part of the plant. However, if the total overhead costs have increased by only a small fraction of the total, an additional effect of the accounting allocation may be that every other product made in the plant will receive a smaller allocation of the continuing overheads. The net difference must be the only true addition.

9.1.3 Depreciation & Taxes

Depreciation is an annual charge against income that distributes the original cost of the fixed asset over its expected service life. It is treated as an expense, but unlike most expenses it does not entail current cash expenditure. This means that in addition to the net reported profit for the period, the company has available additional funds corresponding to the depreciation charge which was included in the profit calculation but did not involve a cash outlay. This is the capital recovery or a partial regeneration of the original cost of the fixed assets and it is available for use as any other accumulated cash.

Clearly, the treatment of depreciation is of great interest to the taxman, because it affects the amount of income tax that goes into the treasury chest. Without taxes there would be no reason for government regulation of depreciation accounting. Indeed, depreciation owes its importance to government regulations of income taxes.

As might be expected, in all the developed countries governments regulate the deduction for depreciation since it directly affects income taxes paid by a company. Depreciation life may or may not have the same meaning as the physical lives of the individual pieces of equipment. In general, tax procedures include guidelines which put a certain type of equipment in a particular industry into a single life category. For example, 11 years is the standard life for chemical plant equipment, except office furniture (10 years), transport equipment (3-6 years) and buildings (45 years). Guideline of lives for other industries are: Petroleum Refining (16 years); Iron & Steel (18 years); Food Products (15-18 years); Engineering (12 years). Salvage values may be considered in setting the life for the equipment. There is no guideline for the life of membrane equipment; it is expected that this will be in line with the standard life for equipment in the industry where they are employed.

There are different methods for spreading the original cost of a fixed asset over its estimated useful service life. The simplest of these is the straight line method which distributes the cost of the asset uniformly over its depreciable life as follows:

$$D = (C - S) / n$$

where C is the cost of the asset; S is the expected salvage value at the end of the asset life; and n is the estimated life in years.

The salvage or liquidation value of an asset decreases sharply in the early years of its life than in later years. In order to show the book value of an asset at its likely salvage value at any time, expenses charged in the earlier years will have to more nearly reflect the annual decrease in value. There are two main methods of depreciation accounting to show this effect. In the Declining Balance Method a constant percentage of the remaining book value is written off each year as follows:

$$D = k(C - D_c)/n$$

Where D_c = cumulative depreciation charged in previous years; k = a constant (from 1 - 2). This is a geometric progression which would never provide a complete depreciation of an asset. The other method which charges depreciation at a decreasing rate is the Sum-of-Years-Digits (SYD) method which is an arithmetic progression which would result in complete charging off the asset cost during the lifetime of the asset. The basic SYD method involves numbering the years of life in order and summing these numbers (years digits). The depreciation rate for each year is obtained by multiplying the asset cost by the numbers of the years in reverse order and dividing by sum of the years digits. This may be expressed as follows:

$$D_y = 2C(n - y + 1)/[n(n + 1)]$$

where y is the age of the asset in years. The graph below illustrates the different methods of depreciation.

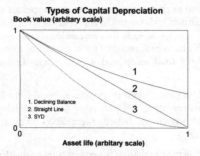

Types of Capital Depreciation
Book value (arbitary scale)

1. Declining Balance
2. Straight Line
3. SYD

Asset life (arbitary scale)

Figure 9.1-1 Variation of capital depreciation over asset life for various depreciation conventions

Membrane cost typically represents 20% of the total investment cost in a membrane plant. While the life of the whole plant may be depreciated over 15 years, membrane must be replaced every 2-5 years. Thus it is appropriate to separate membrane replacement cost from the depreciation equation for the rest of the plant as follows:

$$D = f(C - K, n)$$

$$R = K/m$$

where C is the total cost of the membrane plant; K is the cost of replacing all the membrane; R is the annual membrane replacement cost; f is one of the depreciation functions above; m is membrane life and n is the estimated life of the plant in years. R is also an annual charge against income.

9.1.4   Working Capital

The engineer engaged in cost and profitability estimation must constantly use the concept of working capital, and use it properly in order to estimate the profitability of an investment which is being proposed. Like fixed capital, it is part of the investment on which a return must be earned. Like land, working capital is not depreciated over the life of the project, nor is it tax-deductible as an expense. It is considered liquid and convertible into cash when the project is terminated. Since there is no depreciation allowance on working capital, it must be maintained for the entire life of the project.

For an economic evaluation of the proposed investment, working capital is defined as the funds, in addition to the fixed investment, which a company must contribute to the project. These must be adequate to get the plant operating and to meet subsequent obligations as they come due. Working capital is not a one-time investment which can be known once and for all at the start of a project. Working capital varies with production level and other factors. This definition for working capital is consistent with that used for fixed capital. It is an amount of money actually put aside by the company to get the plant going. There are a number of different types of working capital. A consistent method for valuation of each type of working capital is discussed below:

The easiest element of working capital to estimate a value is stocks. Raw materials, work-in-progress and finished-product should all be considered. Raw materials should, of course, be valued at their delivery cost to the site. However, finished products represent saleable materials to which the value of manufacture has already been added. Although there are arguments in favour of valuation at sale price, working capital as defined here is the cash outlay needed to support the project. The consistent approach is to value finished products only at the cash value of the cash required to produce them. This would exclude non-cash items such as depreciation and profit.

Accounts receivable represent goods which have been sold but for which payment has not been received. Valuation of account receivable, for the purpose of determining the working capital requirement is another problem. The balance sheets show them at sales value. However, on a cash basis the investment by the company in accounts receivable is only the cash cost of the finished products plus freight and any shipping charges. Any valuation between these two may be used since all costs attributable to the product have been spent, but nothing has been received, including the profit. In a prompt paying business, the differences in this range of valuation should not be significant.

Accounts payable represent goods, services, etc., which are available for use (or already used) in the plant but for which payment is not yet been made. Depending on the occurrence of the pay-day for various parts of the human resources, most companies accrue a considerable accounts payable as wages or salaries payable, representing the labour's value added to various products. In as much as labour is expended on making the products before it is paid for, the employees are essentially advancing money to the company. Payment for utilities after the date of use should likewise also be considered a source of funds. If pipe-line water or gas is a raw material, an account pay will also exist, as described above. Such accounts payable represent an offset against accounts receivable. Payable should be included in working capital analyses to find the net amount of company funds required over and above the fixed investment.

Each profitable sale incurs an obligation to pay taxes, but until it becomes due, the cash equivalent of the tax portion received from the customer is usable as working capital by the company. If taxable income is earned uniformly over the year, accrued taxes payable average six month of taxes. As profits accumulate, the capital provided from the tax payable frequently reduces the working capital required by an going project. Any profitable investment adds profits, and the taxes accrued on such profits may in some cases offset all the other items of working capital considered above.

Cash is the balancing factor in any business transaction. Cash is needed only to take care of the non-uniformity of the business cycle (within the month or season). Cash held in excess of this requirement is really idle, not working, capital, held for reasons other than immediate needs. A minimum cash balance must be maintained to cover the expected and also the random variations in receipt of income versus bills.

Thus working capital is the net sum of the five elements: stocks, accounts receivable, accounts payable, taxes payable, and cash.

9.1.5   Risks

The foregoing discussion relates to investment information to enable a better-informed decision and therefore reduce the likelihood of an unsuccessful investment. For simplicity, discussions tend to deal with a combination of best-estimate forecasts for each of the many elements involved and combination of these into a single, deterministic forecast of future conditions. Most companies use such a simple presentation for formal capital sanctions. However, the future is far from certain and many of the forecasts necessary for these analyses will not be precisely achieved.

Supplementary techniques and approximations have been developed to permit a better grasp of the impact of likely deviations from project forecasts. They may be used for decision making, though usually not incorporated into any formal systems.

Risks may be of many different types and nature. The major risk for a membrane plant investment is fouling. Failure to allow a sufficient operating flux margin may result in a serious deficiency in the plant capacity whereas irreversible long term fouling may require more frequent membrane replacement than anticipated. Since the design flux determines the size of a membrane plant, hence the capital cost; an optimistic flux assumption would of course present a very attractive economic case. Thus, when two rival membrane system suppliers compete for the same business, but each quoting a different working flux in his bid, the customer should be aware that a low flux system does not necessary indicate poor efficiency or that a high flux assumption carry greater risk. Rather, the customer should question the validity of the assumptions and ask for evidence such as reliability of any test data or any long term experience with similar applications elsewhere. If, necessary the customer may ask for a revised bid which is based on a greater design flux margin (thereby reducing the risk by paying a higher premium). The treatment of flux data and adoption of an appropriate flux value in system design is discussed further in the design chapter.

Another significant risk, of course, is that profitability will fall below the forecast. Sales volume and price are most susceptible to unforeseen declines (e.g. loss of market share; falling demand for product). By substituting forecasts for certain sensitive elements such as these, the sensitivity of the result to changes in such forecasts can be determined. However, such a technique still may not give a reliable feel for the likelihood of achieving the minimum acceptable rate of return. For a detailed treatment of the subject of risks in capital investment and their evaluation, the reader is referred to specialist literature on the subject [1],[2].

9.2   Capital cost estimation

9.2.1   Types and Accuracy of Estimates

This section explains the techniques which may be used to provide an estimate of the capital cost of a membrane installation which require a minimum expenditure of time and money. It is not possible to produce fully detailed estimates of all equipment, materials, labour, and overhead expense for each new membrane plant proposal. Even if complete drawings and specifications are available, a detailed estimate prepared from them would not necessarily be an accurate one, since 25 to 40 per cent of the total cost of any capital project is for such components as labour, off-site expenses, engineering, and start-up, which cannot be priced precisely. Large differences may also result from the estimator's bias, whether optimistic or pessimistic. Two simple preliminary steps toward more consistent and accurate estimates are:

1. *Check the completeness of the project scope*
2. *Draw on experience of previously proven capital projects to reduce any bias.*

There are many factors which contribute to poor estimates, the most significant is that of not covering the full scope of the proposed project. Initial estimates are often lower than actual project costs. The cause is often the omissions of some significant pieces of equipment, services, or auxiliary facilities rather than any significant errors in pricing.

Estimates fall into a number of categories from a rough guesstimate to a highly detailed procedure based on finished plans and specifications. The process industry commonly employs five kinds of estimates in the evaluation of capital costs for process plants, namely:

- *Order of magnitude estimate*
- *Study estimate*
- *Budget authorisation estimate*
- *Project control estimate*
- *Firm or contractor's estimate*

Order of magnitude estimate is primarily a rule-of-thumb applied only to repetitive types of installations with good cost history. At the other extreme is the firm or contractor's estimate, requiring complete specifications, and site surveys. Time seldom allows the preparation of such estimates prior to a project sanction. Three other basic types of estimates are frequently used. The study estimate is used to examine the economic feasibility of a project before significant spending on piloting, market studies, land surveys, and acquisition. Such an estimate may have an accuracy of only 30 per cent, but it can be prepared with minimum data at a very low cost. Budget authorisation estimate has an accuracy of only 20 per cent but requires more detailed information. Project control estimate is accurate to produce within 10 per cent with the information available as indicated on Table 9.2.1. The limits of error indicated on the table imply variable results depending on available information. There is a greater probability that the actually cost will be more than the estimated cost where the information is incomplete.

First step toward accurate estimating is adoption of a standard list of items for use on all construction projects. A good list should separately identify cost elements useful in the estimation of future capital cost. It should also be functional to provide the most convenient means of cost collection in the field. For instance, the total installed cost for a pumping station, including pumps, piping, instrumentation, foundation may properly records but cannot be used for later estimating a similar unit. Collecting costs functions allow identification of charges for individual items comprising the whole, thus providing information from which the installed plants may be evaluated. Accumulation of detailed cost data of itself, does not ensure high estimating accuracy. Even contractors based on completed drawings and specifications show variations between low and high bids as wide as 30 per cent.

Table 9.2.1 Estimating information guide

Order of magnitude estimate (+/- over 30%)	Study estimate (+/-30%)	Budget Authorisation estimate (+/-20%)	Project control estimate (+/-10%)	Firm estimate (+/-5%)	Required information	
✓					Plant capacity, location & site requirements, Utility & service requirements. Building & auxiliary requirements. Raw materials & finished product. Handling & storage requirements.	Project Scope
	✓				Rough sketches	Process flowsheet
		✓			Preliminary	
			✓	✓	Engineered	
✓	✓				Preliminary sizing & material specifications	Equipment list
			✓	✓	Engineered specifications	
✓	✓	✓	✓		Membrane module type, size & number	
	✓	✓	✓	✓	Pump type, size & number	
					General arrangement	
	✓	✓			(a) Preliminary	
			✓		(b) Engineered	
✓	✓				Approximate sizes & type of construction	Buildings & Structure
	✓	✓			Foundation sketches	
		✓	✓	✓	Architectural & construction	
	✓				Preliminary structural design	
			✓	✓	General arrangement & elevations	
				✓	Detailed drawings	
✓	✓				Preliminary flow sheet & specifications	Piping
			✓	✓	Engineered flow sheets	
				✓	Piping layouts & schedules	
	✓				Rough specifications	Insulation
		✓			Preliminary list of equipment & piping to be Insulated	
			✓	✓	Insulation specifications & schedules	
				✓	Well developed drawings or specifications	
	✓				Preliminary instrument list	Instrumentation
			✓	✓	Engineered list & flow sheet	
				✓	Well developed drawings	
✓	✓				Preliminary motor list with approximate sizes	Electrical
			✓	✓	Engineered list & sizes	
	✓		✓	✓	Substations, number & sizes, specification	
				✓	Distribution specifications	
	✓				Preliminary lighting specifications	
			✓		Preliminary interlock, control & instrument wiring specs.	
			✓	✓	Engineered line diagrams (power & light)	
				✓	Well developed drawings	
✓					Approx. quantities gas, steam, water, electricity, etc.	Utilities Usage
	✓				Preliminary heat balance	
	✓				Preliminary flow sheets	
			✓	✓	Engineered heat balance	
			✓	✓	Engineered flow sheets	
				✓	Well developed drawings	
	✓	✓	✓	✓	Engineering & drafting	Labour
				✓	Labour by trades	
				✓	Supervision	
	✓	✓	✓	✓	Location	Site
		✓	✓	✓	General description	
		✓	✓	✓	Soil bearing	
		✓	✓	✓	Location & dimensions of any infrastructure	
			✓	✓	Site plan	
				✓	Site facilities	

9.2.2   Order of Magnitude Cost Estimates

Table 9.2.2 may be used to produce order of magnitude estimates (accuracy +/- 50%) for both RO and UF plants. This method assumes that centrifugal pumps with 60-80% efficiencies are used. The estimates do not include the costs of the following:

• *Land*
• *Civil Engineering*
• *Buildings*
• *Water intake and treated water pumping stations or storage*

Table 9.2.2 *Membrane (RO & UF) Plant Cost Estimates (+/- 50%)*

Plant Size / Cost[1]	Spirally Wound	Hollow Fibre	Tubular
30-100 m² Plant[2]	20,000 + 1500A	25,000 + 1800A	30,000 + 2500A
101-500 m² Plant[3]	1250A	1600 A	2000A
501-1000 m² Plant[4]	1000A	1400A	1500A

[1]   A = Estimated membrane area; £(1996)

[2]   Assume batch operation

[3]   Assume 3-4 staged plants

[4]   Assume 4-6 staged plants

9.2.3   Scaling From Previous Estimates

This method may be used to adjust one set of estimates to a different design size. Estimates produced by this method may have an accuracy of greater than +/-30 per cent and is useful for the evaluation of the economic feasibility of a project.

It is known that equipment size and cost correlate fairly well by the logarithmic relationship known as the "six-tenth factor". However, the relationship cannot be used if auxiliary facilities are not identical between the old and new estimates. The equation below is a simple form of the six-tenths factor relationship.

$$C_n = r^{0.6}C$$

Where C_n is the new plant cost, C is the previous plant cost and r is the ratio of new to previous capacity.

Estimations of the capital cost for desalination plants using RO (reverse osmosis) and ED (electrodialysis) may be produced from charts given by Glueckstern & Arad [3]. Similarly, capital cost estimates for wastewater treatment plants with RO and UF are given by Rogers[4]. These correlations were based on cost data for plants of 380-380,000 m³/d capacities between 1976-81. As such, all the prices must be adjusted by the appropriate cost index. However, improvements in membrane technology in recent years and the pressure of completion has driven the cost of membrane down substantially which means that the margin of error by the estimates may be considerable.

9.2.4   Costing By Percentage Factors

Table 9.2.3 is a collection of cost data expressed as percentages of total plant installed costs for water treatment plants using microfiltration. It can be used to compile quickly the cost of a new facility from a minimum of data. Given a preliminary list of equipment and rough specifications of structural, electrical, and piping requirements, an acceptable study estimate can prepared (accuracy of 30 per cent). The method requires a reasonably good understanding the process involved to enable the selection of the proper percentage factor within the ranges shown in the table. It is necessary to establish a base by first estimating the cost of the process equipment from the rough flow sheets. Then assume a factor for the installed cost of the equipment as a percentage of total plant cost to arrive at the trial total cost. The estimated costs of other components of the plant are obtained by applying to the trial total cost reasonable percentage factors within the ranges listed in Table 9.2.3. A contingency within the +/-30 % range must be applied to the estimated cost of the plant. During an inflationary period the value selected should be in the positive part, preferably on the high side.

Table 9.2.3 Range of Components for a Membrane Water Treatment Plant		
	% of total installed plant cost	
Component	Range, %	Median, %[1]
Membrane housings	15-20	18.75
Membrane elements	10-15	12.5
Pre-filtration equipment	1-6	5
Tanks and drums	3-6	2.5
Pumps and motors		1.25
Clean in Place (CIP) equipment		6.25
Chemical dosing equipment		6.25
Valves & Piping		10.
Instruments and controls		6.25
Electrical		3.75
Steel structures and platforms (other than buildings)		1.25
Insulation and painting		0.63
Utility equipment (air compressor, space heaters)		1.25
Safety equipment		1.88
Design		6.25
Labour (building & installation)		12.5
Commissioning		3.75
		0
Total		100

[1]   Adjusted percentage factors

9.2.5   Costing by Scaling Plant Components

By combining the two previous methods cost estimates with an accuracy of +/-25% may be achieved.

The equipment size and their cost correlate fairly well in a logarithmic relationship similar to the "six-tenth factor" relationship. However, the value of the power index for each component vary depending on its sensitivity to the size of the equipment. Thus, for example the cost of membrane increases almost linearly with membrane area, but the cost of the control panel hardly changes with the size of the plant. The cost of a new component is given by:

$$C_n = r^i C$$

Where C_n is the cost of the component in the new plant, C is the previous plant component cost, i is the component cost scaling index and r is the ratio of new to previous capacity. Table 6.2.5 shows the cost indices for some of the common equipment components in water treatment (200-6000 m^3/h plant capacity). The total capital cost of a new plant is made up by the summing all the major component costs and dividing it by the sum of the major component costs as percentages of the total plant cost.

Table 9.2.4 - *Cost indices for water treatment equipment (200-6000 m^3/h plants)*

TYPE OF COST	INDICES
Strainers	0.83
Chemical systems	0.36
Rapid gravity filters (RGF)	0.68
NF Membrane systems, ex pumps	0.95
MF Membrane systems, ex pumps	0.5
Pumps & mixers	0.66
Pipework	0.75
Panels & Cabling	0.1

A similar method for Cost Estimation for the Process Industry is described by Woods [5] which has been used by Cote and Lipsky to examine the economic feasibility of a Pervaporation Processes [6].

9.3   Operating cost estimation

Operating cost estimates of various degrees of accuracy are required in order to evaluate a new investment, to evaluate new products for processing with existing (available) equipment; to justify some process changes, and to evaluate the positions of competitors. As with other estimates, accuracy and extent of details depend on the purpose of the estimate.

Estimations of the unit production cost for water from sea water and brackish water using RO reverse osmosis) and ED (electrodialysis) may be produced by the correlation given by Glueckstern & Arad [3]. Similarly, the unit treatment cost for some wastewater by RO and UF are given by Rogers [4]. These correlations were based on operational data for plants 380-380,000 m^3/d capacities between 1976-81. As such, all the prices must be adjusted by the appropriate cost index.

A complete cost estimate must include a large number of items, such as shown in Table 6.3. Estimating effort should be devoted to the most important elements of cost. Operating cost estimates should follow the format normally adopted within one's own company. A typical format for process cost estimates, for example, may have five column headings including items of cost, units of measure, variable costs, fixed costs and total costs. The list of items of cost may be grouped as in Table 6.3 in order to establish the conditions of the estimate and permit estimates to be made at other conditions. Such format permits gathering information on a single table which will constitute both an incremental and a total cost estimate. Estimates of total accounting cost are usually needed by the department concerned in order to see how a specific project will affect their operating statements.

Table 9.3 Check list for operating cost estimation

RAW MATERIALS	Processing materials
	Utilities (power, water, steam, etc.)
	Maintenance materials (cleaning chemicals, preservatives, spares, etc.)
MANPOWER	Direct labour
	Supervision
	Direct maintenance labour
	National insurance & pension contributions
	Sickness & holiday pay
PLANT OVERHEADS	Administration
	Laboratory services
	Legal, Technical & Engineering services
	Purchasing, receiving, and warehousing
	Personnel (staff)
	Safety & Inspection
	Accounting
	Telecommunication
	Canteen facilities
	Insurance & local tax
	Waste disposal and pollution control
DEPRECIATION	Rent and / or depreciation charges
	Membrane replacement cost (per year)
MARKETING & DISTRIBUTION	Warehousing
	Freight
	Advertising and promotional literature
	Technical sales services
	Samples and displays
	Travel and entertainment
	Market research and sales analysis
RESEARCH AND DEVELOPMENT	In house work
	Consultant fees & contract work
FINANCIAL	Interest payments on fixed investment & working capital
	Insurance
	Taxes

For plants already in operation, accounting records provide a major source of information for estimates. The main difficulty is in the transfer of information with proper use of allocated costs and overheads. For new processes, parallels may be drawn with existing operations, and fairly accurate estimates can often be made. When an investment proposal is to be justified estimates may be necessary for both the proposed and base cases. The base case will often be an existing operation and the proposed case will be operation with the expanded or new facilities.

The list of items of costs in Table 9.3 may require some elaboration. The scope of application for some of the more significant items is discussed further in the following paragraphs.

If an expansion to an existing plant is made which will results in an addition to the overhead staff, only the costs which really change should be allocated to the expanded plant in justifying the new investment, regardless of the cost accounting procedures. Likewise, when considering the economics of a proposed process, only the cost of additional requirement for each utility or services should be used. However, when the actual cost accounting takes place the requirements of all the processes will be combined. The accountant will allocate a fair share of overhead to the expanded part of the plant.

Processing materials

As membrane plants are often installed for pollution abatement or product recovery purposes, the cost of processing materials has different meanings depending on the objective of the process and the purpose of the estimates. If a membrane process is being considered as an alternative to another method, for example a volume reduction to lower the cost of transportation or COD removal from a waste stream before discharge to reduce effluent charges then the material cost may be considered to be nil. In the case COD removal the material is considered to be a waste. However, in the case of the volume reduction, the material may or may not be a waste (e.g. transportation of milk and sewage sludge). On the other hand, if it is proposed to use membrane to add value to the material, for example to produce proteins from cheese whey which has traditionally been used as a pig feed supplement, then there is a definite cost which is equivalent to the loss income from selling the whey as a pig feed supplement.

Utilities

The following are usually considered utilities, although in some companies one or more of them may be treated under other categories on the cost sheet: steam, water, cooling water, deionized water, electric power, heating gas, refrigeration and compressed air. Their effect on the cost of the product will of course depend on the process involved. In addition to the cost of the supplies from the utility companies, supply may require treatment or must be transformed and distributed through the plant proper. For a process plant these distribution charges may be significant.

Labour and Supervision

Investment proposals often involve either an addition or a reduction in the labour force. In general, direct labour is more important for batch than for continuous processes, and scheduling is considerably more flexible. If a batch process is not running for part of the time, the men may often be scheduled regularly to operate more than one unit. Different manpower on a given batch operation, operated the same number of shift, can often result in widely differing overall time cycles of production. Conversely, for continuous processes the labour force is almost constant whether the production rate is high or low. Labour becomes more of a fixed cost than in the batch operation. As automation in processes increases, direct labour in the process plant decreases, but is offset somewhat in the need for increased skilled instrument maintenance labour. The minimum labour requirement for a given operation is usually determined by safety considerations rather than the number of duties to which the operators must attend. With regard to indirect labour, there are many possible categories of non-operating labour a plant, but most accounting systems and hence most cost estimates do not allocate them directly to a cost centre or a product but group them in plant overheads as overhead labour.

Fringe Benefits

All companies have labour costs in addition to those paid for hours worked. They include National Health Insurance contributions, sick pay, travel and accident insurance, meal allowances, holidays, pension contributions, etc. These costs are significant in relation to direct wage payments, they may typically add 15 to 40 per cent to the wage payment and must be included in any cost estimates.

Maintenance

The total cost of maintenance will be a function of complexity of the process, material of construction, and the skill level of the maintenance force. Maintenance cost has three components: materials required; manpower to install them, and overhead for supervision and scheduling. These three parts of maintenance cost may vary widely in their proportions of the total cost. However, the most frequent method for estimating maintenance costs before construction of a plant is to include a certain percentage of a the fixed investment as an annual maintenance expense. In general, for investment in a membrane plant, 4 per cent per year of the new investment for maintenance is a good guide. For plants with extensive instrumentation, this may increase to as high as 7 to 10 per cent of the investment. Maintenance costs are not constant over the life of the plant but will increase during later years. Also, accounting treatment varies from company to company in distinguishing between equipment additions that are maintenance and those which are capital additions.

Plant Overheads

All costs on the operating facility which are not chargeable to any particular product or operation are considered plant overheads and must be charged back to the operating cost of the product on some allocated basis. This may includes all the plant overheads items in Table 6.3. Since no one cost centre or product is responsible for a particular overhead cost, the cost must be allocated in the most reasonable manner.

Depreciation

Plant depreciation and membrane replacement cost are covered elsewhere in this chapter (see General investment evaluation).

9.4   Profitability measures

9.4.1   Introduction

There are many methods for choosing between alternative investment projects on the basis of earning potential. In this section the most common methods for measuring return on new investments are discussed in terms of their ability to account for the many factors affecting the future of an investment. The most important differences among the various methods concern their ability to handle the key factors essential to an investment decision, not ease of calculation. Calculation of return on investment itself should be an small part of the expense of developing the valuation of any investment.

9.4.2   Pay-Back Period

The pay-back period of an investment is defined as the length of time required for the stream of cash flows generated by the investment to equal the original cash outlay. It should be noted that this definition requires the earnings to be taken after taxation since tax payments diminish the net cash flow, but before depreciation since the provision for depreciation is purely a book-keeping transaction and, therefore, has no effect on cash flow. The advantages this method are:

- *It is simple to calculate, but also simple in concept and, therefore, easily understood by non-financial executives.*
- *It indicates the timing of the investment being available for re-investment.*
- *By concentrating on the earliest pay-back date it recognises that early returns are preferable to those which accrue later.*

This method fails to give any consideration at all to those earnings which accrue after the pay-back date. In some instances the subsequent earnings can be very high or very low. Furthermore, pay-back ignores the timing of earnings prior to the pay-back date.

9.4.3   *Average Return on Capital Employed*

This method of capital priority ranking measures the average annual return throughout the estimated life of the project. Taxation makes no difference to the ranking of projects but the return will usually be preferred on the basis of profit after taxation since this represents the funds at the disposal of the company. The annual R.O.C.E. is calculated on the average capital employed throughout the life of the asset. The advantages of the R.O.C.E. method include:

- *The calculation is easy to make but (unlike pay-back) it gives consideration to the total earnings throughout the life of the asset.*
- *The average R.O.C.E. provides a reasonable basis of comparison for earnings with alternative investment opportunities*
- *It is possible to introduce a minimum acceptable R.O.C.E as a criteria for project selection.*

The main weakness of R.O.C.E. method is that it fails to recognise the timing of cash flow.

9.4.4   *Net Present Value*

A good method for the allocation of capital funds should take into account the earnings over the life of the project; give due weight to the duration of the earnings; and also make allowance for the timing of the cash flows. This can be achieved by the use of discounted cash flow in a method known as the *Net Present Value* (NPV) method. NPV, also known as Present Worth is a profitability measure in which compound interest factors are used to discount all cash flows to their equivalent value at time zero, using minimum acceptable rate of return as the interest rate. Time zero is taken as the start of operations. It is usual to list cash incomes separately from cash outlays, using discounted values, and to show sums of both as of time zero. NPV is the excess of the present value of the total income throughout the project life over the present value of the investment.

Principle of Discounted Cash Flow

The *Discounted Cash Flow* (DCF) technique recognises that £1 received today is more valuable than £1 receivable in (say) five years' time. The term "present value" is simply a time value of the money. When money is gathering interest the longer it has been invested, the more it is worth today. The present value factor compounds money received or disbursed before zero time from their face value. Similarly, the present value factor discounts or deducts interest from money received or disbursed in the future. The table below shows the value today of £1 to be received or paid after a given number of years.

Present Value Table

The value today of £1 to be received or paid after a given number of years

Number of years	Discount Rate											
	1%	5%	10%	12%	14%	16%	18%	20%	25%	30%	35%	40%
1	0.99	0.95	0.91	0.89	0.88	0.86	0.85	0.83	0.8	0.77	0.74	0.71
2	0.98	0.91	0.83	0.8	0.77	0.74	0.72	0.69	0.64	0.59	0.55	0.51
3	0.97	0.86	0.75	0.71	0.68	0.64	0.61	0.58	0.51	0.46	0.41	0.36
4	0.96	0.82	0.68	0.64	0.59	0.55	0.52	0.48	0.41	0.35	0.3	0.26
5	0.95	0.78	0.62	0.57	0.52	0.48	0.44	0.4	0.33	0.27	0.22	0.19
6	0.94	0.75	0.56	0.51	0.46	0.41	0.37	0.34	0.26	0.21	0.17	0.13
7	0.93	0.71	0.51	0.45	0.4	0.35	0.31	0.28	0.21	0.16	0.12	0.1
8	0.92	0.68	0.47	0.4	0.35	0.31	0.27	0.23	0.17	0.12	0.09	0.07
9	0.91	0.65	0.42	0.36	0.31	0.26	0.23	0.19	0.13	0.09	0.07	0.05
10	0.91	0.61	0.39	0.32	0.27	0.23	0.19	0.16	0.11	0.07	0.05	0.04
11	0.9	0.59	0.35	0.29	0.24	0.2	0.16	0.14	0.09	0.06	0.04	0.03
12	0.89	0.56	0.32	0.26	0.21	0.17	0.14	0.11	0.07	0.04	0.03	0.02
13	0.88	0.53	0.29	0.23	0.18	0.15	0.12	0.09	0.06	0.03	0.02	0.01
14	0.87	0.51	0.26	0.21	0.16	0.13	0.1	0.08	0.04	0.03	0.02	0.01
15	0.86	0.48	0.24	0.18	0.14	0.11	0.08	0.07	0.04	0.02	0.01	0.01
16	0.85	0.46	0.22	0.16	0.12	0.09	0.07	0.05	0.03	0.02	0.01	0.01
17	0.84	0.44	0.2	0.15	0.11	0.08	0.06	0.05	0.02	0.01	0.01	0
18	0.84	0.42	0.18	0.13	0.1	0.07	0.05	0.04	0.02	0.01	0.01	0
19	0.83	0.4	0.16	0.12	0.08	0.06	0.04	0.03	0.01	0.01	0	0
20	0.82	0.38	0.15	0.1	0.07	0.05	0.04	0.03	0.01	0.01	0	0

Internal Rate of Return (IRR)

The choice of the discount rate may be critical under the NPV method since the project priority ranking can differ at varying discount rates. This method, also know as the Internal Rate of Return is based upon the proposition that the best investment is one which yields the highest return from equating the NPV and the capital invested, i.e. the rate at which discounted earnings equal the present value of the capital expenditure. A trial-and-error solution is necessary to calculate the internal rate of return, since the discounting rate is the unknown sought in the calculation. The cash flows are set forth before discounting, in columns showing all cash flows, from start of construction to recovery of land and working capital after the project is shut down. Trial discounting rates are then applied to determine which rate makes the present value of earnings equivalent to the present value of all investments. Note that the interest rate of return is not earned on the original or on the total amount of investment. The discounting procedure corresponds to alteration of the principal invested in the project by the amount of the interest. Interest accumulated over each compounding period is added to the principal at the beginning of the period to obtain the new principal for the next period. The term obligation is used to refer to this changing sum and avoid confusion with the original principal sum. Obligation denotes the total indebtedness of the project to the investor.

Cumulative Cash Position

The cumulative cash position chart is an excellent tool for portraying the realities of the status of an investment in relation to the investor. As its name implies, it shows the cumulative cash effect of the project on the company. This chart (*Figure 9.4*) uses time as the horizontal scale and cash position as the vertical scale. The start of production is chosen as zero on the time scale. The initial zero cash position usually occurs prior to this time, at start of construction. Since the chart shows cumulative

9 - 16

cash position, a single line indicates, throughout the construction period and the earning life, the cumulative cash status of the project at any time. During construction, the line slopes downward from the zero line to indicate increasing negative cash position. Working capital is added instantaneously at time zero; this is a simplifying approximation. As income starts, the line heads up reducing the investment to zero. Where the line crosses the zero line, the point is know as the pay back time. As cash continues to come in, the slope continues upward into the positive cash position (profit) area. At the termination of the project, recovery of land and working capital adds a final portion of cash to establish the final cash position. Note that in some cases the cost of decommissioning a plant or clean up of a site where the land has been contaminated may add a negative value to the cash position rather than producing a positive recovery value.

The dotted line below the cumulative cash line is the discounted cash flow where the discount is rate is merely the cost of capital or a minimum acceptable rate of return. The point where the dotted line crosses the zero line corresponds to the "payback time including interest". The amount of profit made available by this project is assumed to be re-invested at the cost of the capital in other projects, so that the terminal cash position after applying the cost of capital may be greater than the final undiscounted cash position if the plant economic life is sufficiently long. The dotted line above the cumulative cash line shows the relationship between the terminal worth and the present worth discounted at the cost of capital.

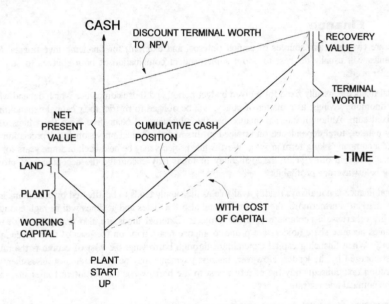

Figure 9.4 Cumulative Cash Position Chart

Whole Life Cost

Where an investment is made in order to combat pollution or to facilitate waste management as the case with a large number of membrane installations, there may be no income stream to speak of. The purpose of the investment, therefore, is to save cost or simply to meet a legal obligation. In such cases, terms which normally associate with capital investments such as profitability or R.O.C.E. do not have real meaning. Instead, terms such as cost of ownership and cost saving become the key measures for investment ranking. There are no widely accepted definitions for these terms. However, cost of ownership is often taken to mean the complete package of costs including the initial purchase cost plus the cost any service and maintenance contract. Cost saving is the amount of cash that is not expended which otherwise would have been expended if the investment was not made. Thus investment ranking by these methods places the proposed investments in order of the size of the estimated cost savings to be made or the cost of the investment plus any annual running costs. Such treatments of investment do not recognise the timing of cash flow or the length of the asset's useful life.

A good method for the allocation of capital funds should take into account the cost savings over the life of the project; give due weight to the working life of the asset; and also make allowance for the timing of the cash flows. This can be achieved by the use of discounted cash flow in the Whole life cost method. It is a cost measure in which compound interest factors are used to discount all cash flows to their equivalent value at time zero. Time zero is taken as the start of operations. It is usual to list cash savings separately from cash outlays, using discounted values, and to show sums of both as of time zero. Whole life cost is the present value of the net operating cost throughout the project life plus present value of the investment.

9.5   Finance

There are two principal sources of capital (internal and external) for financing investments. Most companies will usually do best to adopt some prudent combination of both sources to suit their particular needs.

Internal finance is usually fixed for a given budget period and those companies which depend wholly on autonomous savings, as a matter of policy, will be obliged to restrict their capital expenditure to that fixed sum. Although internal finance has the advantage of being free from the whims of the money market, total dependence on retained profits and the annual provision for depreciation may impede a company's long-term growth. Worthwhile projects may be held back in some years for lack of funds, while in other years the availability of funds may encourage the acceptance of projects earning below average profitability.

External finance is not always readily available as the supply tends to be affected by conditions in the money market. Furthermore, the price of capital also fluctuates and at times will be high enough to disqualify otherwise acceptable investment projects. External funds are also unpopular with some companies because share-holders are prone to impose restrictions on the uses of their funds, albeit indirectly. When financing capital expenditure through borrowing, the cost of capital is the rate of interest charged by the lender. However, interest payments may be off-set against assessments for corporation tax; consequently the effective cost to the borrower is the net interest after tax, rather than the nominal interest rate.

The foregoing discussion assumes that the user of the asset has out-right ownership of itself. However, there are many different financing schemes which allow companies to raise capital to finance an investment (or to provide working capital) by transferring ownership of an asset to a third party. Such schemes include debentures; leasing; BOO & BOOT arrangements.

A debenture is a bond given in exchange for a loan by which the company agrees to pay a stated rate of interest for its use throughout an agreed term. Debentures are normally redeemable on a specified date or in stages within an agreed number of years. Debenture interest is a charge upon the company and must be met in full whether or not any profits are made. Debenture holders always rank for payment in full before shareholders of any class even if unsecured. Debentures may be unsecured loans; that is, a simple promise to pay; or secured by a charge upon the whole or part of the company's assets. Mortgage Debentures are secured on specified assets of the business. The property constituting the security is vested in trustees and held by them for the debenture holders who become the virtual owners of the property. No dealing in the property or change in its constitution can be made without the approval of the trustees. Floating Debentures are secured by a floating charge (or general lien) upon the total assets of the business. In other words, it is secured on all assets in general but on none in particular. A general line does not prevent the company from dealing with its assets in the normal course of business.

In a leasing arrangement the company actually transfers ownership of the property to an investment company and then make regular rental payments to the latter for the use of the property. This type of arrangement is widely applied to office buildings, retail complexes and factories.

BOO (Build Own and Operate) arrangements are particularly suitable for the supply of utility services such as electrical power, gas, oxygen, water, sewerage, etc. In theses instances the costs of installing the infrastructures would be quite prohibitive and/or specialist personnel are required for the operation. By such arrangements the company simply pays for the services it uses as required. A BOOT scheme (Build Own Operate and Transfer) only differs from a BOO scheme in so far as the operation of the asset is transferred to the company personnel after a suitable period of training.

9.6 Trade-offs

To most technical problems there are a multiplicity of solutions. It is important in looking for these solutions one must look at the customers circumstances and preferences i.e. the technical solution for one customer might be quite different to another. In membrane processes one frequently has to trade-off between various features. Some examples are

- *Capital for better treatment versus long term performance*
- *Preheating the feed versus a larger plant*
- *Higher operating pressure versus longer life*
- *More instrumentation and lower supervision.*
- *Capital cost versus operating cost*
- *Tax/finacial incentives*

Several of these are examined briefly in this section

9.6.1 Issues of Scale Changes

The effect of scale on various components is based on the basic design of a plant being independent of scale. It might be thought that the modular nature of membrane technology means that the various factors indicated leave little scope. However, there are some key design concepts which are scale dependent, and are a function of the size of the size of these basic building blocks. This is well illustrated in reverse osmosis plants. For optimum performance on a high recovery plant one would design a tapered plant However, this can only be achieved if there is sufficient building blocks. If the plant is small this means on has to use small elements. As can be seen from figure 9.6.1 there is significant operating cost saving in using large elements. This design conflict is resolved by designing a plant with a reject recycle. The sacrifice is some rejection loss, and an increase in the energy cost.

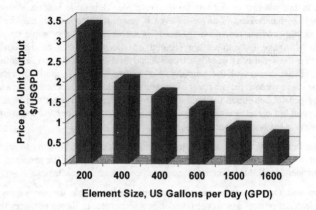

Figure 9.6-1 Variation in price per unit output with element size (measured as production output under equivalent conditions) in 1994.

Another scale cost issue relates to the pressure vessels. The pressure vessels are usually more expensive than the elements, but their depreciation period is much longer. The cost of a pressure vessel is only slightly dependent of length since the major costs are assoocaited with the end caps.

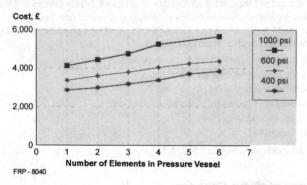

FRP - 8040

Figure 9.6-2 Variation in pressure vessel (fibre re-inforced polymer) cost with number of 8040 elements it contains rated for different operating pressures.

9.6.2 Energy and Exergy

In considering a plant design it is normal to consider the energy required to make the process work. However, if one looks at the system as a whole it is useful to consider the exergy since the concentrate can contain useful energy which can be exploited. The most familiar example of this is in sea water desalination where various energy devices are used to recover energy from the concentrate and return it to the feed. In this way as much as 80 % of the energy can be recovered.

Another interesting example is the use of heat to raise the feed temperature. This is sometime used in reverse osmosis system since the production rate increase 2.5 % per degree. Reverse osmosis plants are designed around an operating flux. Hence, the benefit is realised not in a reduced plant size but a lower operating pressure. Clearly, one can equate the cost of the energy to raise the water temperature versus the reduced pressure and increased capital cost.

9.6.3 Optimum Current Density for Electrodialysis

As with most membrane processes one can go for a more capital expensive solution but lower operating cost or vice versa. In terms of the total cost however there always exist an optimum i.e. the two extremes will usually represent a higher total cost. A simple example of this is electrodialayis, in which costs are taken to be made up of three components

- *Energy cost per unit product*
- *The Capital investment cost per unit product*
- *Cost of added chemical, replacement membrane*

The energy cost per unit product is proportional to the electric current, I. Now as we increase the current on a plant it can process more. Thus the investment cost per unit product is inversely proportional to the current. The cost of additives an replacement membranes is proportional to the amount t of materials treated. Thus, the total cost is given by an equation of the form

$$C = aI + \frac{b}{I} + c$$

It is readily seen that this equation gives a minima total cost at a given current, which can be shown to be occur when the energy costs equal the capital costs, and is given by

$$I_{min} = \sqrt{\frac{b}{a}}$$

This result is illustrated graphically in figure 9.6-3

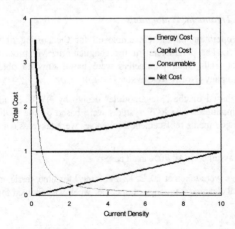

Figure 9.6-3 Trade off between operating cost and capital cost indicates an optimum current density.

9.6.4   Tax and other Financial Incentives

Investment decisions are governed as much by internal company need / policy as is by external influences. In times when there is a shortage of capital for investment or in a business which has already entered the decline phase, any investment would only be geared toward a short term goal. Thus, projects which show a low capital requirement would be favoured even if the operating costs are high. Conversely, many environmental projects are concerned with pollution abatement or conservation of resources. In such cases capital expenditure take priority over any possible operating cost savings. Indeed, governments often encourage projects of this nature through generous capital allowance or punitive levey on the use of certain resources.

9.7   Examples

The following three examples illustrate the use of the cost estimation techniques and investment appraisal approaches discussed in this chapter for different situations: a comparison of the economic viability of a new coagulation-plus-MF process versus more conventional rapid gravity filtration (RGF) for the removal of Natural Organic Matter (NOM); use of NPV in producing budget authorisation estimate for sludge thickening in an anaerobic digester; and a real life investment decision choosing between alternative processes for the recovery of paint from wastewater.

Example 1.   This example examines the economic feasibility of the combined Coagulation / MF process for NOM removal. The process is based on results from a pilot trial programme. The analysis was based on the capital and annual operating costs for plants at two capacities: 50 and 150 Mld (2080 and 6250 m^3/h). The costs presented here were based on the full design and construction of the treatment plants. Study estimates are considered to have an accuracy of +/-30% only.

In order to compare the cost of the combined Coagulation / MF process for NOM removal to an alternative process, the costs of another treatment process with the same plant capacities are also presented. The alternative process in this case was Rapid Gravity Filtration (RGF).

Process Parameters & Treatment Assumptions

The basic design and operating parameters employed for the costing of the various plants are presented in Table 9.7.1. The parameters for the coagulation / MF process were based on data obtained from the pilot trial. Other parameters were based on data obtained from experience elsewhere within the company (North West Water Group).

The MF systems were based on the "Megamodule" design by NWW Acumem Ltd with 0.7 mm hollow fibre in 40 m^2 modules. The RGF systems were based on the traditional type filters with sedimentation. Membrane / media replacement frequencies were taken as 6 and 10 years for MF and RGF respectively.

Pump efficiencies were assumed to be 70% in all cases.

Average chemical dosage were taken at 20, 2, 12 and 2 mg/l for lime, ferric, acid and polyelectrolyte respectively. A chemical cleaning frequency of once per week was allowed for MF process.

Table 9.7.1 - Design, Operating Parameters Assumptions

Description	Units	PLANT CAPACITY			
		50 Mld		150 Mld	
		MF	RGF	MF	RGF
STRAINERS					
rating	µm	100	100	100	100
pressure drop	bar	0.2	0.2	0.2	0.2
filtration duration	hours	0.5	0.5	0.5	0.5
backwash duration	seconds	30	30	30	30
recovery	%	98	98	98	98
CHEMICAL SYSTEMS					
ferric sulphate		yes	yes	yes	yes
sulphuric acid		yes	yes	yes	yes
lime		yes	yes	yes	yes
polyelectrolyte		no	yes	no	yes
MEMBRANE SYSTEMS					
Working flux	LMH	110		110	
working pressure	barg	0.8		0.8	
Cross flow velocity	m/s	NA		NA	
Filtration duration	hours	0.17		0.17	
Backwash or purge duration	seconds	5		5	
Recovery	%	95		95	
Membrane life	years	6		6	
Replacement cost	£/m^2	65		65	
RAPID GRAVITY FILTERS					
Bed height	m		2		2
Sand particle size	mm		2		2
Working pressure	barg		0.8		0.8
Filtration rate	m3/m2/h		8		8
Filtration duration	hours		10		10
Backwash	hours		0.5		0.5
Recovery	%		92		92
COST PARAMETERS		ANNUAL USAGE			
Amortisation period, years	20				
Interest rate, %	10				
Labour rate, £/hour	10	8,000	8,000	8,000	8,000
Electricity cost, £/kWh	0.05	80	80	241	241
Ferric sulphate, £/ton	56	36.5	36.5	109.5	109.5
Sulphuric acid, £/ton	42	219	219	657	657
Lime, £/ton	56	36.5	36.5	109.5	109.5
Polyelectrolyte, £/ton	910	0	36.5	0	109.5
Cleaning chemicals, £/ton	2,000	0.5	0	1	0

Costing Assumptions

The following were not included in the analysis:

- *Land*
- *Civil Engineering*
- *Buildings*
- *Water intake and treated water pumping stations or storage*
- *Wastewater treatment*
- *Telemetry*

It should be noted that membrane systems have smaller space requirements and minimal civil engineering requirement compare to the non-membrane processes. Although these costs were not included in the estimates, they would ultimately reduce the total capital costs of membrane systems more significantly than compared to the non-membrane systems. In all cases, a 100% plant operating factor was assumed.

Process Economics Discussion

Capital cost estimates for the various plants were based on the actual cost data for components in water treatment plants of capacities in the range 4 20 Mld and component cost scaling was the method used for scale up cost estimation as described in section 9.2.5.

Table 9.7.2 - Process Cost Estimations

Description	Unit	PLANT CAPACITY			
		50 Mld		150 Mld	
		MF	RGF	MF	RGF
CAPITAL COST BREAKDOWN £000 (1996)					
Strainers		66	66	166	166
Chemical systems		294	355	438	532
Rapid gravity filters		0	2,713	0	5,726
Membrane systems, ex pumps		2,019	0	5,057	0
Pumps & mixers		83	83	183	183
Pipework		210	280	524	591
Panels & Cabling		310	310	330	330
TOTAL		2,982	3,807	6,698	7,528
OPERATING COST BREAKDOWN £000 (1996)					
Depreciation		149	190	335	376
Finance		298	381	670	753
Labour		80	80	80	80
Power		35	35	106	106
Chemicals		14	47	42	140
Membrane /media replacement		205	52	616	156
Maintenance		78	78	185	161
TOTAL		860	863	2,032	1,772

Operating costs for the two process options at 50 Mld and 150 Mld capacities are produced according the assumptions in Table 9.7.1. These estimates are shown in Table 9.7.2 .

At 50 Mld capacity the capital costs of the two options vary by only +/-14% with the RGF being more costly, MF being less. On the other hand at this level the MF and RGF show very similar operating costs. The unit treatment cost is 4.7 p per m³ of treated water for both processes.

The capital costs for the two processes appear to increase at a similar rate to each other up to the 150 Mld capacity. However, there is a significant departure in the pattern of operating costs at this level. The unit treatment costs are 3.7 and 3.2 p per m³ of treated water for the MF and RGF processes respectively. Membrane replacement costs account for the increase in the treatment cost.

It appears from this analysis that as far as MF and RGF are concerned the capital cost of the latter is always greater than the former (particularly if land and building costs are taken into account). However, the unit treatment cost favours an MF option for plants with capacities under 100 Mld and for greater capacities RGF would be more favourable.

EXAMPLE 2. This section summarises the economic case for a membrane system for biomass retention in an enhanced anaerobic digestion process and involves the preparation of budget authorisation estimates and net present value calculations (NPV).

Background:

An effluent treatment works operates two anaerobic digesters, each has a volume of 2250 m³. The digesters convert the organic matters in the sludge from the primary settling tanks and waste activated sludge from the aeration basins to methane gas for the CHP unit on site. Currently, The digesters operate with an HTR of 13 days at maximum hydraulic loading comparing with the 12 days required by regulations. The one day operating margin does not provide sufficient safety under abnormal loading conditions. Indeed, There has been instances of excursions.

The Works management is keen to explore options and implement to most effective one for raising the digesters retention times.

Options

Options for raising the anaerobic digesters retention times at the Works include:

1. *Take no action*
2. *Install extra digester capacity*
3. *Introduce sludge thickening to reduce the volumetric loading to the digesters*
4. *Decouple SRT and HTR for the digesters using a membrane system.*

Since the loading on the works is likely to increase not decrease in the future, option 1 is really not tenable. Option 2 would require a major capital expenditure and as such would only be considered as a last resort.

A number sludge thickening techniques involving solid liquid separation equipment such as centrifuges, drum thickeners, etc. have been proposed as option 3. Irrespective of the capital requirement for investment in such processes, there are inherent drawbacks which are common with these processes including:

- *A supernatant stream of high BOD is returned to the aeration basins adding to the load on the activated sludge process at the same time reducing the gas yield.*

- *Although gas yield would improved because of the longer retention times, the biomass in the digester is not enhanced, therefore, the rate of conversion is not improved.*
- *Difficult to provide an anaerobic condition for their operation.*

Mechanical thickeners require expensive chemical dosing and have a high energy consumption. Equipment such as centrifuges also have critical components which may required specialist maintenance fitters. They are notorious for mechanical failure.

Membrane Option

The membrane option is technically the most versatile with the following main features:

1. *Capability to decouple the SRT from the HRT for the digester*
2. *Enhanced biomass which increases conversion rate resulting in a much better gas yield than any pre-thickening methods and reduce sludge volume generation.*
3. *Simple anaerobic operation.*

Ultrafiltration (UF) is a membrane process that has established itself as a unit separation process for over 20 years finding favour particularly in the dairy industry for milk concentration and protein recovery from cheese whey (traditionally a wastewater). Although it is not widely used in anaerobic digestion hitherto because of cost, however, the economic climate is changing and membrane may find an important place in the future of wastewater treatment. Indeed, a number of large scale use of membrane for anaerobic digestion have been operating successfully. In house trials on digested sludge have also been carried out. Excellent performance was achieved with tubular UF membrane over a range of digested sludge solids concentration of interest (2-5.4%).

The present proposal involves the use of proven commercial tubular UF membrane (1" tube) to remove water from one of the digester.

Proposed Operation

The proposed operation is shown in the above diagram. In simple terms sludge is removed from the heat exchanger loop and passed to the UF unit where a part of the water is removed. The concentrated sludge is returned to the heat exchanger. The water being free of any suspended solids is sent for further treatment in the Activated Sludge Process.

It is envisaged that the additional load on the Activated Sludge Process represents only 0.1% its normal load therefore will have no noticeable effect. However, the reduction in the sludge production rate will increase the retention time in Digester No. 1 from 13 days currently to 19 days. The increased retention time represents a substantial increase in the operating margin for the digester beside any savings that might accrue from the reduction in the costs of sludge disposal.

Process Economics Assumptions

The following assumptions were made in the calculations:

- *Inflation rate 3.5% pa*
- *Interest rate 11% pa*
- *Tax rate 13% (special concession for water industry)*
- *Depreciation linear over 25 years*

The major items of running costs are power (43%) and membrane replacement (35%). Cleaning chemicals, maintenance and manpower account for the remaining 22%.

The impact on the true cost saving is difficult to predict at this point since it is not directly proportional to the reduction in sludge volume. Increased gas production is also not quantifiable. However, to facilitate an assessment it is assumed that cost saving will be the same at the current disposal cost (£4.56 per m³ of sludge).

The financial performance of the proposed investment and an economic summary is shown by Tables 1 and 2 respectively. Cumulative cash flow is also shown by the chart below. The dip in cash flow in year 25 is the effect of asset renewal when the proposed life comes to the end of its economic life.

Performance

Financial performance of this investment depend on the interest rate, the savings from sludge disposal costs and the flux performance of the membrane.

The proposed operation is based on a membrane flux performance level of 30 $Lm^{-2}hr^{-1}$ (LMH) because the in house trials were not sufficiently long to predict long terms membrane performance with absolute certainty (6 years is the assumed membrane life). However, theoretical calculations indicate a long term performance level of 62 LMH.

If theoretical performance is achieved, it will be possible to use this membrane plant to process sludge from the 2nd digester also, on a time share basis (one hour alternately). Such operation would more than double the NPV of the investment (or reduce the pay back time by half). The plant performance will be monitored closely in the first year to see if this could be achieved.

Table 9.7.3 - Financial performance

	Interest rate %	Assume Flux 30 LMH		Assume Flux 60 LMH	
		5 yrs	30 yrs	5 yrs	30 yrs
Net Present Values		£000	£000	£000	£000
High	15.0%	(40)	61	91	265
Medium	11.0%	(26)	118	119	365
Low	7.0%	(9)	203	152	509
Payback (years)		4.3		2.2	

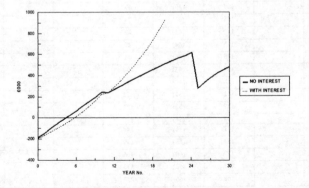

Figure 9.7-1 Cumulative cash flow chart with and without interesr payments

Table 9.7.4 - Economic summary

ECONOMIC SUMMARY			Economic life: 25 years
			Depreciable life: 25 years
Investment period: 30 years			Rate of return: 11 %
Prepared by: MSL, June 21, 1996			

All figures in £000

INCREMENTAL INVESTMENT & EXPENSE IN FIRST YEAR

Fixed Capital (1)	Project requirement: £150	Expenditure (3)
	Base case requirement: £0	Installation: £5
	Net requirement: £150	Start up: £25
Working Capital (2)	Project requirement: £0	
	Base case requirement: £0	Base case requirement: £0
	Net requirement: £0	Total Incr. expenditure: £30
Total Incr. Investment: £150		**Total Requirement (1+2+3): £180**

INCREMENTAL CASH FLOW SUMMARY

Annual Period	Cash Income	Gross Savings	Total Expenses	Profit After Tax	Fixed Investment	Working Capital	Net Incr. Cashflow
current		20	(53)	(41)	(150)		(183)
1		62	(24)	35			38
2		64	(25)	19			50
3		67	(26)	24			40
4		69	(26)	28			40
5		71	(27)	29			41
6		74	(28)	34			42
7		76	(29)	38			42
8		79	(30)	43			43
9		82	(31)	48			44
10		85	(32)	4			44
11		88	(34)	44			(6)
12		91	(35)	47			32
13		94	(36)	51			32
14		97	(37)	56			32
15		101	(39)	62			32
16		104	(40)	67			31
17		108	(41)	73			31
18		111	(43)	80			30
19		115	(44)	87			29
20		119	(46)	94			28
21		124	(47)	104			27
22		128	(49)	113			26
23		132	(51)	122			24
24		137	(53)	132			23
25		142	(54)	119	(354)		(334)
26		147	(56)	149			53
27		152	(58)	159			49
28		157	(60)	166			41
29		163	(62)	175			34
30		168	(65)	253			28

EXAMPLE 3. In this real life example the manufacturer of an emulsion paint used a payback ranking method in the economic analysis to compare three options for dealing with the factory effluent. The

effluent resulted from the washing of the paint manufacturing equipment. It has been shown through a pilot study that the effluent could be concentrated by evaporation or ultrafiltration to produce a saleable paint product. Additionally, the membrane permeate could be re-used for cleaning.

Options:

1. *No treatment, carry on with landfill (current position)*
2. *Membrane separation and recycle/ re-use*
3. *Evaporation and recycle/re-use*

Table 9.7.5 Comparison of effluent treatment cost at an emulsion paint producer

£000 (1994)	Landfill	UF			Evaporation		
		Sensitivity analysis			Sensitivity analysis		
		case 1	case 2	case 3	case 1	case 2	case 3
Raw Materials cost	60	60	60	60	60	60	60
Land Fill	75						
Cost To The Factory	135	60	60	60	60	60	60
Operating Costs Pa Including Standard Service Contract		43	43	43	50	50	50
Total Costs Pa (Raw Materials/ Operation/ Landfill)	135	103	103	103	110	110	110
Recoverable Value Of Paint Concentrate, Pa	0	15	60	120	15	60	120
Net Costs Pa (Total Costs[1] Less Recoverable Value)	135	88	43	+17	95	50	+10
Net Saving (Current Landfill Cost less net costs	0	47	92	152	40	85	145
Total Capital Cost	0	220	220	220	220	220	220
Payback Period For Capital Investment (Years)	N/A	4.7	2.4	1.4	5.5	2.6	1.5
Ranking Based On Payback[2]		1			2		

[1]   A (+) indicates an actual income rather than cost.

[2]   The landfill option is the base case which involved no capital outlay.

Assumptions:

1. *Raw material losses to effluent per annum costs approximately £60K*
2. *Operating costs for Membrane system include cleaning fluids, replacement membranes, service visits, emergency service, electricity and labour.*
3. *Operating costs for Evaporation include minor maintenance, electricity, steam and labour.*
4. *Recoverable value of paint concentrate: (i) Case 1 - minimum value @ £100/t (ii) Case 2 - recovery of raw material costs only (iii) Case 3 - recovery of raw material costs and its incorporation into the standard paint product, which will therefore include a margin (assumed 50% margin).*
5. *Permeate from the membrane process is re-used at no cost. Evaporation has no effluent.*
6. *Capital cost - This includes considerable site work carried out by the manufacturer including the installation of storage tanks, bunds, and paint transfer equipment. The costs for the membrane system and the evaporator were firm quotations from the respective suppliers.*

Discussion:

Evaporation and membrane separation options were both economically more sensible than straight landfill. This was because with evaporation and membrane separations a payback on capital investment was possible due to the savings made on raw materials. However, the membrane option gives a more rapid return on the capital investment in all situations in the sensitivity analysis. It should be noted that this particular factory steam was available at relatively low cost. The producer also had other manufacturing sites which did not have steam, so that a new boiler would add substantially the capital requirement. There was a realisation, too, that landfill cost was rising rapidly which would change the scenario decisively in favour of a treatment process. All these factors led the paint producer to make an investment in the membrane process.

9.8   References

1. Hess and Quigley, Chem. Eng. Progr., **59** (1963), Symp. Ser. 42, pp55-63.

2. Mallory, J.B., Chem. Eng. Progr., **67**, 10, 68 (1971).

3. Glueckstern, P. and Arad, N., *"Desalting Brackish and Sea Water"*, in *"Economics of the Application of Membrane Processes"*, p479-507.

4. Rogers, N. , *"Wastewater Treatment, in Economics of the Application of Membrane Processes"*, p509-546.

5. Woods, D.R., *"Cost Estimation for the Process Industry"*, McMaster University, Hamilton, Ontario, 1983

6. Cote, P. and Lipsky, C., in Proceedings of the 4th Int. Conf. on *"Pervaporation Processes in the Chemical Industry"*, Bakish, R. (ed.), Bakish Corporation, NJ, 1989.

7. K S Spiegler *"Principles of Energetics "* Spinger-Verlag

Chapter 10

SURFACE WATER TREATMENT

Contents

10.1   Objective of surface water treatment

Water treatment plants are designed to produce potable water consistently. While quite exacting, the potable quality standard allows a large degree of freedom for changes in composition. In areas that have historically received constant supplies it can come has a major shock when there is a significant variation. In contrast industrial customers design their processes to accommodate the variations, where changes are frequent. Variations in water are particularly common in surface water supplies. To allow for these variations, the design of a water treatment plant may include a number of processing steps.   As in all engineering design, the most judicious combination will be sought from both the technical and the economic viewpoints (capital cost and operating costs).

One measure of raw and product water quality is turbidity. Turbidity is caused by the scatter of light due to the presence of mainly sub-micron sized particles that are dispersed in the water. Colour on the hand is due to the absorbance at certain wavelengths by organic impurities (mainly humic and fulvic acids) in the water. *Natural organic matter* (NOM) in water may comprise of many naturally occurring substances including carbohydrates, polypeptides, polyhydroxyl aromatic compounds, and others. Generally they have a MW range of 100-10,000. In addition to causing the colour problem, poor taste and odour, organic matter provides the precursors to the formation of disinfection by-products (dbp) which may pose a health risk.

Water quality standards are essentially concerned with the level of a number of specified parameters including micro-organisms, suspended solids, organic substances and other micro-pollutants (e.g. pesticides). In the UK the *Drinking Water Inspectorate* (DWI) is the agency responsible for enforcing such quality regulations. Similarly, in the US the equivalent body is the *United States Environmental Protection Agency* (USEPA).

The amendments to the 1986 *Safe Drinking Water Act* (SDWA) required the USEPA to promulgate several new drinking water regulations. *The Surface Water Treatment Rule* (SWTR), promulgated in June 1989, was developed to ensure the adequacy of Giardia and virus removal or inactivation for surface waters by specifying treatment requirements for filtration and disinfection. In addition a *Disinfectant / Disinfection By-Products* (D/DBP) Rule promulgation is expected soon. The organic compounds targeted for regulation include *total trihalomethanes* (THMs), *haloacetic acids* (HAAs), and chloral hydrate.

In Europe, the 1989 EU Directive on drinking water quality confirmed, in legislation, quality standards for potable water covering microbiological and chemical parameters. These standards have been incorporated into UK legislation as the *prescribed concentration or values* (known as PCVs). The PCV for coloured material in drinking water was set at 20 degrees hazen (°H). Chlorination of *natural organic matter* (NOM) forms a series of by-products, commonly known as trihalomethanes, and a PCV of 100 µg/l of total THM at the customers tap is the current UK standard. More recently this limit has been reviewed by the EU and standards for individual THMs are to be set. Also, a new set of disinfection by-products created through the chlorination of humic substances such as haloacetic acids (HAA) are likely to be the subject of future legislation. The regulations are being tightened in line with improved water treatment methods and improved analytical techniques and the completion of *World Health Organisation* (WHO) toxicological assessment of individual chlorination by-products.

10.2   Traditional approach to water treatment

Figure 10.2-1 illustrates the main treatment process combinations commonly used in water treatment, which may be supplemented by additional special treatment processes made necessary by the presence of specific unwanted substances in the raw water (fluorine, nitrates, calcium, etc.). The quality of the raw water dictates the type of treatment required. For instance, for clean, unpolluted raw water only disinfection is used to achieve the required micro-biological quality. Where the water contains a small quantity of colloids, or has a more pronounced colour, in-line coagulation, settlement and filtration will solve the problem. If the quantity of coagulant required to remove the colloids or reduce the colour is too high the floc formed will be large and will rapidly clog the filter creating the need for frequent washing. It is therefore essential to provide a floc separation stage that uses settling or flotation techniques prior to filtration.

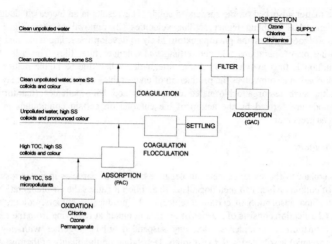

Figure 10.2-1 Main treatment processes for surface water

Roughing treatment:

Depending on the type of water available, the first treatment stage may be a roughing treatment, which is designed to remove coarse particles likely to affect the subsequent processes. Such pre-treatment may include:

- a screening unit,

- a straining unit, also known as macro straining, which is required if the water contains grass, leaves, plastic debris, etc.,

- a grit chamber, located either upstream or downstream of the straining unit,

- a micro-straining unit if the amount of plankton is limited and no settling stage is planned,

- a surface de-oiler

- a preliminary sedimentation unit if the level of suspended solids in the raw water (silt, clay, etc.) is very high,

Pre-oxidation:

- Aeration may be necessary if the water is low in oxygen. Aeration results in oxidation of ferrous and manganous ions, and increase in oxygen content. As a result the water has a better taste. Aeration may also be necessary if the water contains undesirable gases. For example hydrogen suulphide (H_2S), which imparts a very unpleasant taste, is easily removed by atmospheric aeration. Some carbon dioxide is removed by aeration at atmospheric pressure. The extent of carbon dioxide removal will vary according to the mineral content of the water.

- Chemical treatment includes the use of chlorine and its derivatives, ozone, and potassium permanganate. Pre-oxidation has the advantages of: (1) improved flocculation by acting on

10 - 3

organic matter adsorbed on the suspended solids. This results in an increased sludge cohesion coefficient, thus enabling higher settling velocities (2) removal of most algae and other organisms (zooplankton and phytoplankton) likely to develop in settling tanks and filters, thus facilitating plant operations (cleaner settling tanks, longer filter life). The risk of anaerobic fermentation is thus avoided (3) colour attenuation when due to humic matter. However, pre-chlorination also involves the production of undesirable compounds which may be harmful to health, such as organochlorinated compounds and halo-methanes (haloforms). These compounds are formed by the action of the chlorine on certain compounds in the water, known as precursors.

Water Clarification:

Adding a coagulant to the water cancels out negative charges of particles in water resulting in the coagulation of colloids which can then flocculate. In addition it causes the co-precipitation of certain organic matter and adsorption of certain disenfection by-product precursors, colour, etc. on the formed floc. Clarification consists of a series of operations aimed at removing from the raw water by adsorption, formation of complexes, etc., any suspended solids, together with any pollutants (organic or inorganic) associated with these solids. Depending on the quality of the raw water, large quantities of coagulant may be required. It is therefore necessary to remove the larger part of the resulting floc by either by settling or flotation before filtration. Without this preliminary operation, the filter would clog rapidly, requiring frequent washing and resulting in unacceptable water consumption for backwashing. If the required coagulant treatment rate is low, it may be possible to use direct filtration following addition of the coagulant and any necessary coagulant aid. The latter process is referred to as in-line coagulation and can also be used where partial coagulation produces water of a satisfactory quality. If the floc formed after complete coagulation contains a high proportion of hydroxide (high ratio of hydroxides to suspended solids in the raw water) then the density of the floc will be low and it is more advantageous to employ flotation (DAF) rather than settling for the separation stage.

Direct rapid gravity filtration:

Essentially, the filter comprises a packed bed of a medium such as fine sand, anthracite or granular activated carbon (GAC). The filter bed provides a plug flow reactor in which numerous interactions between the medium and floc particles take place. Floc attaches to the filter medium by a number of mechanisms i.e. gravity settlement, hydrodynamic forces, diffusion, attractive forces. As more and more floc is deposited on the filter medium the resistance to filtration or headloss increases and a detachment of floc can occur, this phenomenon is known as a breakthrough. The filter should be cleaned prior to a breakthrough. Cleaning is achieved by a back washing process which is performed once a day, typically. Upon returning to service the filter goes through a maturation or ripening period before optimum performance is achieved.

Removal Of Organic Matter:

Organic matter (OM) covers a wide range of compounds, which feature differing physical and chemical properties. Each stage of the treatment process contributes to the removal of a certain part of the OM. Certain volatile substances are removed by aeration, which can be introduced at the head of the treatment works (e.g. by stripping techniques, cascading or spraying). If the organic pollution in the water is due only to volatile compounds, then they can be removed by stripping techniques. In line coagulation removal rate for OM is limited to 10-30%. With settling or flotation this rate may be increased to 40-60%. Adsorption with activated carbon[1] is particularly effective for OM with high molecular weight or low polarity e.g.. surfactants, phenols, pesticides, etc. Oxidation with ozone considerably improves the organoleptic quality of water (OM levels, colour and taste). The conditions under which ozone is used in drinking water treatment processes generally preclude complete oxidation of organic matter. But transforming organic matter by ozone produces compounds with lower molecular weights and higher polarity, and with higher biodegradability. Ozone is particularly effective for removing phenols, detergents, polycyclic hydrocarbons, certain pesticides, such as aldrin and can under certain condition reduce the level of haloform precursors. On the other hand, ozone will not remove other pesticides, such as lindane. Combined coagulation and *powdered activated carbon* (PAC) may yield an OM removal rate up to 70%.

Disinfection:

Disinfection is the final stage of treatment before the drinking water is distributed. Disinfection removes all pathogenic micro-organisms from the water. Some harmless germs may remain in the waters as disinfection does not mean sterilisation. The disinfection of water comprises two important stages corresponding to the two different effects of a given disinfectant: the bactericidal effect or the capacity to destroy germs at a given rate; and the remnant effect or the ability of a disinfectant to remain in the distribution system and guarantee the bacteriological quality of the water. Disinfection is only effective when applied to good quality water. The concentration of suspended solids should be as low as possible (under 1 mg/l). Bacteria and other micro-organisms can agglomerate on the suspended solids, which protect them against the effect of disinfectants. The OM content, the TOC and above all the content of assimilable organic carbon (AOC) must be as low as possible. If these parameters are too high, the water will consume the residual disinfectant, thus allowing for the possible resurgence of bacteria. It would also be difficult, if not impossible, to maintain a constant residual disinfectant level in the distribution network. Applying disinfectant to water often leads to reactions producing by-products which must be minimised as far as possible.

10.3   Membrane approach to water treatment

Membrane processes are essentially new unit operations that are potentially useful in water treatment. As in the traditional approach, the quality of the raw water dictates the type of treatment required. Membrane processes may be used by themselves or in combination with one or more of the more traditional processes. During recent years, use of membrane technology has become widespread for the clarification of turbid waters with low concentrations of organic matter, such as water from shallow underground aquifers influenced by runoff waters. These water resources are characterised by a total organic carbon (TOC) ranging from 1 to 2 mg/l and turbidity lower than 1 NTU, which can reach 100 NTU and higher after a rainy period. During such an episode the total organic carbon (TOC) can increase up to 10 mg/l. Membrane technology has the capability of removing a wide variety of contaminants ranging from large particles to dissolved solids from water, is a promising alternative to conventional treatment processes. Membrane treatment is a physical process that separates components into a permeate (clean water stream) and a concentrate, which is rejected by the membrane with no addition or minimal addition of chemicals. Figure 10.3-1 shows

---

[1]   Granular activated carbon (GAC) & Powdered activated carbon (PAC)

the some of the most probable water treatment schemes which include at least one membrane process. Considerations of the application of membranes to some of the most common water problems are given in the following paragraphs:

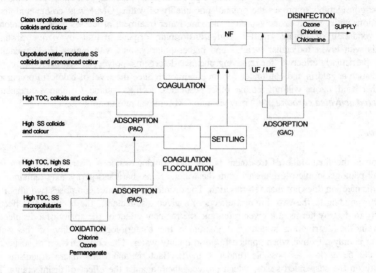

Figure 10.3-1 Application of membrane processes in surface water treatment

Membranes In Water Softening And Organic Removal:

Some water resources with low turbidity do not require a clarification step. On the other hand they can present hardness, salinity or colour problems that are often accompanied by high concentrations of organic matter which causes by-product formation during the final chlorination step. Furthermore, some waters may be contaminated by organic micro-pollutants. Treatment of brackish underground waters necessitates the resolution of one or more of these problems. Taylor and co-workers [1] have shown that membranes with a molecular weight cut-off lower than 300 to 500 Daltons were necessary to respectively remove hardness and 90 % of the *trihalomethane precursors* (THP). On the other hand a membrane cut-off between 500 to 1,000 Daltons was sufficient to remove natural humic acids. The solutions to problems of hardness, micropollutants, THP and colour residue require the implementation of membranes with MWCO from 300 to 1000 and 100,000 Daltons which correspond to nanofiltration and ultrafiltration respectively. In the latter case a combination with adsorption on powdered activated carbon is necessary.

Membranes In Colour Removal

Some surface waters naturally contain high concentrations of organic matter (humic substances) that cause highly coloured water. For example, some lake waters have colour values of about 60 mg/L Pt-Co, a TOC content at least of 8 mg/L and turbidity of 1 NTU. However, even if the colour could be removed by nanofiltration then this treatment should not be applied if the waster does not present

10 - 6

a hardness problem. In fact, nanofiltration for organic matter removal associated with softening makes the water aggressive and requires soda or lime addition to restore equilibrium before distribution. Furthermore, a MWCO of 1,000 Daltons is apparently sufficient to remove colour. Hollow fibre ultrafiltration membranes (500 to 800 MWCO) have shown their ability to remove 95% of the colour and 80% of trihalomethane precursors, while only slightly modifying the salinity and hardness of the water [2]. On the other hand, high MWCO ultrafiltration membrane, even in conjunction with PAC adsorption does not achieve more than 40% colour removal in most cases. Low MWCO ultrafiltration membranes present three advantages in comparison to nanofiltration :

- no need for pre-treatment
- production of high quality water that does not need to be re-mineralised
- lower operating costs and investment costs

Membranes In Micropollutant Removal

Karstic underground waters do not generally have dissolved organic pollution problems, but they may occasionally be contaminated by micropollutants. An evaluation of a number nanofiltration membranes for micropollutants removal has been carried out by KIWA in the Netherlands [3]. It was found that the Hydranautics PVD1 membrane removed over 90% of all pesticides (Atrazine, Diuron). However, in the same test, three other membranes showed only 50% removal efficiency for Diuron and between 65 to 80% for Atrazine. Thus, nanofiltration would be useful as a treatment for underground waters contaminated with micropollutants, but it has the disadvantage of associated partial de-mineralisation of the water. Furthermore karstic underground waters are also influenced by surface water run-off during rainy periods which could present important turbidity surges that makes a clarification step necessary as a pre-treatment for nanofiltration. Ultrafiltration membranes alone do not remove the micropollutants. However, pre-dosing the raw water with PAC will allow the removal of these pollutants by ultrafiltration. The combined PAC/UF process is applicable regardless of the turbidity of the raw water and does not require a clarification pre-treatment step. For water with a TOC level lower than 2 ppm, a PAC dose of 5 to 10 mg/l PAC would be sufficient to ensure a total removal of any adsorbable pesticides.

Membranes For The Treatment Waters With High Organic Contents

Regardless of pore size membranes cannot be directly applied to a surface water that has a high organic matter concentration for potable production. Typically, the composition of this type of surface water may comprise a turbidity of 10 NTU with peaks of up to 50 NTU, a presence of a high bacteria count, a high concentration of organic matter (TOC from 3 to 6 mg/l, and a THMPF from 150 to 300 μg/l), some micropollutants such as high concentrations of pesticides or chlorinated solvents, taste and odour problems and often associates with a high ammonia concentrations (over 0 .5mg/l). Membrane processes used as a single unit operation in this case are inadequate because of either poor flux or poor rejection of organic compounds or both. Therefore, membranes can only be used in combination with the more conventional treatments such as oxidation, flocculation-coagulation or powdered activated carbon adsorption. They are applied either directly to raw water or as polishing treatment after a clarification step.

10.4 Nanofiltration

Brackish water treatment

Originally nanofiltration (NF) membranes were developed in order to remove sulphate hardness and total hardness related to calcium and magnesium concentrations in brackish waters. They were designed to reduce the overall salinity (50 to 70%) as well as a significant proportion of the divalent ions (up to 95% rejection possible), i.e. mainly calcium and magnesium that contribute to water hardness. Their applications were limited to water resources without suspended matter and a low fouling index (SDI < 5). For such waters pre-filtration at 5 μm would normally be adequate for nanofiltration. The addition of a chelating agent and an acid may be required so as to prevent scaling. In Florida, a number of water softening plants with a total treatment capacities over 280,000 m3/day using nanofiltration have been operating successfully for many years [6]. These water resources contain small amounts of suspended solids, but they have high total hardness, colour and THMFP values. Total hardness is reduced by about 90% (i.e. from 316 to 24 mg/l as $CaCO_3$). The softening is accompanied by a partial de-mineralisation as the total dissolved solids (TDS) are reduced. Nanofiltration is now regarded as a good alternative to the more traditional use of reverse osmosis for softening and partial salt removal, it operates at a lower pressure (5 to 10 bars) and gives a better water recovery rates (up to 90% compared with only 80% for RO) [1].

Treatment of river waters

A comprehensive evaluation of Nanofiltration for surface water treatment has been undertaken by the AWWA [4]. In this study four NF membranes were used: two hollow fibre NF and two spiral wound NF with three different sources of river waters. The spiral wound membranes had a MWCO of 200-300 Daltons and were operated with UF pre-treatment. The two hollow fibre NF membranes had a MWCO from 400 to 800 Daltons. The spiral-wound NF membranes removed

- 94 % removal of sulphate
- 50-80 % removal of calcium
- 7-35 % removal of bromide
- 55-83 % removal of TOC
- 71-97 % reduction in UV 254 absorbance

The lowest removals occurred with the hollow fibre NF membranes which was treating a river water; this finding was consistent with gel permeation chromatography, which indicated that the water had a higher percentage of low molecular weight materials than the other two waters. The highest TOC and UV254 absorbance removals were achieved by the spiral wound membranes. Average THM precursor removals by NF ranged from 31 to 94 percent, depending on the membrane and source water evaluated. In general, as MWCO decreased, removal of precursor material increased. The flux rates for the hollow fibre NF membranes ranged from 10 to 20 gfd (17 to 34 LMH). Flux rates for the spiral wound NF membranes ranged from 17 to 31 gfd (29 to 52 LMH). It was noted that hollow fibre NF could be employed on surface waters of varying quality without pretreatment. Additionally, it appeared that once the membrane was acclimatised to the feed water, membrane flux was fairly steady irrespective of feed water recovery and time periods of

10 - 8

operation. Although the spiral wound membranes had higher specific flux rates than the hollow fibre NF membranes, they required pretreatment. In this case, UF was employed as the pretreatment, but in other similar applications microfiltration or conventional methods have been used [5]. Using UF as a pretreatment should always ensure that feed water turbidity for the NF stage was less than 0.1 NTU; in comparison, during the testing of the hollow fibre NF membranes, feed water turbidity ranged from 7.6 to 23 NTU.

A long term pilot trial with Trisep TS80 nanofiltration modules for water quality improvement on a surface water source in Melbourne, Florida has been carried out over a period of 3800 hours. The results of the trial are summarised in Table 10.4.1 below:

Table 10.4.1 Summary of the NF treatment for the Melbourne pilot plant (adapted from [7])

Element model:	4040-TS80-TSA
Tube length:	3 element
Feed pressure:	75-100 psig
Recovery:	45-65%
Flux:	10-15 gfd
Pre-treatment:	GAC pre-filter and acid dosing

Parameters	Raw water analysis	Permeate water analysis
TOXFP[1], as ppb Cl	7,164	143
THMFP, ppb	625	25
Colour, cpu	210	not detectable
TDS, ppm	413	41
Total Hardness, ppm CaCO$_3$	147	6
Alkalinity, as ppm CaCO$_3$	81	5.5
Na, ppm	54	15
Fe, ppm	263	5
Cl, ppm	99	16

[1]   Total organic halide forming potential

As can be seen from Table 10.4.1, the TS80 nanofiltration membrane gave very good inorganic ion rejections and trihalomethane formation potential (THMFP) reduction. During the course of the test, the average THMFP in the permeate was less than 25 parts per billion. The membrane also exhibited very high water flux rates. Normalised to a pressure of 100 psi, this equates to 18 gfd (31 LMH). It was noted that the membrane operated with a minimum of pre-treatment and chemical cleanings.

Treatment of water from lakes and Highlands

Common fresh water sources are rivers, lakes and low salinity ground water. In Norway there is an abundant supply of fresh water. However, as common with the waters in other places with similar climate (such as in the Highlands of Scotland) the waters are often polluted by bacteria and significant amounts of dissolved organic matter, mostly brown plant residuals, or humus. The presence of such substances in the water is seasonal being at its worst during the winter months as the summer vegetation decay. Humic substances include a wide range of molecular weights, and carry several different chemical groups, some of them ionic. As they include a range of molecule sizes, one should expect a gradual reduction of colour retention with increasing pore sizes of the membrane, and similarly in relation to TOC retention. Indeed, this was found to be true [8], but the retention of TOC was less, mainly because the larger humus molecules contribute most to colour. Norway probably has the highest concentration of NF plants for colour removal in the world. Jansen and Thorsen [8] reported that currently there are at least 14 such plants in operation. The first plant has been in service since 1989. According to these authors humic substances are most troublesome as they tend to foul the membranes easily. Membrane deposits from one site in had 15% ash content (dry weight) mainly of iron and calcium oxides, the rest was organic. Analyses from other work [9] showed an ash content of more than 60% (surface water) and 70% (ground water). Since the nature of the fouling deposit will be different in different sites, one therefore has to expect that measures taken to combat fouling may have to be different for different plants. The role of the feed water composition should be investigated before a membrane installation. The majority of the Norwegian plants use 50 μm cartridge filters as a pre-treatment. Fouling is controlled by operating with a membrane flux under 20 LMH plus the use of effective cleaning chemicals that can be used for short rinsing at regular intervals. Several different types of membranes are used in these plants including: polyamide and a proprietary thin film type, but most are variants of CA. Experiences with them suggested that hydrophilic types are best for long term performance.

10.5   Ultrafiltration

Treatment Of Waters From Shallow Underground Aquifers

During recent years much interest has focused on ultrafiltration membranes as a means for clarification of turbid waters with low concentrations of organic matter. One such developments has been undertaken by Lyonnaise des Eaux-Dumez in France. A cellulosic ultrafiltration membrane (hollow fibre type with a nominal pore size of 0.01 μm) which has been developed by the Lyonnaise des Eaux-Dumez is employed in at least 20 plants representing an overall production capacity of 2600 m3/d [11]. A large number of the plants are used to treat waters from shallow underground aquifers. The largest plant on this type of resource is the plant at Sauve (Gard) which has a capacity of 2000 m3/day. In all cases the treated water turbidity never exceeded 0.1 NTU whatever the raw water turbidity. On the other hand the plants typically remove only 20% of the TOC and trihalomethane precursors (THMFP). It was claimed that the ultrafiltration plants completely remove all micro-organisms whether they were bacteria, protozoa cysts or viruses. It was also claimed that the process has the advantage of being able to operate in a dead end mode or under a low recirculation rate (lower than 1 m/s) which limits energy consumption to 100 to 300 Wh/m^3).

Treatment Of River Waters

Frequent quality problems associated with river waters include the following:

- waters with variable turbidity and microbial contamination problems
- coloured waters
- hard waters

- specific micropollutant problems (pesticides, chlorinated solvents, etc.) that may occur in any one of the above situations.

While MF is suitable for treating problems relating to particulate matters, in the case of hard and /or coloured waters, nanofiltration can provide the process solution. Thus, in these first three categories of water, membrane processes are sufficient for drinking water production. On the other hand, in cases where the waters present the additional problem of either chronic or accidental organic pollution, or in the even more complicated cases of surface waters, membranes alone do not provide a satisfactory solution. In such cases it becomes necessary to combine the membrane process with complementary processes (adsorption, oxidation, etc.) or to include the membranes in a complete water treatment scheme. In most cases, ultrafiltration can be combined with adsorption in order to produce a treated water complying with drinking water standards. In this way, PAC dosage is lower than when applied directly on the raw water (e.g. 10 to 20 mg/l instead of 30 to 40 mg/l for raw water treatment). For specific cases of organic problems, an oxidation step with 2 mg/l ozone can be applied prior to the addition of PAC.

Unlike in the combination with microfiltration where PAC is used as a supplementary treatment for occasional pollution (pesticides, chlorinated solvents, etc.) or for specific organic problems which are not removed by coagulation, oxidation and MF, the PAC and UF combination is designed to remove the major part of the dissolved organic matter not removed by ultrafiltration membranes alone. PAC dosages used in order to reach a pre-determined permeate quality in terms of dissolved organic carbon, UV light absorption or THMFP is also often adequate for easily adsorbable micropollutant removal. In a study [13] on Seine river water using UF with PAC (20 mg/l) complete Atrazine (2 μg/l) removal has been reported. The use of PAC in an ultrafiltration loop may be considered as completely stirred tank reactors (CSTR), which made it possible to reach high adsorption efficiencies for short contact times. A mathematical model combining hydraulic with kinetic and thermodynamic parameters for the adsorption PAC, established by the University of Illinois [14] makes it possible to predict the water quality for a given ultrafiltration plant (as a function of the type of water, PAC type and dosage and plant capacity) or to design a treatment plant for a desired permeate quality from a given water. The advantages of the PAC/UF process include a consistent hydraulic production rate, since it limits fouling and therefore cuts down on regeneration frequency and the PAC also increases the backwashing efficiency by making the deposits occurring on the membrane surface much more porous. By the same token, PAC makes it possible to maintain higher fluxes than those obtained without PAC (up to 25% higher).

Different studies comparing ultrafiltration membranes have been carried out, notably in the USA. In fact, new regulations concerning potable water treatment (Surface Water Treatment Rule) have provided an impetus for an increased interest in membrane technologies: Jacangelo et al [12] and Cabassud et al [16] compared different types of ultrafiltration membranes, including cellulosic, acrylic and ceramic membranes in hollow fibre and spiral wound configurations with coagulation as a pre-treatment. These authors found that for an equivalent permeate quality the cellulosic membrane produced the best flux performances. For example, CA hollow fibre provided a flux that was two or three times higher than other spiral wound polymeric membranes for an energy consumption two to three times lower. The spiral wound membrane also experienced fouling problems related to its design and required chemical cleanings twice per week). Although they produced a high flux (300 LMH), ceramic ultrafiltration membranes combined with coagulation at optimum coagulant dosages require high energy consumption as well as frequent chemical cleaning (once a week). Thus, they would incur operating costs twice as high as with cellulosic hollow fibre membranes as well as higher investment cost.

10 - 11

In another study with 3 different river waters the use of alum was compared with PAC as a pre-treatment for UF [4]. It was reported that with in-line alum coagulation organic removal was enhanced compared to treatment with UF only. For the Mokelumne water, 40-45 percent of TOC was removed at an alum dose of 15 mg/l. For the delta water the removal was approximately 50 percent at an alum dose of 50 mg/l. The addition of 50 mg/l of alum to the Delta water was found to be more effective in removing TOC than any PAC dose employed. Although the use of in-line alum coagulation as a pretreatment to UF increased DBP precursor removal, it was found to foul the membrane readily in both the Mokelumne and the Delta waters. When 50 mg/l of alum was used for the Delta water, the transmembrane pressure increased from approximately 6 to 21 psi (0.4 to 1.5 bars) in less than a week of operation. For the Mokelumne water, the increase in head loss through the membrane was apparently due to a coating of the inner lumen of the hollow fibre membrane with the coagulant. Consequently, alum did not appear to be a feasible pre-treatment for ultrafiltration. The addition of PAC as a pre-treatment did not cause fouling in any of the three waters examined in the study. In the case of the Ottawa water the addition 10 to 30 mg/l of PAC as a pre-treatment to UF appeared to retard but not completely inhibit fouling by this water, which had a relatively high organics content (7.4 mg/l TOC).

UF Membrane As A Disinfection Technique

Although conventional water treatment processes with rapid sand filtration provides several lines of defence, chlorination of the treated water has been imperative when treating grossly polluted supplies. However, in case of a power failure, such systems may be susceptible to contamination of the treated water. Also, increasing concern over transmission of viral diseases had prompted a study of the ability of a membrane possessing high flux to retain virus. Ultrafiltration membrane offers a simple, one-step process which provides a positive barrier between the contaminated supply and the treated water. As a part of this study [15], tap water was inoculated with E. coli phage in water suspensions containing 10^7 to 10^9 organisms per ml. The final phage concentration in the feed water was between 3.5×10^4 to 4.7×10^4 per ml. Although E. coli phage have no sanitary significance, the size and ease of quantitative measurement suggested that they would give a good indicator for viruses. Bench scale trial with UF membrane produced a permeate in which no virus could be detected. To demonstrate the effectiveness of ultrafiltration membranes in treating an actual water supply, a 500 gpd pilot plant was installed near the Mianus River in Greenwich, Connecticut. The plant was operated for a period of two months. It was reported [15] that a substantial flux (35-55 LMH) was achieved with no chemical use and very little attention was required. To demonstrate coliform rejection, their concentration in the permeate was measured during the 2 month trial period. The following results were obtained:

Table 10.5.1 - Coliform count in UF permeate (adapted from ref. 15)

Coliform Organisms/100 ml	Percent time equal or less than
0	50
2	83
2.2	90
6.2	97

The positive results were thought to be due to contamination during sampling or possible growth in the permeate channel.

10.6 Microfiltration

10.6.1 Introduction

Microporous membranes were first developed in Germany during World War II for military uses in the detection of microbiological contamination of water supplies. They were first commercialised by the Millipore Corporation after the War. Traditionally, the most significant applications of microfiltration (MF) are to be found in cartridge filtration for beverage clarification (0.45 µm), cold sterilisation for pharmaceutical preparations (0.2µm) and ultrapure water for semiconductor processing (<0.1 µm). The most applicable membrane for potable water uses is the 0.2 µm membrane. MF is often preferred over UF for clean water applications because of it very low pressure requirement (typically 0.2-1 barg compared to 1.5 to 6 barg for UF). Since water is a commodity which must be delivered in large volumes at minimum cost, the hollow fibre which offer high membrane area to volume ration and can be manufactured in high volumes at little cost, has emerged as a clear winner in this application.

Memtec of Australia (also known as Memcor Limited in the UK) is no doubt the pioneer of large scale water treatment using hollow fibre MF. The MEMCOR® Continuous Microfiltration (CMF) process is based on microporous hollow fibre membranes which provide effective barrier filtration to 0.2 micron. The process operates in direct flow (also known as deadend) filtration thereby ensuring low energy consumption. An illustration of this process is shown by Figure 10.6.1. Filtration rates are maintained by the combination of a patented gas backwash sequence and chemical clean-in-place system. Memcor claims more than 300 plants using this process currently installed world-wide. When applied to the treatment of surface or ground waters either as a single-stage treatment or in combination with other treatments where the feed water parameters exceed requirements, the addition of chemical coagulants can be applied upstream. The process provides effective turbidity removal and primary disinfection, irrespective of variations in feed water quality. Primary disinfection occurs without the addition of chemicals and includes the removal of chlorine resistant micro-organisms such as *Giardia lamblia* and *Cryptosporidium*. The patented gas pulse system is a key feature of the MEMCOR® CMF process. Gas, usually clean compressed air (or CO_2 or N_2 for food products), is forced into the lumens of the hollow fibre. The sequential opening of downstream valves creates a sudden reduction in pressure through the membrane which results in a rapid flow of gas through the membrane matrix, taking with it any contaminants lodged in the matrix and on the surface. This gas pulse provides total removal of contaminants within seconds which repeated regularly ensuring the highest performance possible.

The membrane employed in the MEMCOR® CMF process is made from polypropylene which gives rise to a number of process limitations. Cumulative exposure to strong oxidising agents, particularly chlorine, causes membrane embrittlement and eventually failure. The hydrophobic nature of the polypropylene tends to encourage membrane fouling by the adsorption of oil, fats, proteins and humic substances. The membrane also requires a wetting agents for re-wetting if it is allowed to dry out. In recent years new hydrophilic HF microporous membranes have become available in large volumes and they are beginning to displace the hydrophobic type. New membranes of particular interest include the X-Flow HF (a modified polyethersulphone blend) and the Lyonnaise des Eaux-Dumez HF (very open UF made from a cellulose acetate blend). These membranes can be backflushed with clean water (permeate) only.

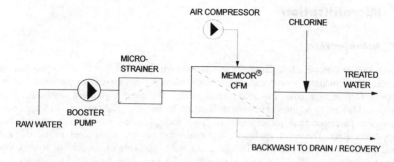

Figure 10.6-1 Schematic of a typical MEMCOR® CMF process for water treatment from a clean source

10.6.2 MF in Colour/ Turbidity treatment Without the Use of Chemical Coagulants

Over the past four or five years a large number of pilot trials and full scale implementations of the MEMCOR® CMF process for the production of potable water have been carried out, mainly in Australia and North America [17, 18, 19, 20]. These processes have been used as alternatives to conventional water purification plants for the removal of taste, odour and colour from various surface water sources. It is claimed [17] that the water quality emanating from the CMF systems achieves or betters the relevant current Australian and international guidelines for drinking water. Generally, the physical parameters show the filtrate to have a true colour of less than 15 Platinum Cobalt Units (PCU) and typically 5 PCU or less in most cases. The turbidity of the filtrate is less than 0.2 NTU even when the feed water varies from 5-500 NTU. Additionally, iron levels are typically reduced to less than 0.2 mg/l. Taste and odour were also removed in tests on a site affected by algal blooms. Table 10.6.1 shows the results of the water quality test from the Fishing Creek Reservoir trial with the MEMCOR® CMF process without the use coagulant. Clearly, the CMF system performed well as a solid-liquid separation device. Most beneficial was its high efficiency in removing turbidity and micro-organisms. With no chemical coagulation or disinfection, microfiltration removed all measurable coliform bacteria and Giardia cysts and reduced MS2 bacteriophage levels by 1 to 2 logs. As expected, microfiltration did not remove significant amounts of soluble organics or DBP precursors. However, the excellent removal of micro-organisms might allow reduced chlorination resulting in lower DBP formation.

Table 10.6.1 - *Water quality from the Fishing Creek Reservoir trial with the MEMCOR® CMF process without coagulant. (Adapted from ref. [17]*

Water Quality Parameters	Raw water	Permeate
Turbidity , NTU	0.5-12	<0.2
pH	6.8-7.41	6.6-7.4
Alkalinity , mg/l	2.1-4.5	2.1-4.4
Particle count, No. / ml	400-1000	<50
Particle size distribution, μm	0.5-15	<0.5
Total coliform per 100 ml	10-100	<1
Heterotrophic plate count, cfu	10-1000	1-100
MS2 bacteriophage (Seeded) log (PFU / 100 ml)	5.69-6.82	4.02-6.08
Total organic carbon, mg/l	1.3-5.4	0.8-4.4
Dissolved organic carbon, mg/l	0.9-3.9	0.78-3.5
Trihalomethane formation potential, μg/l	58.6-634.7	42.3-400
Giardia Muris (Seeded) , 10^5 / 100 ml	3.5-5.0	nil

Operation with the CMF systems has produced variable experiences. The small domestic water systems were found to be simple and easy to use [19]. However, it was reported [20] that the design of a 5 MGD (20,000 m^3/day) plant was quite sophisticated and mechanically complex; so much so that about 40 hours of training was necessary to enable the operators to perform the basic functions and diagnose the remedies for specific alarms. Plant start up was very labour demanding. A significant effort was expended on resolving the programmable logic controllers (PLC) software for data communications in the control of various plant functions (e.g. normal filtration, backwash, membrane re-wet, membrane integrity test, and CIP operation). A large number of PLCs were involved as each individual CMF unit is controlled by its own PLC which is linked to a master PLC and a remote telemetry system. System performance is maintained by the use of an air backwash controlled by the PLC. Although compressed air (at 90 to 100 psig pressure) is used to effect the backwash, a portion of the water must be wasted in the sludge (backwash water). For high quality raw water sources (turbidity < 5 NTU), daily backwash volume average about 1 percent of the feed water. During high turbidity events, this can increased to over 7 percent. Backwashes do not fully restore membrane performance and there is a gradual build up of transmembrane pressure, so periodic chemical cleaning of the membranes is required. For clean feed water (turbidity < 5 NTU) chemical cleaning would be needed every 4-6 week. Chemical cleaning removes biological and organic fouling.

10.6.3 Combined Coagulation And Microfiltration

The use of coagulants in conjunction with MF processes is not new. Coagulants such as alum, ferric salts and poly-aluminium chloride have been previously applied in industrial wastewater applications as part of the precipitation chemistry necessary to remove insoluble metals. This treatment has been evaluated as a flux enhancing agent, to reduce colour, or to address phosphorous reduction in wastewaters [21]. The concept of using a coagulant follows a rational that coagulant use will improve the filterability of the water, therefore facilitating the operation and backwashing of a system. Coagulant may also provide additional removal of organic matter which may cause membrane fouling. Advantages of this treatment combination include:

- Higher NOM removals are possible because a membrane filter has higher particulate removal efficiency.

- Lower coagulant dosing are possible because a full and settleable floc is not required prior to filtration.

- Higher NOM removal because the coagulation chemistry for organic removal and coagulant selection can be optimised specifically for NOM removal as particle removal is guaranteed.

The methodology for coagulant addition is determined by the rate at which the reaction occurs between the coagulant and the soluble constituent. The kinetics of coagulant reactions are very fast and are normally completed within the first few second after coagulant addition. This clearly illustrated by figure10.6-2 which shows the rate of growth of coagulant floc particles sized 2-5 μm. Coagulant addition may be as simple as flash mixing which may occur at the pump suction.

No. of particles (2-5 μm)

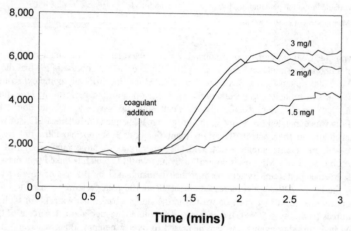

Figure 10.6-2 Rate of growth of ferric sulphate floc particles 2-5 μm (Adapted from ref. 24) on addition to an upland water (pH 5.2-5.3)

The use of MF in conjunction with coagulation processes for the production of drinking water is described in a report by Vickers and Cline [25]. They indicated that without coagulation MF only achieved a nominal removal of 10% for *disinfection by-product* (DBP) precursors, but with coagulation typical TOC removal was between 5 to 70% depending on the quality of the water. However, Weisner et al [23] claimed that TOC removal by MF alone was 15-20% whereas with coagulant (aluminium sulphate or ferric chloride) it was increased to 50-75%. Additionally, flux was enhanced by 100% (from 100 LMH to 200 LMH using a 0.2 μm TiO_2 membrane at 3-5 m/s cross flow velocity and 2-3 barg pressure). High cross flow velocity and high pressure use would of course involve additional capital expenditure in pumping equipment and put an increased burden on the operating costs. Recently, Hillis et al [24] described a pilot trial involving coagulation and microfiltration in a deadend mode for the removal of *natural organic matter* (NOM) from an Upland water source in potable water treatment. Fig. 6.6.3 shows a schematic of the pilot process.

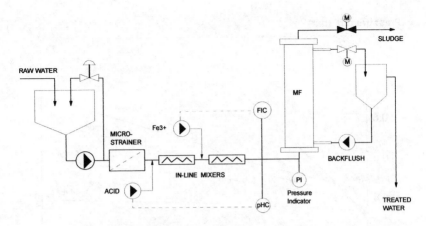

Figure *10.6-3 Process schematic of pilot plant using coagulation and microfiltration for NOM removal from an Upland water source*

The pilot plant has a design capacity of 70,000 l/day. It was comprised of a raw water holding tank, a booster pump, a 100 μm strainer, two in line mixers, a 10 m² polyethersulphone 0.2μm hollow fibre module and a permeate holding tank all in series. The through put rate was set by a bypass valve on the booster pump. Sulphuric acid was added prior to the first mixer. Coagulant dosing rate was controlled by a flow controller. The hold up volume of the system allowed a delay of 20 seconds between coagulant addition and entry into the MF module. Two solenoid valves and a backflush pump enabled the membrane module to be backwashed, while allowing the wastewater (sludge) to be diverted to drain. The contents of the permeate overflowed to drain.

In operation, water from the raw water holding tank, at a constant flow, was passed through the strainer before being mixed with coagulant (ferric sulphate). The pH controller maintained the water at pH 5.1 to pH 5.5 to allow the coagulation process to start before it reaches the MF module. As water passed through the membrane any suspended solids including any micro-organisms were retained by the membrane. Build up of solids in the membrane module caused the inlet pressure to rise with time as exhibited by figure 10.6-4. In order to reduce the rate of flux decline the membrane module was backflushed typically every 10 minutes with up to 5% of the permeate and then purged with 1% of the raw water. Cleaning was carried out once per week with either sodium hypochlorite or sulphuric acid. This was achieved by backflushing the cleaning solution through the membrane and allowing it to steep in the module for about 30 minute prior to backflushing with permeate and purging with raw water.

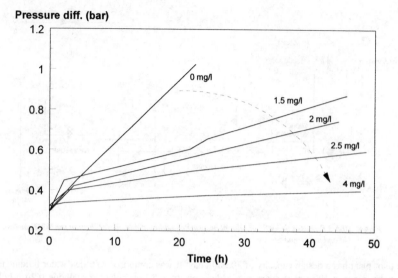

***Figure** 10.6-4 Pressure rise of a microfiltration unit being operated at a constant filtration rate (110 LMH) on an Upland water, and illustrates the effect of various ferric doses on the rate of pressure rise.*

The quality of the treated water from this pilot trial is shown in a comparison with the raw water quality parameters in Table 10.6.2 While the turbidity and the iron level in the permeate were fairly constant (low), the colour and THMFP reduction improved as the rate of coagulant dose was increased. A dosing rate of 2.5 to 3 mg/l of ferric was necessary to ensure good removal of organics for this Upland water.

The largest plant of this type with (2,000 m3/day capacity) for which ceramic membranes (0.2 μm) are combined with coagulation (using alum dosing) has been operating in France since 1988 [22].

***Table** 10.6.2Comparison of water quality parameters before and after coagulation/microfiltration (Adapted from [24])*

Ferric dose (mg/l)	Iron (mg/l)		Colour (Hazen)		Turbidity (NTU)		THMFP (μg/l)	
	Raw	Permeate	Raw	Permeate	Raw	Permeate	Raw	Permeate
0	79	40	25	23	0.7	0.2	458	427
1	35	21	11	4	0.3	0.2	477	150
1.5	73	37	11	<2	0.5	0.2	393	ND
2	73	29	11	<2	1.4	<0.2	322	82
2.5	51	35	15	2	0.9	0.2	371	63

Recently, North West Water has commissioned a number of plants which uses fine hollow fibre (hydrophilic polyethersulphone, 0.7 mm ID, 0.2 μm pore size). The largest of these treat water from the river Dee after it has been treated with alum to coagulate the silts and humics. The process has a capacity of 80 Mld (80,000 m^3 / day) with GAC polishing for pesticide removal prior to chlorination.

North West Water have a number of other large scale microfiltration plants (see table 10.6.3). Interestingly, there are a wide variety of business drivers which the membrane plants are helping to resolve.

Table 10.6.3 Large-scale microfiltration plants being operated by North West Water

Location	Capacity ML/day	Water Source	Drivers
Huntington	80	River	Emergency supplies
Winnick	16	Bore-hole	Cryptosporidium removal
Cornhow	32	Lake	Cryptosporidium removal Waste minimisation
Whitebull	45	Upland Surface	THM precursor removal

10.6.4 Experience Of Microbial Removal By MF

Microbial challenges are necessary for membrane technology demonstrations. They are required where the removal mechanism is not fully understood. The challenge must be conducted under the most severe conditions: a clean membrane, maximum transmembrane pressure and flux, maximum disinfectant dose, etc.. Virus removal efficiencies can only be determined by a viral challenge whereas particulates may be used as substitutes for larger micro-organisms. The **Surface Water Treatment Rule** (SWTR) requires a three log (99.9 percent) removal or inactivation of Giardia and a four log (99.99 percent) removal / inactivation of viruses. The Enhanced SWTR may require some water suppliers to provide even greater removal of these organisms, depending on source water quality. However, to date there has been little information in the peer reviewed literature on the efficacy of membrane for removal of these micro-organisms.

Jacangelo and Adham [27] have performed some study of the removal efficiency of both UF and MF membranes against a number of microbial agents including viruses. Each membrane was challenged with *Cryptosporidium parvum, Giardia muris, E. coli, Pseudomonas aeruginosa*, and MS2 bacteriophage (a virus). Initial experiments focused on determining the membrane rejection capabilities under a worst case conditions (distilled water, deadend mode at the highest manufacturer recommended trans-membrane pressure). Results indicated that both UF and MF (polymeric membranes) can provide an absolute barrier to protozoa and selected bacteria. The membrane removal mechanism of these micro-organisms is most likely to be physical sieving. These authors expected that both UF and MF membranes tested can meet the current SWTR requirements of 3 log removal of Giardia. Further, these membranes should also meet the enhanced SWTR requirements that may cover other micro-organisms such as *Cryptosporidium*. The UF membrane with a MWCO of 100k Daltons was able to reject all the seeded MS2 virus achieving more than 6 logs reduction. Thus, UF membranes with MWCO under 100k Daltons should easily meet the current SWTR requirements of 4 log removal of virus.

According to Hultquist [26] tests in California with the MEMCOR® CMF process has credited it with a 3 log removal of Giardia cysts and a 0.5 log virus removal. A 1.5 log virus removal was demonstrated, but with a large uncertainty. A 1 log virus removal factor of safety margin was applied as called for in the California SWTR guidance in such situation. The MEMCOR® CMF process has been approved for used on most surface waters in California at a specified trans-membrane pressure and flux. Approval is granted or under consideration for at least 12 such plants. Only two other specific membranes have been approved in California so far.

10.7   Some technical and economic considerations

Nanofiltration can be applied without fouling problems on good quality waters (Silt Density Index < 5). The advantage of such a process is the production of an excellent and constant water quality in terms of the removal of dissolved organic carbon, UV light absorption, THM precursors, colour. However, some micropollutants or coloured compounds or biodegradable organic compounds are not completely removed by this kind of membrane. Furthermore, since nanofiltration membranes reject divalent ions, remineralisation must often be implemented on the permeate in order to produce potable water.

Particularly interesting is the use of the PAC and UF combination as a polishing step. Permeate fluxes are very high as compared to those obtained by direct ultrafiltration of river water or settled waters without PAC addition.

Post treatment of membrane-filtered water is an important consideration when designing and selecting a membrane process. Water from the UF with or without PAC pretreatment, as well as the loose (400 to 600 MWCO) UF / NF membranes, would probably not require measures in addition to those that would normally be applied to the source water for corrosion control. In fact, corrosion control with UF and loose NF may be reduced as compared to conventional treatment because no coagulant would be added. However, if the tighter NF membranes (200 to 300 MWCO) are employed, it appears that pH adjustment or other methods of corrosion control would need to be implemented.

The chlorine demand of a water is an important water quality parameter that influences DBP formation as well as disinfectant costs. As chlorine demand is reduced or removed, less disinfectant is needed for primary disinfection or for maintaining a residual in the distribution system. UF alone removed between 11 and 25 percent of the chlorine demand, depending on the water. The addition of coagulants should remove 25-65 % of chlorine demand. Also chlorine demand decreased with increasing PAC addition. At dosages commonly employed in water treatment (5 to 30 mg/L PAC), chlorine demand is reduced by up to 30%. In general, NF is very effective in reducing chlorine demand from 38 to 87 percent, depending on the water and membrane employed.

Odours is another a post membrane treatment consideration. UF membranes can reduce the threshold odour number (TON) in some water from 9 to 5. Addition of 40 mg/L of PAC as a pretreatment to UF can further reduce the TON to 2.

As far as the treatment of ground and surface waters with low concentrations of organic matter is concerned, hollow fibre cellulosic ultrafiltration appears to be the best reliable technology for the production of clarified and disinfected (including viruses removal) waters without the use of any chemicals. Even though operating costs are similar for cellulosic UF membranes and polypropylene MF membranes one must keep in mind that resulting treated water qualities are different. Nanofiltration can remove hardness and organic matter as a single unit operation. On the other hand for coloured water low MWCO ultrafiltration hollow fibres (500 to 1000 Daltons) are the best compromise between economic and water quality. For all the groundwaters which exhibit a chronic or erratic pesticides pollution as well as regular turbidity episodes, the combined treatment of PAC addition with cellulosic ultrafiltration allows clarification/disinfection as well as pesticides adsorption in a single unit operation.

Surface waters with high organic matter concentrations are quite difficult to treat using membrane technology as single operation. For small treatment plants with a production capacity lower than 5000 m^3/d, the treatment comprising a combination of oxidation, PAC adsorption and cellulosic ultrafiltration is a viable alternative to conventional complete treatment lines (e.g.. clarification, oxidation, GAC adsorption) as far as treated water quality and costs are concerned. However, for economic reasons, membrane processes should only be implemented for plants with over 25,000 m^3/d capacity if a polishing step is required as in this case, the only technologies which can produce a better water quality than conventional treatment scheme are the combined PAC/UF or nanofiltration processes. However, at this scale, nanofiltration still remains much more expensive than either conventional (ozone, GAC filtration) or ultrafiltration based polishing.

For surface water treatment involving coagulation followed by either MF or rapid gravity filtration (RGF) it appears that as far as MF and RGF are concerned the capital cost of the latter is always greater than the former (particularly if land and building costs are taken into account). However, the unit treatment cost favours an MF option for plants with capacities under 100,000 m^3/d and for greater capacities RGF would be more favourable (see example 1, chapter 11).

10.8   References

1. Taylor, J.S.; Thompson. D.M.; Carswell, J.K. *"Applying Membrane Processes to Groundwater Sources for Trihalomethane Precursor Control"*, Journal of the American Water Works Association 79 (8), (August 1987).

2. Demers, J.F., KYRIACOS, *"Utilisation des membranes de nanofiltration pour l'enlevement de la couleur et des matieres organiques naturelles dans la production d'eau potable"*. Info EAU, Avril 1991.

3. Kruithof, J.C. and Meirjers, R.T. *"Removal of Pesticides with Nanofiltration"*. Emerging Technologies conference, Vienna, 1992.

4. Jacangelo, J.G. and Laine, J.M., *"Evaluation of Ultrafiltration Membrane Pretreatment and Nanofiltration of Surface Waters"* AWWA Research Foundation and American Water Works Association, 1994.

5. Reiss, C.R., and J.S. Taylor. *"Membrane Pretreatment of a Surface Water"*. In Proc. of the American Water Works Association in Membrane Technology in the Water industry Conference. Denver, Colo. (1991) AWWA.

6. Conlon, W.J. and McClellan, S.A., *"Membrane softening: A treatment process comes of age"*. Journal of the American Water Works Association, 81 (11) Nov. 1989, p48.

7. Trisep Corporation, *"Nanofiltration pilot plant studies"*. In Filtration & Separation, November 1993, pp 632-624.

8. Jansen, K. and Thorsen, T., *"Treatment of water containing humic substances - Status and Prospects"*, in Proceedings of the membranes in water treatment workshop, Paris, 1995.

9. Mallevialle, J., Anselme, C. and Marsigny, O. *"Effects of Humic Substances on Membrane Processes"*. In *"Aquatic Humic Substances"*, Am. Chem. Soc., Washington 1989.

10. Thorsen, T., Krogh, T. and Bergan, E. *"Removal of Humic Substances with Membranes - System, Use and Experiences"*. Membr. Tech. Conf, AWWW, Baltimore, 1993.

11. Chevalier, M.R. and Anselme, C., "*Ultrafiltration in drinking water treatment long term estimation of operating cost and water quality*", paper presented at the International Membrane Technology Conference, IMSTEC 92, Sydney, 1992.

12. Jacangelo, J.G., Laine, J.M., Carns K.E., Cummings, E.W. and Mallevialle, J. "*Low pressure membrane filtration for removing Giardia and microbial indicators*". Journal of the American Water Works Association, 83 (9) Sept. 1991, p97.

13. Anselme, C., Baudin, I., Mazounie, P. and Mallevialle, J. "*Production of drinking water by combination treatments: ultrafiltration and adsorption on powdered activated carbon*". Paper presented at the American Water Works Association Annual conference, Vancouver, June 1992.

14. Adham, S.S., Snoeying, V.L., Clark, M.M. and Bersillon, J.L. "*Predicting and verifying organics removal by PAC in an ultrafiltration system*". Research and Technology, Journal of the American Water Works Association, Dec. 1991 p81-91.

15. Smith, C.V. and Di Gregorio, D., "*Ultrafiltration Water Treatment*" in Membrane Science and Technology, 2nd edition, J.E. Flinn (ed.), Plenum Press, N.Y. 1970.

16. Cabassud, C., Anselme, C., Bersillon, J.L. and Aptel, P., "*Ultrafiltration as a non-polluting alternative to traditional clarification in water treatment*", Proceedings of the Filtration Society, Filtration & Separation (May / June) 1991.

17. Morris, R.A., Watson, I. and Tstsaronis, S., "*The use of microfiltration to remove colour and turbidity from surface waters without the use of chemical coagulants*", In *"Effective Membrane Processes - New Perspective"*, R. Paterson (ed), Mechanical Engineering Publications Ltd, London, 1993.

18. Yoo, S.R., Pardini, J., Bentsan, G.D., Schreiber, P.J., Daniel, P. and Brown, D., "*Making the Move to Microfiltration: Experiences at a 5 MGD Microfiltration Drinking Water Treatment Plant*". Paper presented at the *Microfiltration For Water Treatment Symposium*, Irvine, CA, August 1994.

19. Coffey, B., Stewart, M.H. and Wattier, K., "*Evaluation of Microfiltration for Metropolitan's Small Domestic Water Systems*". Paper presented at the *Microfiltration For Water Treatment Symposium*, Irvine, CA, August 1994.

20. Kostelecky, J.D. and Ellersick, M.C., "*Full-Scale implementation of Microfiltration at Metropolitan's Colorado River Aqueduct Pumping Plants*". Paper presented at the *Microfiltration For Water Treatment Symposium*, Irvine, CA, August 1994.

21. Kohl, H.R., Lozier, J., Bedford, D. and Fulgham, B., " *Evaluating Reverse Osmosis Membrane Performance on Secondary Effluent Pre-treated by Membrane Microfiltration*". AWWA National Conference, San Antonio, TX, June 1993.

22. Bourdon, F., Bourbigot, M.M. and Faivre, M., "*Microfiltration tangentialle des eaux souterraines d'origine karstique*". In L'eau, l'Industrie, les Nuisances, September 1988.

23. Weisner, M., Clark, M. and Mallevialle, J, "*Membranes filtration of coagulated suspensions*". In Journal of Envi. Engin. American Society of Civil Engineering, 1 (1989) 115.

24. Hillis, P., Le, M.S. and Padley, M., "*Removal of Natural Organic Matter (NOM) by coagulation and Microfiltration in potable water treatment*", Paper presented at the Workshop on Colloid Science in Membrane Engineering, Toulouse, May 1996. To be published in The Journal of Colloid and Surfaces.

25. Cline, G. and Vickers, J., "*Drinking water treatment using Microfiltration in conjunction with coagulation processes*". Paper presented at the *Microfiltration For Water Treatment Symposium*, Irvine, CA, August 1994.

26. Hultquist, R., "Microfiltration and the question of compliance: Surface water treatment and Reclamation criteria". Paper presented at the *Microfiltration For Water Treatment Symposium*, Irvine, CA, August 1994.

27. Jacangelo, J. and Adham, S., "*Comparison of Microfiltration and Ultrafiltration for Microbial Removal*". Paper presented at the *Microfiltration For Water Treatment Symposium*, Irvine, CA, August 1994.

Chapter 11

MEMBRANES IN BIOLOGICAL
WASTEWATER TREATMENT

Contents

11.1   Wastewater characteristics and need for treatment

Wastewater is produced whenever and wherever water is used then discarded. The usual sources of wastewaters are households, businesses and industries together with any ground or surface water.

If untreated wastewater is allowed to accumulate, the decomposition of the organic materials it contains leads to the production of large quantities of malodorous gases. In addition, untreated wastewater usually contains numerous pathogenic, or disease-causing micro-organisms that derive from the human intestinal tract or that may be present in certain industrial waste. Wastewater also contains nutrients, which can stimulate the growth of aquatic plants, and it may contain toxic compounds. For these reasons, the immediate and nuisance-free removal of wastewater from its sources of generation, followed by treatment and disposal, is not only desirable but also necessary in an industrialised society.

The most appropriate processes for wastewater treatment depend not only on the nature of the feed, but the size of the flow, the location of the treatment works and the availability of land amongst many other factors. Figure 11.1-1 shows a summary of the unit operations commonly found in wastewater treatment.

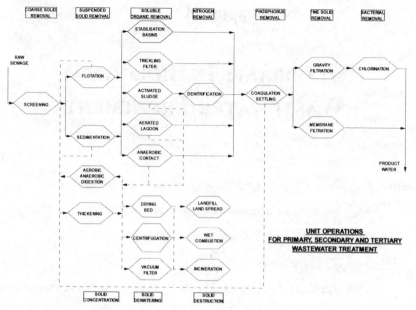

Figure 11.1-1 Unit operations in wastewater treatment

Wastewater is characterised in terms of its physical, chemical, and biological composition. The principal physical properties and the chemical and biological constituents of wastewater and their sources are reported in Table 11.1.1. It should be noted that many of the parameters listed in Table 11.1.1 are interrelated. For example, temperature, a physical property, affects both the biological activity in the wastewater and the amounts of gases dissolved in the wastewater.

TABLE 11.1.1 Physical, chemical, and biological characteristics of wastewater and their sources

Physical properties:

Colour	Domestic and industrial wastes, natural decay of organic materials
Odour	Decomposing wastewater, industrial wastes
Solids	Domestic and industrial wastes, soil erosion
Temperature	Domestic and industrial wastes

Organic constituents:

Carbohydrates	Domestic, commercial, and industrial wastes
Fats, oils, and grease	Domestic, commercial, and industrial wastes
Pesticides	Agricultural wastes
Phenols	Industrial wastes
Proteins	Domestic, commercial, and industrial wastes
Priority pollutants	Domestic, commercial, and industrial wastes
Surfactants	Domestic, commercial, and industrial wastes
Volatile organic	Domestic, commercial, and industrial wastes
Others	Natural decay of organic materials

Inorganic constituents:

Alkalinity	Domestic wastes, domestic water supply, groundwater infiltration
Chlorides	Domestic wastes, domestic water supply, groundwater infiltration
Heavy metals	Industrial wastes
Nitrogen	Domestic and agricultural wastes
pH	Domestic, commercial, and industrial wastes
Phosphorus	Domestic, commercial, and industrial wastes; natural runoff
Priority pollutants	Domestic, commercial, and industrial wastes
Sulphur	Domestic water supply; domestic, commercial, and industrial wastes

Gases:

Hydrogen sulphide	Decomposition of domestic wastes
Methane	Decomposition of domestic wastes
Oxygen	Domestic water supply, surface-water infiltration

Biological constituents:

Animals	Open watercourses and treatment plants
Plants	Open watercourses and treatment plant

Protists:

Eubacteria	Domestic wastes, surface-water infiltration, treatment plants
Archaebacteria	Domestic wastes, surface-water infiltration, treatment plants

Viruses   Domestic wastes

The important contaminants of concern in wastewater treatment are listed in Table 11.1.2. Secondary treatment standards for wastewater are concerned with the removal of biodegradable organics, suspended solids, and pathogens. Many of the more stringent standards that have been developed recently deal with the removal of nutrients and priority pollutants. When wastewater is to be reused, standards normally include requirements for the removal of refractory organics, heavy metals, and in some cases dissolved inorganic solids.

TABLE 11.1.2 Important contaminants of concern in wastewater treatment

Contaminants	Reason for importance
Suspended solids	Suspended solids can lead to the development of sludge deposits and anaerobic conditions when untreated wastewater is discharged in the aquatic environment.
Biodegradable	Composed principally of proteins, carbohydrates, and fats, organics biodegradable organics are measured most commonly in terms of BOD (biochemical oxygen demand) and COD (chemical oxygen demand.) If discharged untreated to the environment, their biological stabilisation can lead to the depletion of natural oxygen resources and to the development of septic conditions.
Pathogens	Communicable diseases can be transmitted by the pathogenic organisms in wastewater.
Nutrients	Both nitrogen and phosphorus, along with carbon, are essential nutrients for growth. When discharged to the aquatic environment, these nutrients can lead to the growth of undesirable aquatic life. When discharged in excessive amounts on land, they can also lead to the pollution of groundwater.
Priority pollutants	Organic and inorganic compounds selected on the basis of their known or suspected carcinogenicity, mutagenicity, or high acute toxicity. Many of these compounds are found in wastewater.

Refractory organics	These organics tend to resist conventional methods of wastewater treatment. Typical examples include surfactants, phenols, and agricultural pesticides.
Heavy metals	Heavy metals are usually added to wastewater from commercial and industrial activities and may have to be removed if the wastewater is to be reused.
Dissolved inorganic	Inorganic constituents such as calcium, sodium, and sulphate are added to the original domestic water supply as a result of water use and may have to be removed if the wastewater is to be reused.

11.2 Microbiology of wastewater treatment

11.2.1 Introduction

Biological treatment is likely to be suitable for treatment of wastewaters of animal or vegetable origin in so far as such treatment is essentially an acceleration of the natural processes of decay, biological stabilisation and mineralisation. Biological treatment can also be used to bring about oxidation or reduction of nitrogen, sulphur, phosphorus, iron and other elements associated with living matter including effluents from industrial processes.

11.2.2 Micro-organisms Involved In Aerobic Wastewater Treatment

At present, largely undefined, heterogeneous cultures of micro-organisms are used in biological wastewater treatment processes to convert organic pollutants into environmentally acceptable forms. This situation differs from other biological treatments where specific enzymes or defined cell cultures are employed as the biocatalysts to mediate biological reactions. Such a difference is not because the nature of biological reactions involved in the two cases are fundamentally different but due largely to practical reasons. Unlike most industrial raw materials that have a defined composition, industrial and domestic wastewaters often have complex and variable compositions. The metabolism of this requires a versatile and flexible combination of enzymes, that at present could only be obtained from a mixed culture of bacteria. In addition, organic pollutants are often present in a much diluted form compared with industrial raw materials, so a high cost could be incurred in the treatment processes if stringent conditions were to be satisfied to maintain a specific group of enzymes or a defined culture of bacteria in the treatment plant.

Two major categories of biological wastewater treatment processes exist, which are known as aerobic and anaerobic processes. In aerobic metabolism dissolved oxygen is required as an electron acceptor, while in anaerobic metabolism the existence of oxygen is not permitted. Some bacteria however, do not use dissolved oxygen as the electron acceptor during metabolism, but use oxygen in its oxidised forms instead. The most important examples are the nitrate reducers and the sulphate reducers. The nitrate reducers are grouped together with aerobic organisms, while the sulphate reducers are grouped with the anaerobic ones because the former are more closely associated with aerobic treatment and the later with anaerobic treatment.

In aerobic processes, the microbial community consists of both micro and macro-organisms that can metabolise organic pollutants in wastewaters mainly to carbon dioxide and water in the presence of dissolved oxygen. A large variety of bacteria species are found in aerobic treatment processes, but the dominating species may differ from one plant to another. In general, the microbial community in suspended growth systems show a lower diversity than that in fixed-film systems. This is mainly due to the absence of Annelida, Insecta and Arachnida. The diversity and distribution of the

11 - 4

micro-community in aerobic treatment plants depends on the selective pressure of the environment, which may arise from the design of the reactor the mode of operation and the nature of the wastewater. Table 11.2.1 shows a typical make up of the microbial population in activated sludge.

Table *11.2.1 Typical make up of an activated sludge*

Type	Function	Examples	
Heterotrophic	carbonaceous oxidation denitrification phosphorus removal	*Pseudomonas Flavobacterium Chromobacterium Mycillus Alcakugenes*	*Micrococcus Staphylococcus Bacillus Acinetobacter*
Autotroph	Nitrification	*Nitrosomonas Nitrobacter*	
Filamentous	Carbonaceous oxidation	*Type 021 N M. Parvicella Type 1701 S. Natans*	*Nocardia sp . N. Limicola Beggiatoa sp.*
Protozoa	Prey on bacteria and algae, both attached and free swimming bacteria	*Ciliatea Rhizopodea Phytomastigophorea Actinopodea*	
Fixed film systems			
Bacteria	Metabolism of organic pollutants, nitrification and denitrification.	*Achromobacter Escherichia coli Myxobacterium Nitrosomonas*	*Nitrobacter Pseudomonas Streptococcus Chromobacterium*
Protozoa	Prey on bacteria and algae, including both attached and free swirnming.	*Ciliatea Phytomastigophorea Rhizopodea*	*Zoomasdgophorea Actinopodea*
Nematode	Prey on bacteria, other nematodes and rotifera.	*Diplogasterinae Rhabdidnea*	*Non-Rhabditida*
Rotifera	Prey on bacteria and algae.	*Adinetidea Philodinidae Brachionidae*	*Notommatidae Testudinellidae*
Annelida	Prey on bacteria and other suspended organic materials.	*Aelosomatidae Enchytraeidae*	*Lurnbirculidae Lumbricidae*
Insecta	Grazers feeding on biofilm or its components.	*Collembola Coleptera*	*Hymenoptera*
Arachnida	Prey on insects and worms.	*Linyphudae Astigmata*	*Mesostigmata*

11.2.3 Micro-organisms Involved In Anaerobic Wastewater Treatment

Anaerobic digestion is a complex microbial process involving many species of bacteria. Its primary objective is the stabilisation of organic matter through its conversion to carbon dioxide and methane in an oxygen free environment. The overall process is often subdivided into four separate processes as follows:

Hydrolysis - Organic polymeric material cannot be utilised by micro-organisms unless it is broken down into soluble compounds, usually monomers or dimers, which can then pass across the cell membrane. Therefore hydrolysis is the first stage in the anaerobic degradation of complex polymeric organics. Three groups of organics are generally considered these being carbohydrates, proteins and lipids.

Acidogenesis - Following hydrolysis of carbohydrates, lipids and proteins fermentative processes occur yielding a variety of intermediates of which volatile fatty acids (VFA) form the major constituents. During the fermentation of carbohydrates a variety of reduction products are produced, including ethanol, lactose, propionate, formate, butyrate, succinate, caproate, acetate, butanol, propanol, and acetone. Carbon dioxide and molecular hydrogen are also produced.

Acetogenesis - This results in acetate and a range short chain volatile fatty acids and alcohols. Short chain volatile fatty acids are important intermediates in anaerobic environments. Propionate and butyrate account for approximately 15 and 5%, respectively, of the methane produced in anaerobic digesters. The acetogenic bacteria are acetate forming organisms that exhibit two different types of metabolism. These are termed acetogenic dehydrogenations and acetogenic hydrogenations. Several obligate proton reducers are involved. *Syntrophomonas wolfei* oxidises even numbered carbon chain fatty acids to acetate and hydrogen while odd numbered carbon chain fatty acids are oxidised to propionate, acetate and hydrogen. *Syntrophobacter wolinii* oxidises propionate to acetate, carbon dioxide and hydrogen.

Methanogenesis - Methanogens are responsible for the production of methane from a limited number of single carbon compounds. Different species often specialise in the number of substrates from which they can produce methane. Some species utilise their substrates as both the carbon and energy source, while other organisms require additional carbon compounds. Unique co-enzymes are involved in the production of methane from acetate and single carbon compounds. The methanogenic bacteria belong to the group *archaebacteri*. These are distinguished from the *Eubacteria* by the absence of muranic acid in the cell walls, and the presence of unusual lipids particularly polyisoprenoid ether linked lipids. A limited number of substrates are used by the 47 species of methanogenic bacteria known. Two major groups have been identified. Group I contains 33 species belonging to the families: *Methanobacteriaceae, Methanothermaceae, Methanococcaceae, Methanomicrobiaceae, Methanoplanaceae*. These species reduce carbon dioxide with hydrogen or utilise formate in the formation of methane. Group II includes 14 species belonging to the family *Methanosarcinaceae*. These species utilise acetate, methylamines and/or methanol. *M. barkeri* and *M. vacuolata* are most versatile as they use all known methanogenic substrates except formate. All Methanogens obtain energy for growth from the formation of methane.

11.3 Types of bioreactors

11.3.1 Aerobic systems

Suspended Floc Systems

These systems may be regarded as utilising more intensively the natural purification processes that occur in streams and other water bodies. In such processes micro-organisms are cultivated in the form of flocs suspended in a tank which must then be separated from the wastewater to produce a clarified effluent. Also sufficient quantities of micro-organisms must be maintained in the reactor so that continuous and efficient removal of pollutants can be sustained.

The simplest form is the aerated lagoon, but the activated sludge process is that most widely used. Numerous variants exist of the activated sludge process differing in their hydraulic flow patterns, relative sizes and shapes of tanks, method and intensities of aeration, and materials and costs of construction. In recent years, the process has been intensified by increasing biomass concentrations, intensification of oxygen transfer systems, increasing substrate transfer to the biomass, changing the type of biomass, and modifications for nutrient removal.

Activated sludge process

This process was developed for the treatment of domestic sewage by Ardern and Lockett in 1913 under the guidance of W.J.Fowler. It is now also widely used for treatment of industrial wastewaters. The process, often preceded by primary sedimentation of the wastewater, normally involves two stages.

    1. Aeration of the settled wastewater in admixture with activated sludge.

    2. Secondary sedimentation in order to separate activated sludge for re-use in stage 1.

The aeration tank contains an aerobic suspension of many different micro-organisms which under the right operating conditions can produce the required degree of treatment. The wastewater to be treated enters the activated sludge tank and is mixed with the microbial suspension and supplied with oxygen. Organic material in the wastewater is oxidised and there is a corresponding growth of microbial biomass. This biomass is separated from the treated effluent in a settling tank and the thickened sludge is recycled back to the aeration tank inlet. Effluent is discharged from this final settling tank. A small proportion of the recycled flow is removed from the system as surplus sludge. This fraction is an important process variable and its value not only determines the average concentration of the **mixed liquor suspended solids** (MLSS) in the aeration tank but also the overall rate of treatment. Continuous plants use one or more aeration tanks usually of square cross-section, in which the contents are completely mixed. When a single tank is used, or when the tanks are operated in parallel, the plant is known as a "completely-mixed" activated-sludge plant (see figure 8.3.1). This has advantages in load balancing where rapid changes to the influent occur, and is more resistant to toxic shock.

The design variables for the activated sludge process are used to describe the process variants and can be used to compare performance. They include: wastewater retention time; sludge loading rate; sludge age; sludge production rate; aeration requirements; solids liquid separation. The latter is a major disadvantage of all suspended floc systems. Poor settling sludges are common and can make operation unsustainable. This has resulted in the development of a number of process variants to overcome the problem.

Fixed Film Systems

Fixed film systems have developed from the early practice of treating sewage by spreading on land. Essentially the processes involve flow of the wastewater to be treated as a thin film over the surfaces of an inert medium exposed to the air. A bacterial film forms on the surfaces of the medium over the course of a few weeks from growth of bacteria present in the wastewaters, or from those derived from air and soil. The bacteria utilise oxygen dissolved from the air to oxidise organic matter in the wastewater and obtain energy for growth. In time protozoa and later metazoa (worms, fly larvae, springtails, etc.) populate the film until there is a balanced community of organisms and the accumulation of film on the medium is balanced by the loss of humus solids in the flow of water.

The treatment processes covered include: conventional biological filters, *biological aerated filters* (BAFs and BAFFs), *rotating biological contactors* (RBCs), *biological fluidised beds* (BFBs) and reed beds.

The main requirements for the basic design of a fixed-film system are the dimensions of the fixed bed of medium, and final settling tank, if it is used. For BAFs and fluidised support systems the design should specify the amount of process oxygen to be supplied. For conventional biological filters and reed beds natural ventilation is usually sufficient.

Biological Fluidised Beds (BFBs)

A fluidised bed, Figure 11.3-2, differs from the biological aerated filter in that the bed of media expands until the gravitational and lifting forces on the particles are equal. The media remains suspended in free motion in the upward flow of water. In the presence of nutrients, derived from the wastewater, an adherent biological film develops on each particle, reducing its density and increasing its resistance to flow. The bed expands further and the least dense particles migrate to the upper part of the bed. In time they overflow with effluent and can be separated by settlement. A fluidised bed may be used as an anoxic reactor (dissolved oxygen absent and nitrate used as oxidant), or it may be used with oxygenated or aerated feed as an alternative to biological filtration or activated sludge treatment.

In fluidised beds recycled effluent is mixed with influent wastewater and passed through an oxygenator to supply the oxygen required for treatment. The ratio of the recycled flow to the influent flow is set at a minimum value so that the oxygen supply matches the requirements by the BOD and ammonia loads.

The recycled flow must also be sufficient to fluidise the bed. A suitable superficial velocity is in the range 20 to 50m/h. These constraints lead to recycle ratios between 2:1 and 100:1 depending on the total oxygen demand of the influent.

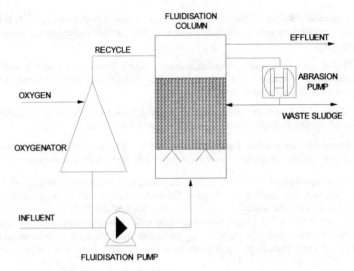

Figure 11.3-2 Biological fluidised beds

11.3.2 Anaerobic systems

Anaerobic digestion is one of the oldest processes used for the stabilisation of sludges. It involves the decomposition of organic and inorganic matter in the absence of molecular oxygen. The major applications have been, and remain today, in the stabilisation of concentrated sludges produced from the treatment of wastewater and in the treatment of some industrial wastes. More recently, it has been demonstrated that dilute organic wastes can also be treated anaerobically.

In the anaerobic digestion process, the organic material in mixtures of primary settled and biological sludges is converted biologically, under anaerobic conditions, to a variety of end products including methane (CH_4) and carbon dioxide (CO_2). The process is carried out in an airtight reactor. Sludge, introduced continuously or intermittently, is retained in the reactor for varying periods of time. The stabilised sludge, withdrawn continuously or intermittently from the reactor, is reduced in organic and pathogen content and is non-putrescible.

In recent years a number of different anaerobic processes have been developed for the treatment of high strength organic waste. The more common processes now in use include the following:

1. *Unmixed anaerobic digestion (also know as standard-rate anaerobic digestion). In this process, the contents of the digester are usually unheated and unmixed. Detention times for the process vary from 30 to 60 days.*

2. *Completely mixed anaerobic digestion (also know as high-rate anaerobic digestion). In this process (see Figure 11.3-3), the contents of the digester are heated and mixed completely. The required detention time for high-rate digestion is typically 15 days or less.*

3. **Anaerobic Filter** *(AF)*. *Biomass retention is achieved by both attachment of the bacteria onto packing materials, and enhancement of the internal sedimentation through these packings.*

4. **Upflow anaerobic sludge blanket process** *(UASB)*. *Biomass retention is achieved by cultivation of sludge which have high settling velocity, and can be retained within the reactor by a simple internal settler.*

5. **Fluidised bed process**. *Biomass retention is achieved by attachment of bacteria onto small particles which is fluidised in the reactor to provide mixing.*

6. **Anaerobic sequencing batch reactor** *(ASBR)*. *Biomass retention is achieved by alternating the operation mode of the reactor either for reaction or for sedimentation.*

Selection of the appropriate process is critical for successful anaerobic treatment of industrial wastewaters and warrants detailed consideration. Decision should be based mainly on the technical possibility of achieving the essential engineering objectives which include: maintaining a high concentration of biomass in the reactor, achieving a long *solids retention time* (SRT) independent of the *hydraulic retention time* (HRT), achieving favourable mass transport conditions, minimising inconsistency of performance, preventing accumulation of inert suspended solids within the bioreactor.

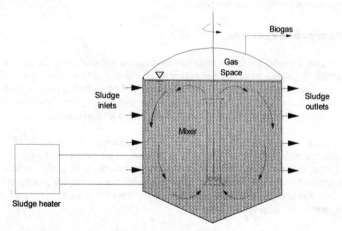

Figure *11.3-3 Completely mixed anaerobic digester*

11.4 Applicability of Membrane in Biological Wastewater Treatment

11.4.1 Introduction

Membrane processes may be useful in situations where a traditional sedimentation or concentration process is inadequate to provide the desired effluent. In particular, membrane processes can accept influents with a much higher solids concentration than conventional settlement equipment. Solid concentration may be in line with the upstream aeration tanks which permits a reduction in the volume of the tanks compared to conventional clarification systems, resulting in a significantly smaller footprint for the overall process.

11.4.2 Opportunities for membrane use

With the rising demands on industry to deliver effluents to higher standards and more reliable quality, there is a need to develop and deliver processes that will meet legislative requirements more cost effectively. With its versatile separation capability membrane technology is making an impact on a number of wastewater treatment areas including:

- *tertiary solid capture,*
- *enhanced clarification and biological process intensification,*
- *sludge thickening.*

In every case, membrane technology offers a reliable, high rate filtration process with absolute removal for sub-micron sized particles of suspended solids, biological materials and coliform bacteria. The low footprint requirement ensures minimum land space usage, and this provides an excellent scope for retrofitting existing wastewater treatment works. Potential exists for the development of disinfection processes for effluent where conventional clarification is the limiting step in secondary treatment. Advances in the technology is required in order to reduce the high operating costs at present.

Consider the case of sludge. Sludge is a suspension with water concentrations varying between 92 and 98%. Sludges result from the removal of solid pollutants during all types of water and wastewater treatment. The main sources of sludge are from primary and storm settling tanks and humus or surplus activated sludge. Secondary sludge is difficult to treat and is usually mixed with primary sludge, sometimes by re-settling them in primary tanks. The treatments to be carried out on the sludge are dictated by the final disposal methods. Such methods include: sea dumping, landfill, agricultural uses, and incineration. Sludge treatment can be by anaerobic digestion, aerobic digestion (composting), chemical or thermal methods. Such methods are often followed by a dewatering technique such as evaporation and or filtration. Sea disposal usually involves no treatment, but landfilling and incineration often require a semi-solid sludge. Sludge for agricultural uses is normally composted to ensure nitrification and destruction of harmful anaerobes.

Recently new legislation has come into effect making a number of disposal routes illegal or economically prohibitive. Notably, the availability of landfill site is rapidly diminishing; incineration cost is escalating and the practice of land spreading is becoming unpopular. The impact of these changes is not only altering the economic equations of sludge, but is also forcing operators to be more innovative in their approach to the whole issue of sludge management. If, for example, sludge could be concentrated, say by a factor of 2X, the cost of any subsequent treatments, e.g. heat, chemical conditioning, or transportation, could be reduced by up to 50%. Even with composting or anaerobic digestion the reactor volume required for treatment is reduced by half, saving both capital and operating costs. The biological processes would also be more efficient with a higher overall conversion.

In the case of the activated sludge process, the operation is restricted to a fairly low biomass concentration because of the final settlement limitation. Consequently, the required reactor volumes are large (high capital cost) and the conversion rates are necessarily very low (high operating cost).

Over recent years attempts have been made to increase the working biomass (e.g. BAFFs, BFBs). These developments have brought improvement, but there is still plenty of scope for highly efficient systems to be developed, particularly with the use of membranes for biomass retention. Membrane systems have the potential for a very high sludge age which generally means a reduction in the surplus sludge volume.

11.4.3 How membrane can be used with bioreactors

The simplest membrane bioreactor type is termed the recycle bioreactor (see figure 11.4-1a). Typically, a reaction vessel, operated as a stirred tank reactor, is coupled in a semi-closed loop via a suitable pump to a membrane module. In operation, the reaction vessel is filled with the substrate and enzyme or micro-organisms. The contents of the reaction vessel are continuously circulated through the membrane module and recycled back to the reaction vessel. At steady state, water and product molecules small enough to permeate through the membrane are removed from the system while the enzyme or microbial cells are returned to the reaction vessel for further reaction. The total volume of the system is maintained constant by matching the feed rate with the permeate flux. If the substrate molecule is too large to pass through the membrane it would be returned to the reaction vessel, otherwise it would pass into the permeate. This type of reactor operates very near to a completely mixed condition and for practical purposes may be modelled as a simple *continuous stirred tank reactor* (CSTR). In effect, the concentration of product in the reactor and the membrane module will be essentially the same at any given time. In the case of low molecular weight substrates that are freely permeable, their concentration in the reaction vessel will be the same as in the product stream. Thus CSTR type bio-reactors are more suited for substrate inhibited reactions than product-inhibited reactions, when the conversion rate is high. The high productivity associated with membrane bio-reactors is usually due to higher enzyme or cell concentration and its continuous re-use.

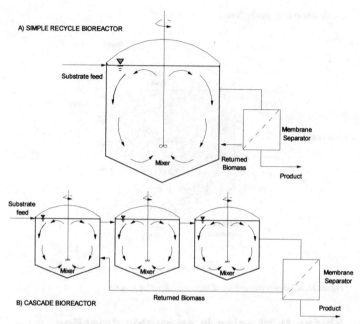

Figure 11.4-1 Membrane bioreactor configurations.

The completely mixed condition of the CSTR is a disadvantage for most enzymatic or microbial conversion processes since they have a positive order of reaction, i.e. the rate of conversion increases with increasing substrate concentration. These reactors will operate at very low conversion rates, because the substrate concentration in the system is very low. Consequently, the reactor volume must be significantly larger than for a plug flow reactor for the same conversion. To overcome this limitation, a number of CSTR bio-reactors may be put in series in a cascade fashion (see figure 11.4-1b). The total volume of the system should be less than if only one stage is used, since all stages, except the last one will operate at a higher conversion rate.

Early examples of the use of membrane bio-reactors for wastewater treatment were in conjunction with anaerobic digestion. Not surprisingly, where the treatment of high strength, low volume waste is concerned, the relatively high cost of membranes is not significant. Figure 11.4.2 shows a schematic of a basic anaerobic membrane digester. This design is essentially the same as the simple recycle bio-reactor described above, except that the reaction vessel is enclosed and has a head space for gas collection and removal. The membrane module is conveniently placed in series with the sludge heat exchanger. The key benefits of the use of membrane for biomass retention in anaerobic digestion include:

1. *Capability to decouple the SRT from the HRT for the digester*

2. *Reduce capital cost with smaller digester volume or*

3. *Enhanced biomass and increase retention time resulting in greater gas yield reduced sludge volume.*

4. *Simple anaerobic operation.*

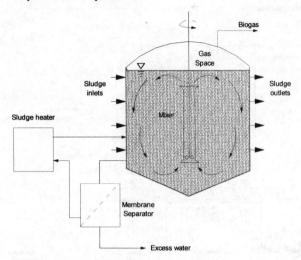

Figure 11.4-2 Membrane-enhanced anaerobic digestion process.

11.5 Sludge thickening in anaerobic digestion

Some of the earliest membrane development work in sludge thickening for anaerobic digestion was attributed to Dorr-Oliver, the company that patented the commercial MARS *(membrane anaerobic reactor system)* process [1]. The use of membranes as biomass separators in anaerobic digester systems treating industrial effluent was pioneered by Choate et al. [2] and Saw et al. [3]. They also proposed a variation of the anaerobic bio-reactor design in which the process was split into two stages in order to take advantage of the different conditions demanded by acidogenic and methanogenic processes.

Ultrafiltration and anaerobic digestion are two very complementary and interdependent processes; anaerobic digestion breaks down organic materials that would otherwise foul the filter membranes, while the latter for their part serve to retain the biomass some of which would otherwise be lost in the effluent of conventional digester systems. The requirement for the prevention of membrane fouling is the symbiotic self-cleaning action of the anaerobic bacterial population. Influent macromolecules and particulate organics are rejected at the membrane surface and are selectively retained in the digester by the membrane unit until metabolised by the micro-organisms.

A successful large scale application of ultrafiltration in sludge thickening for anaerobic digestion has been described be Ross et al. [4] for the treatment of an effluent from a maize processing factory. The composition of the maize effluent varied according to the processing conditions, but generally consisted of soluble salts, proteins, carbohydrates, fibres, etc. The plant configuration was similar to that shown in figure 11.4-2. Ross et al. established that the membrane life was significantly extended by operation at lower pressures. Operation of membranes with inlet pressures as low as 175 kPa has been made possible by halving the parallel flow path of the tubular membranes (e.g. a flow path of

40 tubes in series was reduced to 2 flow paths of 20 tubes or 4 flow paths of 10 tubes). In each case the feed flow per module was doubled to obtain the same linear flow velocity inside the tubes (2 m/s typical) and the pressure drop was halved for each change in flow. Since the flux in an anaerobic sludge process was relatively insensitive to the feed pressure it was more economical to pump additional volume at reduced pressure than to pump less volume at higher pressure. Such development in the hydrodynamics of tubular membrane modules provided a more reliable low-cost system as there was a much greater scope in pump selection and prices when operating at the lower pressure. Operation at the lower pressure also resulted in higher volumes being pumped (at the same energy level) which further improved the mixing of the digester contents. The operation and the process condition of the anaerobic membrane digester are summarised in Table 11.5.1 below.

Experiences with full scale studies indicate that membrane separation technology offers a viable solution to problems of effluents with very high COD loads. The combination of anaerobic digestion with ultrafiltration achieves the desirable effects of maintaining reasonably high biomass concentrations and long sludge retention times, while producing a good quality colloid-free effluent.

Table 11.5.1 Summary of the UF-coupled anaerobic digestion process for the treatment of a maize effluent (adapted from ref. 4)

Throughput, m³/d	500	Membrane MWCO	20,000 to 80,000
Feed COD, mg/l	15,000	Tube dia, mm	9
Permeate COD, mg/l	400	Membrane area, m²	668
Reactor volume, m³	2,610	Cross flow vel, m/s	1.6 - 2.0
HRT, d	5.2	Feed pressure, kPa	175-450
MLSS, mg/l	21,000	Working flux, LMH	8-37
Space Loading Rate kg COD m⁻³d⁻¹	2.9	Cleaning frequency (using EDTA)	once per month
Temp., °C	35	Membrane life expectancy	6 years

11.6   Intensification of activated sludge process

Much of the early interest in the use of membranes for biomass retention in an activated sludge process occurred in Japan. This stems from the very high cost of space in most Japanese cities, that gives great incentive in minimising the process space requirement. Membranes have been used with both suspended growth and fixed film type activated sludge processes. In suspended growth systems, the membranes obviate the need for both primary and final settlement tanks thereby resulting in considerable saving in space. Additionally, the high level of *mixed liquor suspended solids* (MLSS) effectively achieves nitrification and denitrification without the need for extended aeration. In fixed film processes, the solid retention mechanism of the packed bed is often inadequate to meet local discharge or water re-use requirement, therefore, the membrane is applied to the effluent to remove any residual solids or micro-organisms in order to improve its quality (see section 7 below).

A study of the performance of hollow fibre membranes (0.03 and 0.1 μm) in a nitrifying/denitrifying activated sludge process has been described by Chiemchaisri et al. [5]. In this study the hollow fibre bundle was immersed in the bioreactor; permeate was withdrawn by the application of a vacuum. Sludge build-up around the fibres was minimised by agitation of the liquid. Stable operation was achieved over a period nearly 1 year. Table 11.6.1 shows a summary of the hollow fibre membrane assisted activated process.

Table 11.6.1 Summary of the hollow fibre membrane assisted activated process. Adapted from ref. 5

Feed COD, mg/l	150-450	Membrane cut off, µm	0.03, 0.1
N contribution to COD	0.3-10%	Area, m²	0.3
Permeate COD, mg/l	20.8	stirring speed, rpm	290
Reactor volume, L	62	Feed pressure, kPa	30-70
HRT, days	0.5-1	Working flux, LMH	8-29
MLSS, mg/l	4,000		
N removal	80-90%		
Temp., °C	25-29		

A more conventional type of recycle membrane bio-reactor for activated sludge treatment is described by Magara et al. [6]. These authors showed that the investment and operational cost of the membrane activated sludge process (biological denitrification) treating 100 m³/d collective night soil was less than or equal to a conventional system but requiring less land and supervision. The results of their year long pilot trial is summarised in Table 11.6.2.

Table 11.6.2. Summary of the pilot trial for collective night soil biological denitrification, using tubular membrane in an activated sludge process (adapted from [6]).

Throughput, m³/d	3.8	Membrane MWCO	20,000
BOD load kg/m³/d	1.66	Tube dia, mm	12
Permeate BOD, mg/l	<10	Cross flow vel, m/s	0.5-3.0
TN load kg/m³/d	0.63	Feed pressure, kPa	100-300
HRT, d	4.7	Working flux, LMH	65-85
MLSS, mg/l	15,000-20,000	Area, m²	
Space Loading Rate kg BOD m⁻³d⁻¹		Cleaning frequency (using EDTA)	
Temp., °C	35	membrane life expectancy	

11.7   Kubota submerged membrane bio-reactor

Figure 11.7-1 shows a schematic diagram of the Kubota submerged membrane bio-reactor system. This system is an example of the most successful design of a membrane enhanced activated sludge process.   A large number of full scale treatment works in Japan already employed the Kubota membrane system.   A key feature in this process is the use of membranes as an integral part of the bio-reactor, obviating the need for both primary and final settlement stages. The filtration membrane has a pore size of 0.2-0.4 µm. The membrane cartridges are submerged in the aeration tank and the treated effluent is separated from the mixed liquor under the effect of gravity or negative pressure (suction). The membrane is used to separate the biomass from the effluent exiting the process.

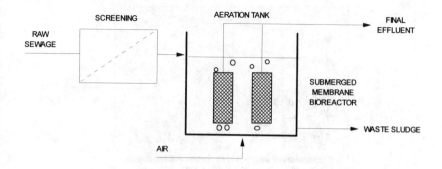

Figure 11.7-1 Kubota Submerged Membrane Activated Sludge Process.

In operation, air is sparged from the bottom of the module unit. This keeps the bacterial mass in suspension, provides oxygen for bio-oxidation processes, and provides a gentle scouring of the membrane. The latter helps to keep the pores of the membrane free from clogging. The liquid under treatment is drawn through the membrane, under gravity or vacuum suction, and is pumped out of the system. The biomass is retained within the system and thereby increasing the efficiency of the bio-degradation process. The process is effective where small quantities of screened sewage or organic laden industrial effluents, with a low suspended solids matter are required to be treated in a confined space and in a controlled manner. It can be used for both secondary and tertiary treatment (e.g. N and P removal). The process appears to be effective in obviating the need for bulky units in sludge settling and recycling. However, it is primarily applicable to effluents containing soluble organics with a low suspended solids concentration. The specific capital cost per volumetric throughput) of the units may be low for small works but the running cost tends to be high. The advantages of the Kubota submerged membrane bio-reactor process include:

- *high MLSS (15 ,000 - 20,000 mg/l),*
- *small footprint,*
- *no primary or secondary settlement requirement,*
- *low operating costs for small works,*
- *excellent final effluent quality, BOD removal, nitrification and clarification in one,*
- *high sludge age, low sludge production rate,*
- *easy retrofit of existing works.*

Most of the systems in use so far have been installed in Japan. However, keen interest world-wide has been shown this process because of its superior technical advantages. At the present, in the UK, a number pilot trials with Kubota submerged membrane bio-reactor are being carried out or planned. The general operational performance of the process is indicated by in Table 11.7.1. In an actual pilot trial (at 80 m³/d) with screened domestic sewage, conducted by North West Water Limited, it was found that the effluent quality over a 6 month period was remarkably consistently (effluent BOD < 5 mg/l, see Figure 11.7.2) despite large variations in the influent quality.

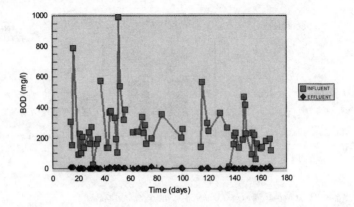

Figure 11.7-2 Variations in the influent / effluent quality - submerged membrane bioreactor trial.

Table 11.7-1 Summary of the operational performance of the Kubota submerged membrane
bio-reactor process

Parameters	operating range
MLSS, mg/l	12000 - 18000
BOD loading, kg/m3d	0.32 - 0.63
F/M ratio, per d	0.025 - 0.042
Hydraulic retention time, h	5-11
Effluent BOD, mg/l	<5
Effluent Total Nitrogen , mg/l	<10
Sludge age (d)	25 - 40
Temp., °C	18-22

The type of micro-organisms that develop in the Kubota submerged membrane bio-reactor system
suggest that the process suffers oxygen limitation, probably due to the high level of MLSS and the
coarse bubble aeration technique used in the process. Figure 11.7.3 illustrates the differences in the
microbial population demography in traditional and the membrane bioreactor activated sludge
samples. The high level of free flagellates in the membrane bioreactor sample (mostly of the genus
Trepomonas) was indicative of chronic oxygen starvation. Since the sludge was very high in MLSS
and viscous, poor oxygen transfer was not unexpected. A high level of dissolved oxygen is required
to encourage a population of grazing ciliate protozoa in order to reduce sludge production.

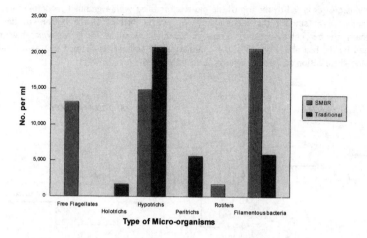

Figure 11.7-3 *Comparison of microbial population in traditional activated sludge (Ellesmere Port WwTW - SS: 2,500 mg/L) & Kubota membrane (SMBR) activated sludge (SS: 15,000 mg/L).*

11.8 Tertiary solid capture

The demand of water for industrial, domestic and irrigation purposes is increasing in many countries [7, 8]. There are already a large number of existing and planned projects, particularly in Japan, the U.S.A. and the Middle East to reuse wastewater for these various ends [9]. Membrane technologies (RO, UF or MF) combined with other physico-chemical (activated carbon adsorption, ozonation) or biological processes allow the water to be reuse for non-drinking applications. In both Japan and the U.S.A., some buildings now include water treatment plants, using membrane bioreactors, to allow the water to be recirculated and used in toilet flushes [10]. Olsen [11] and Olivieri [12] have investigated the possibility of using membrane filtration for improving the quality of municipal secondary effluent. In the UK and in Europe the requirements of the shellfish industry (Fishery Directives), and the development of recreational zones (designated Bathing Water), are making it increasingly necessary to disinfect effluents prior to release, particularly in French coastal zones [13]. The same applies to effluents released into living rivers. The French Public Health Authorities recommend that recycled effluents used to irrigate market gardens (for vegetables that are consumed raw), should contain no more than 1 helminth egg and 1000 thermotolerant coliforms per 100 ml (Health Authority recommendation, 1991). However, all the disinfectants commonly used have the disadvantage of being extremely sensitive to the effluent quality. It is not realistic to reduce the faecal level by 3 or 4 logarithmic units when treating poor quality effluent (COD > 90 mg/l, BOD > 30 mg/l and suspended solids > 30 mg/l). It is against this background of economic drive, public health concern and legislation that there has been an upsurge in interest in the use of membrane for secondary effluent disinfection or tertiary solid capture in recent years.

The MEMBIO™ process (a trademark of Memcor Ltd) is a new fixed film activated sludge process (a BAFF), which combines secondary effluent treatment with solids separation and disinfection by continuous microfiltration. Figure 11.8.1 shows the process flow diagram for the MEMBIO process.

Settled wastewater is fed to the top of the bio-reactor filled with a granular medium (coarse sand), which serves as a site for the bacterial growth. Air is injected at the bottom of the reactor. Periodically, the bed is back-washed to remove accumulated solids. The solids which have not been entrapped by the bed are removed in the continuous MF unit. It is claimed that as a result of the membrane, disinfection of the final effluent is accomplished.

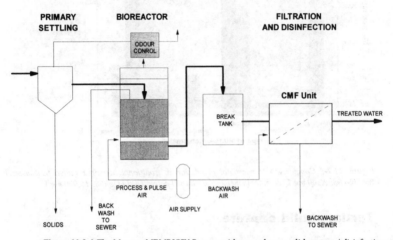

Figure 11.8-1 The Memcor MEMBIO™ Process with a membrane solid capture / disinfection unit.

Langlais et al. [14] reported the pilot trial results of a tertiary solid capture system using a variation of the MEMBIO process. In their trial they employed a commercial Biocarbone ® BAFF (fixed film activated sludge filter by Le Carbone of France) and a membrane unit (0.2 μm polypropylene HF) supplied by MEMCOR   They reported that in a process similar to one shown by Figure 11.8.1, with filtration cycles exceeding 72 hours and a permeate flow of approximately 80 l/m2/h (LMH) was maintained. Despite the poor effluent from the BAFF (COD was between 100 - 400 mg/l; BOD 30 -150 mg/l and TSS 15 - 90 mg/l), the membrane was capable of producing a tertiary effluent containing no faecal coliforms or tenia and ascaris eggs. Subject to confirmation, it was also capable of eliminating all free amoeba cysts. Moreover, the final effluent was free from any turbidity. COD and BOD reductions of 60 and 70 % were achieved which were sufficient to satisfy "level e" requirements (in France the maximum level recommended for recycled effluents, viz. COD < 90 mg/l, BOD < 30 mg/l and suspended solids < 30 mg/l).

In a similar development by North West Water Limited a 0.2 μm HF membrane was coupled to the BFB to provide an effective solid capture mechanism for the process (Figure11.8.2). The TSS in the BFB effluent was in the range 15 - 100 mg/l. There was no detectable turbidity in the membrane permeate. It was found that the membrane flux performance was primarily determined by the applied pressure and the frequency of backwash. Increasing the cross flow velocity also increased the flux, but the effect was not significant. The most favourable operating condition seemed to be at a pressure in the range of 0.4 to 0.9 barg with a backwash rate of 6 per hour (Figure 11.8.3). Under such a condition an average flux in excess of 60 LMH was sustainable with a weekly clean using 500-1000 ppm sodium hypochlorite.

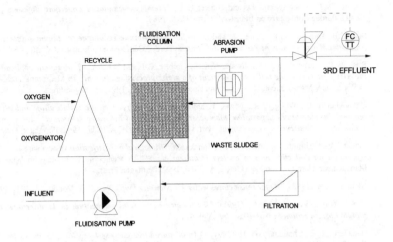

Figure 11.8-2 Tertiary solid capture with a membrane in the Biological Fluidised Bed system.

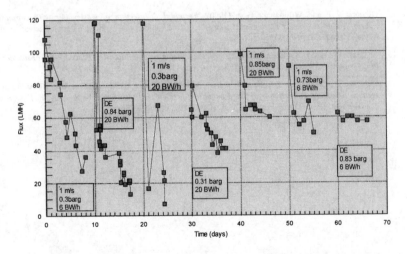

Figure 11.8-3 Microfiltration (0.2 μm HF) performance under various condtions in tertiary solid capture.

11.9   References

1. Stavenger, P.L., *Chem. Eng. Prog.*, No. 67, Vol. 3, P 30, 1971.

2. Choate, W.T., Houldsworth, D. and Butler, G.A., *"Membrane-enhanced anaerobic digesters"*, 37th Annual Purdue Conference on Industrial Waste, USA, 1983.

3. Saw, C.B, Anderson, G.K., James, A. and Le, M.S., *"A Membrane Technique for Biomass Retention in Anaerobic Waste Treatment Process"*, 40th Annual Purdue Conference on Industrial Waste, USA, 1985.

4. Ross, W.R., Barnard, J..P, Strohwald, N.K.H., Grobler, C.J. and Sanetra, J. *"Practical application of the ADUF process to the full scale treatment of a maize-processing effluent"*. In Membrane Technology in Wastewater Management, Hart, O.O. and Buckley, C.A. (Eds), 1992, Permagon Press.

5. Chiemchaisri, C., Wong, Y.K., Urase, T. and Yamamoto, K.Y. *"Organic stabilisation and nitrogen removal in membrane separation bioreactor for domestic wastewater treatment"*. In Membrane Technology in Wastewater Management, Hart, O.O. and Buckley, C.A. (Eds), 1992, Permagon Press.

6. Magara, Y., Nishimura, K., Itoh, M., and Tanaka, M. *"Biological denitrification system with membrane separation for collective human excreta treatment plant."* In Membrane Technology in Wastewater Management, Hart, O.O. and Buckley, C.A. (Eds), 1992, Permagon Press.

7. Nichols, A. B. (1988). *"Water reuse closes water - wastewater loop"*. W.P.C.F., Vol.60, 11, 1931-1937.

8. Thanh, N. C., Visvanathan, C. (1991). *"Wastewater reuse gains momentum in Mediterranean and Middle Eastern regions"*. Wat. Wast. Int., Vol. 6, 1 ,19-26.

9. Hagadorn,.R. E., Chaudhary, D. H. (1991). *"An advanced treatment system with zero discharge"*. Wat. Env. Tech., 16-17.

10. Audic, J.M., Fugita, Y. and Faup, G.M. (1986). *"Le couplage boues activees - membrane: une realite au Japon"*. T.S.M., Vol. 6, 297-300.

11. Olsen, O.J. and Haagensen, U.H. (1983). *"Membrane filtration for the reuse of city wastewater"*. Desalination, Vol. 47, 257-265.

12. Olivieri, V P and Willinghan, G A (1990) *"Terminal disinfection of wastewater with contiuous microfiltration to protect recreational waters"*. Oxford conferences, Bathing waters - Recreation and Management (Dec. 10-11), Bristol, U.K.

13. Langlais, B. (1988). *"La desinfection des effluents rejetes en zone littorale. Procedes de substitution a la chloration"*. L'eau, l'Industrie, les Nuisances, Vol. 118, Avril, 31-32.

14. Langlais, B., Denis, Ph., Triballeau, S. Faivre, M. and Bourbigot, M.M. *"Test on Microfiltration as a tertiary treatment downstream of fixed bacteria filtration"*. In Membrane Technology in Wastewater Management, Hart, O.O. and Buckley, C.A. (Eds), 1992, Permagon Press.

Chapter 12

OILY WASTE-WATER TREATMENT

Contents

12.1   Sources of oily wastes

Motorists in the Western world discard about 2,000,000 m^3 of oil every year by changing their engine oil. Despite the fact that these are among the highest quality products refined from our best crude oils, only about 10% of the waste oil reach recyclers to be re-processed into clean lubricants. The rest is burned as fuel or disposed of as waste. Indeed, a conservative estimate is that over 1,500,000 m^3 pa of used lubricating oil are injected directly into the environment through landfills, burning, and other disposal methods. The incentive for recycling these lubricants flourished when oil was scarce. During world war II the German industry collected and recycled its dirty lubricating oil to help reduce oil imports. Since then, subsidies and legislation have made the concept a permanent fixture in Germany. It is estimated that even today some two thirds of German waste crankcase oil is recycled. Clearly, the rest of the world could follow the example set by Germany.

Stable oil-water emulsions are generated in many diverse industries. Figures 12.1-1 and 12.1-3 show some major industrial manufacturing processes where large volumes of oily wastewaters are generated. Metal working operations use water-soluble coolants, cutting and grinding oils, and lubricants for machining. Metal cleaning tanks and alkaline degreasing baths generate an oily wastewater. Rolling and drawing operations use oil lubricants and coolants. Food processing has waste streams with natural fats and oils from animal and plant processing-particularly vegetable oily wastes. Peel oil is another example from the citrus industry. Textile manufacturing will produce natural oils from wool scouring or fabric finishing oils. Metal cleaning and wool scouring wastes are illustrative of the diversity of oily wastewater treatable by membranes. Metal cleaning operations normally precede painting or plating operations with the objective being to remove dirt and grease.

During the recovery of oil from on-shore and off-shore production facilities substantial volumes of brine are co-produced with the oil. This oilfield brine must be treated before it may be discharged or re-use for injection into the oilfield. The daily production volume from the North Sea operation alone is of the order 400-10,000 m³/d [1].

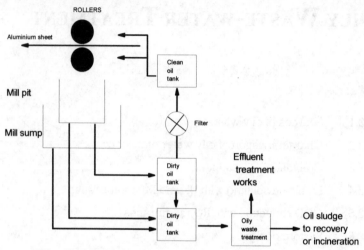

Figure 12.1-1 *Use of oil emulsion in an aluminium hot rolling process*

Oilfield brines comprise of the total water that is co-produced with oil and may include various treatment chemicals (i.e. corrosion inhibitors, scale inhibitors, and emulsions breakers) added during the oil recovery process. The characteristics of oilfield brine vary substantially depending on the geological formation, the oil production operation and the age of the field. The level of oil in the untreated oilfield brine ranges from 3 to 1570 mg/L. The major organic components include phenol, alkyl phenols, carboxylic acids, and esters of aromatic and aliphatic acids. Because of the use of steam to stimulate production of the heavy oil, the temperature of the oilfield brine could reach over 80°C in some cases.

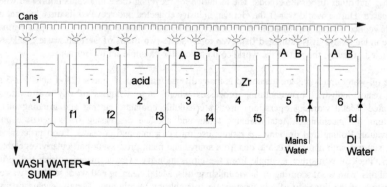

Figure 12.1-2 *Oily wastewater generation from in an aluminium can washing process*

The large flow rates, variability in the oilfield brine characteristics and the nature oil dispersion itself (presence of hydrophobic organics, inorganic colloids, surfactants, etc.) represent a significant challenge for the application of membrane technology to oilfield brine treatment. The ultimate fate of the oilfield brine will dictate the level of oil and suspended solids removal required. In offshore oil production, the oil concentration must be reduced to meet regulatory requirements before disposal by overboard discharge. In various offshore regions the current requirements generally range from 30 to 72 mg/l total oil and grease, but increasingly stringent regulatory levels are anticipated in the future. Alternatively at both onshore and offshore facilities, the oilfield brine may by re-injected into an underground formation for disposal or pressure maintenance. In this case there is no legal requirement for treatment, however, it is generally acknowledged that reduction of both suspended oil and suspended solids to low levels would be desirable in order to minimise the potential for plugging of the formation.

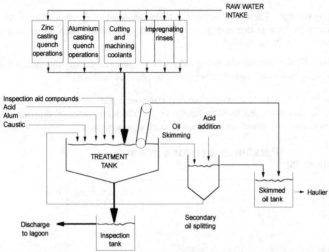

Figure 12.1-3 Use of cooling and lubricating oil in aluminium & zinc die casting process

12.2   Characterisation of oily water emulsions

12.2.1   Classification of oil dispersions

Industrial oily wastewaters are often placed into three broad categories according to the distribution of the oil phase as follows:

- Free-floating oil
- Unstable oil-water emulsions
- Stable oil-water emulsions

A more theoretical approach to aspects of oil water separation will be of great assistance in the application and operation of equipment. In considering the impact of these theoretical aspects of oil separation, it is also of interest to note the nature of oil/water emulsions. Oil may be combined with water in any of the following forms:

1. **Primary Dispersion**: Droplets above 50 micron diameter. These have a significant rising velocity in water and will be separated by gravity devices.

2. **Secondary Dispersion**: Droplets in the range of 3-50 micron diameter. These have progressively smaller rising velocities and may only be separated by special media arrangements designed to affect the surface energies of the droplets.

3. **Mechanical Emulsions**: Droplets below 3 micron diameter. These may be formed by high energy input and are unlikely to be separated by physical processes. Separation may be facilitated by the addition of certain chemicals.

4. **Chemical Emulsions**: Droplets less than 2 micron diameter on which surfactants are adsorbed and create a potential barrier inhibiting the coalescence process. Separable only by chemical means.

5. **Microemulsions:** Thermodynamically stable colloidal dispersion. Coalescence would produce an increase in free energy, and hence does not occur. Such dispersions are created by using an ampithatic molecule.

6. **Dissolved Hydrocarbons** (including oils): Some oils, particularly the lighter fractions will be soluble in water and cannot be separated by physical methods.

Oil dispersions are often not homogenous, but may comprise a mixture of several of the above types of dispersions. This is clearly illustrated by figures 12.2-1 which shows the oil drop size distributions in a paraffin dispersion with water with different amount of a surfactant.

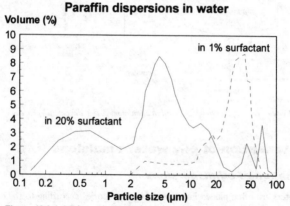

Figure *12.2-1 Oil drop size distribution in an oily water dispersion*

In a coalescing separator such a media filter there may be three major mechanisms involved in providing the basis for separation:

- Stokes Law - In the first instance larger oil droplets will readily rise to the surface by Stokes Law and collect as a surface layer.

- Flocculation - Smaller droplets with a random motion in the fluid are subjected to dampening of their motion by the media which causes the droplets to collect and join together.

12 - 4

- Coalescence - Small droplets are entrained on the surface of the media and combine to form larger droplets which rise to join the surface layer. The presence of any oleophilic material also promotes inter-droplet coalescence of the flocculated droplets.

12.2.2 The physics of an oil droplet in water

Oil and water are immiscible fluids. That is, when mixed they do not go into solution or form a third and unique chemical compound. Rather, each retains its own individual chemical characteristic despite exposure to the other. When a small amount of oil is mixed with a much larger amount of water, a portion of the oil is suspended in the form of small droplets. By assuming that these droplets are spherical, rigid, non-interacting and suspended in a vortex free fluid their behaviour may be described by simple equation known as Stokes' Law.

According to Stokes' Law when a droplet of oil is placed in water, the buoyancy of the lighter oil causes it to rapidly accelerate upward. As the droplet accelerates, the drag force also increases rapidly. These forces are so large that the droplet quickly reaches a terminal velocity. For all practical purposes, the acceleration distance and time are so small that they can be ignored. Of course, exactly the same behaviour may be expected of a water droplet suspended in oil. The only difference is that the water droplet will want to go down instead of up. The terminal velocity is given by the following expression::

$$V = K(S_w - S_o)\, d^2 / U,$$

where

K = a constant,
S_w = the specific gravity of water,
S_o = the specific gravity of oil,
d = the droplet diameter,
U = the kinematics viscosity of the continuous medium.

As an oil droplet rises in water, the drag force on the droplet is proportional to the viscosity of the water. Similarly, as a water droplet falls in oil, its terminal velocity will be affected by the viscosity of the oil. To the extent that a real suspension approximates the ideal conditions upon which Stokes' Law is based, the equation may be used to predict the behaviour of the suspension. Specifically, the distance between the droplets should not be less than 10 times the diameter of the droplets. An accuracy of 0.5% can be obtained so long as the Reynolds number does not exceed 0.4.

An emulsion contains very small droplets, much smaller than the droplets of free oil in water. The correlation between droplet size and oil content depends on the statistical distribution of droplet size within the mixture. So long as that distribution is favourable, mechanical separation can be made.

Mechanical emulsions are formed by centrifugal pumps and other high shear devices which break up the small oil droplets into even smaller droplets. These droplets are generally below 3 micron in diameter and are unlikely to be separated by mechanical processes. Separation may be facilitated by the addition of certain chemicals. To further illustrate the use of Stokes' Law, consider the operation of a parallel plate coalescer. The throughput rate of a typical parallel plate coalescer may be estimated by the following equation [2]:

$$Q = (S_w - S_o)d^2 D^2 / 51000U \text{ (in USGPM)},$$

where

D = tank diameter (inch),
G = (1- 0.94)(2)^2(93)^2 / 51000 (0.81) = 0.05 USGPM (0.2 L/min),

for a tank with a diameter of 93 inches and an average oil droplet size of 2 microns. Any flow rate greater than 0.2 L/min through this separator will cause the mechanical emulsions to be passed through with the water. Clearly, mechanical separators are not a practical proposition for breaking oily emulsions.

A typical chemical emulsion might contain one per cent oil with a droplet size of 0.1 micron. Clearly, these droplets will not be separated by mechanical processes. Therefore, emulsification destroys the theoretical basis upon which mechanical separators which rely on Stokes' Law for their operation. Other methods must be relied upon for emulsion breaking.

12.3 Conventional treatments for oily wastewaters

As the foregoing has shown, free oil is readily removed by mechanical gravity separation devices. Unstable oil-water emulsions can be broken mechanically or chemically. It is the stable oil-water emulsions which are most difficult to break but which at the same time are most amenable to treatment by membranes. Chemical coagulation/flotation or contract hauling are usually the more expensive alternatives. Furthermore, all chemical treatment methods produce a sludge in which the dirt, floc and trapped water remain in the oily phase. The most common method for oily emulsion treatment uses sulphuric acid to dissolve the sludge. The oil phase is further cleaned by sand filtration.

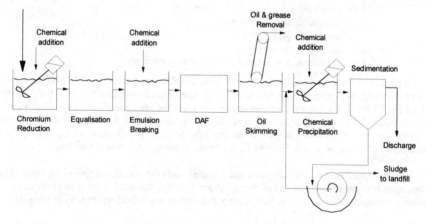

Figure *12.3-1 Traditional treatment of wastewater in can manufacture.*

Membrane is still a relatively new method for oil emulsion breaking and it is often hailed as better and simpler, but on many variable streams the claim has been refuted. Traditional technologies include acid cracking and *dissolved air flotation* (DAF), and they are still viable. In fact, acid is often used on the concentrate from membrane plants. It does not work well on dilute streams. Chemical treatment methods are fairly well established. In larger installations, where proper staffing is available, conventional methods could provide a very cost effective treatment. DAF is probably the major competitive process to membrane filtration. Figures 12.3-1 and 12.3-2 show how the traditional treatment methods are combined in the process schemes for the treatment of effluents from aluminium can manufacture and an industrial laundry. So far, membranes are apparently uneconomic for the process flow rates over 30 m³/h. Perhaps surprisingly, the economics of DAF may be changing faster than the economics of membrane filtration. The capital cost of very new DAF systems for streams of about 5 m³/h are now comparable to that of an UF system. However, this still leaves much scope for membrane processes in smaller scale applications.

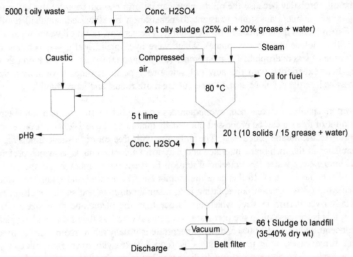

Figure 12.3-2 Traditional treatment of industrial laundry wastewater.

12.4   Emulsion splitting with UF and MF membranes

Oil water separations were seen as an attractive market for Ultrafiltration membranes as early as 20 years ago. Abcor, Inc. now part of Koch International was perhaps the first company to realise that wastewater from the washing of tankers used in petroleum transportation was a major potential for tubular Ultrafiltration. Abcor devoted considerable effort to develop this business, unfortunately legislation governing the discharge of tanker waste at seas have always been difficult to enforce and tanker operators saw no incentive to change from their traditional practice of dumping washwater at seas. Although that venture was not a commercial success, Abcor's pioneering effort in introducing UF to oily emulsion breaking was eventually rewarded by success with applications in the metal working industries. Our understanding of membrane behaviour and performance is due in parts such people and other more academic workers. A review of the various uses of membrane for oily wastes treatment is given by Porter [3]. The mechanisms of oil water separation by UF and MF membranes are examined in greater detail by Tanny and Heisler [4] and Anderson et al [5]. The treatment of offshore produced water by membrane process is examined by Jackson et al [6] and Zaidi et al [1].

Membranes in general do not work well with unstable suspensions. But that is of little consequence, since conventional technologies such as API separators work very effectively with such suspensions, and have a large economic advantage over membrane processes. In fact, there is an excellent combination of conventional technology for the roughing separation, and membranes for the stable emulsions.

Ultrafiltration membranes often work much better than Microfiltration membranes, even though the oil droplets in even the finest emulsions are large enough to be excluded by the microfiltration membranes. There is an exception when micelles are present, because MF will permit most of these to pass, and with them their oil content. The real reason appear to be that MF membranes tend to be wetted by the oil, probably because the oil droplets can deform enough to enter the MF pores. UF pores, being much smaller, are quite resistant to penetration by stabilised, emulsified oil. As a corollary, UF membranes do not need to be oleophobic. The early work on oil water separation was carried out with cellulose acetate membranes, which were hydrophilic, and it was believed that the second generation of less hydrophilic UF membranes (polysulphone) would not work on oily wastes. In fact, long field experience with the new polysulphone type membranes have shown that they perform just as well as the cellulose acetate type but are more robust and more durable.

Oily waste waters suitable for treatment by membranes contain 0.1 to 10% oil in a stable emulsion. A limited amount of free oil can be processed but usually quantities above 1-2% must be removed by a skimmer prior to membrane processing. The difficulty with free oil or unstable emulsions is that the oil accumulates at the membrane interface and may form a continuous layer which preferentially wets the membrane over water (the interfacial tension between the membrane and the oil is lower than that of membrane / water). In this case the membrane will pass oil and retain the water. The secret of successful oil water emulsion splitting is to maintain discrete and stable emulsion particles of oil (generally over 0.1µm in size) which are larger than the membrane pore size (0.01µm or below). When this is the case oil in the permeate will generally be less than 5-20 mg/l. Typically the oil can be concentrated up to 40-60% and the concentrate is usually recovered by further processing with acid, in some cases, it is used in fuel oils (concentrate with over 50% oil can support combustion by itself). The permeate is usually discharge but it may be pure enough to be re-used or may require further treatment, for example with reverse osmosis prior to re-use.

In the concentration polarisation process in membrane separation, the oil-water emulsions generally behave like colloidal suspensions showing typical gel concentrations of 70 to 80% (close packing of spheres) and a flux independent of pressure in the gel polarised region. Low lubricity emulsions (with a high synthetic oil content) are usually characterised by high flux and low fouling rates. Oils containing large amounts of tramp oil with moderate to low lubricity show lower flux and require more frequent membrane cleaning. High lubricity oils (containing natural fatty materials) exhibit low flux and tend to foul the membrane more easily.

In the presence of poorly stabilised oil emulsions, cellulosic membranes significantly outperform medium surface free energy membranes (polysulphones) which, in turn, significantly outperform low surface free energy membranes (e.g. polyolefins). Non-cellulosic membranes can become wetted with oil more easily and loses their water permeability and rejection for oil [5]. If the chemical nature of the oily wastewater requires a more chemically resistant hydrophobic membranes it is important that a high crossflow velocity be maintained.

Figure 12.4-1 typifies the performance of a hydrophilised polyethersulphone membrane in an oil water separation; the flux declines significantly during the first hour or so of the operation then tends to be fairly stable thereafter. This seems to suggest that there is some degree of surface fouling initially which is then stabilised as the flux is reduced (and concentration polarisation is brought within control by the cross flow). Evidence of normal concentration polarisation is exhibited by figure 12.4-2. which shows that as the oil concentration is increased the flux is reduced accordingly.

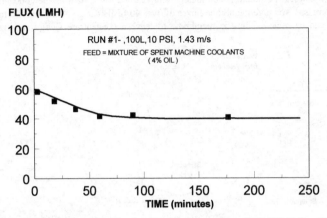

Figure *12.4-1 Initial fouling effect by oil on an UF membrane (150k PES Hollow Fibre/ OWS 501 System).*

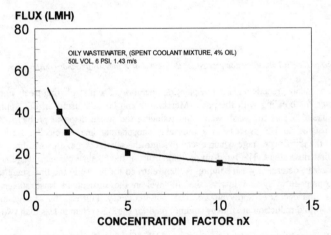

Figure *12.4-2 Effect of oil concentration on the performance of a HF Ultrafiltration system (OWS 5010 system performance).*

It is estimated that there are in excess of five hundred ultrafiltration plants are currently employed for the cleaning of oily wastes with capacities ranging from 1 to 400 m³/day. A survey by Cartwright [7] of over 200 operating systems in the metal finishing industry indicated that operating costs were between £0.3 - £1 per m³ of wastewater treated. In the same survey it was found that the membrane life in these applications were typically between 1.5 and 3 years. Figure 12.4-3. shows the schematic diagram of an ultrafiltration process for the treatment of an industrial laundry wastewater. Most Ultrafiltration plants operate in a semi-batch mode. Typically, the oily waste is first screened to remove large solids, then skimmed to remove any floating oil; sometimes an emulsifier is added to prevent de-stabilisation. In the laundry application oily waste is fed to the plant during a cycle that can last up to a week. Permeate is continually removed and discharged as the concentration of the oil sludge increases. Average membrane fluxes of over 40 LMH can be achieved. Up to 98% of the wastewater can be removed by the membrane to leave a concentrate sludge which can then be incinerated.

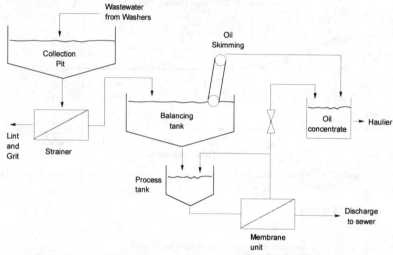

Figure 12.4-3 An ultrafiltration process for the treatment of Industrial laundry wastewater.

Metal cleaning normally precede a plating or painting operation. The phosphating treatment prepares the metal surface for bonding with the paint. Membranes can be very useful in this application in extending the useful life of the wash water and reducing the waste disposal problem. Porter [3] describes the use of an UF process in a pre-paint phosphating line. Typically, fresh water is introduced into the third rinse stage which overflows into the second stage, etc. Alkaline / detergent is used in the first rinse stage. The build-up of oil and dirt in the alkaline cleaning stage is removed by UF; the hot clear, detergent wash solution is then returned for re-use in the first stage tank. This closed loop counter-flow system lowers the oil level in the subsequent rinse stages, reduces detergent costs, and decreases spray nozzle clean out frequency. Further, the savings in disposal charges alone (30 fold reduction in effluent volume) paid for the UF system in less than two years.

Turbie et al [9] describe an industrial application of Ultrafiltration in the treatment of rinse water during the scouring of raw wool. In the wool scouring process, aqueous detergent solutions are used to remove contaminants from raw wool (mainly wool grease and suint or sheep sweat with smaller amounts of soil and faecal matter). During scouring, the wool grease is emulsified by the detergent solution, the suint dissolves, and the mineral particles become suspended. The waste stream is highly polluting with a typical COD of 80,000 mg/l. UF has been effective in reducing the COD to less than 15,000 ppm as well as decreasing disposal costs fresh water consumption was reduced by half.

For spent lubricating oil recovery application, UF has the potential of removing all contaminants, but the flux is very low at room temperature. Inorganic UF membranes operating at 300 °C and a pressure of 7 barg are capable of processing the oil economically. In one plant in Europe, where the spent lubricating oil is pre-treated with thermal shock and centrifugation at 180 °C, the long term flux is reported to be stable between 40-80 LMH. The plant is cleaned only twice annually.

12.5   COD reduction with RO membranes

RO has been found favourably at a pilot scale trial for concentrating cutting oil wastewater streams containing 1.5% oil to 30-60% with the resultant concentrate having a calorific value of about 5000 Btu/lb and a permeate with COD level < 500 mg/l [10]. However, the use of RO for oil water separation has not been a commercial success since in most cases it has to compete against UF which is a more cost effective process, because of the latter's low operating pressure and better throughput performance. However, RO polishing is used to process UF permeate where there is a low COD requirement for discharge into a sensitive water course or for water recycle applications. COD from the UF permeate of a chemical oil water emulsion is due mainly to the soluble surfactants, some residual oil, and trace amounts of anti-microbial agents, corrosion inhibitors and tracer dyes. They have a molecular weight range between 100-750 MW with an average about 250 MW. The surfactants themselves are often a blend of ionic and non-ionic surfactants.

Paulson and Comb [9] showed that oily rinse waters generated from aluminium can manufacturing can be recycled. Following UF, at 110 m³/day of the permeate is re-used as rinse water and the concentrate is further treated by RO before being tankered off-site for disposal. The pay-back period for the membrane plant was reported to be under one year including saving on effluent charges.

During the aluminium rolling process *(see Figure 12.1-1)* an oil water emulsion is used as a coolant for the rolls, a lubricant to reduce frictional losses and as a cleaning solution. Wastes arising from these various uses have an oil concentration of between 0.5 - 5%. Senksen et al [11] describe a combined Ultrafiltration / Reverse Osmosis process to treat an effluent stream of about 400 m³/day to produce de-ionized quality water for re-use in the aluminium rolling process.

Figure 12.5-1 shows how membranes may be used (UF for oil removal followed by RO for COD removal and desalination) for water recycle in an aluminium can washing process.

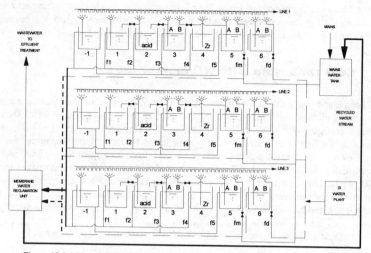

Figure 12.5-1 Use of UF & RO for water recycle in an aluminium can washing process

12.6 System Design

A typical process flowsheet for oily waste treatment by Ultrafiltration is shown by *Figure 12.6*. In general, spent metal working fluids, cleaners, and other wastewaters are collected from the sources and pumped to a balance tank. This tank serves a number of purposes. Firstly, it allows free, or tramp oil and sludge to be separated from the emulsion. Free oil will float to the surface while settleable solids collect at the bottom of the tank. This allows the separation of two troublesome components which have adverse effect on the performance of the membrane. Very often, an oil skimmer is used to removed the free oil at surface. A coalescing filter or API (plate) separator may be used to enhance the performance of the oil skimmer. Secondly, the balance tank tends to even out any variations in the feed flow and the waste composition. Without this equalisation, the membrane system would have to be designed on the worst case basis which would be an inefficient design requiring extra membrane area and make the system more costly.

Stable emulsion is drawn from the balance tank through a pre-filter or strainer to remove any residual solids and pumped to the process tank. Typically, this tank has 3-6 hours capacity (see *12.4.3 Batch Cycle Time*). The process tank level remains constant during this top up mode of the batch run. Any permeate removed from the membrane system during this period is made up by an equal rate of waste transfer from the balance tank to the process tank. If the wastewater has an unstable tendency then pH adjustment and/or surfactant addition can be designed into the process as shown.

From the membrane system, the feed is recirculated back into the process tank. As permeate is removed from the membrane system with each pass of the feed, the process tank will have a slightly higher oil concentration. The permeate leaving the membrane system is essentially oil-free and may be directed to drain if appropriate. Alternatively, post-treatment in the form of reverse osmosis or activated carbon may be used for further polishing if a de-ionised water quality is sought. In the case of synthetic metal working fluids, the permeate will be clear but will also contain most of the original synthetic components and therefore, the permeate may be recycled into the fresh synthetic coolants.

As the batch process continue with fresh emulsion fed to the process tank at the rate equal to the permeate removal from the system, the concentration of oil and grease slowly builds in the process tank. At a certain concentration, typically between 10-15% oil, the feed from the balance tank is stopped and the membrane system is allowed to continue operation. The process tank concentrates or batch down as permeate continue to be withdrawn from the system until a final oil concentration of 25-50% is reached. In this way, the original emulsion (feed) volume is reduced by 80-95%. Once the final concentration is attained, the batch run is terminated and the concentrate is transferred to a concentrate holding tank. The concentrated oil may be hauled away for appropriate disposal, recycled or used as supplementary fuel.

Cleaning is performed to remove surface foulants and metal precipitates from the membrane and pipework. Typical foulants are organic (oil, grease, free oil, fats, waxes, bacteria) and inorganic scale formers (metal oxides, hydroxides, carbonates, sulphates). Cleaning is normally done with an alkaline detergent (see also *12.4.5 Cleaning*) at an elevated temperature (60 °C) for 1-2 hours. For hollow fibre systems, the concentrated cleaning solution can be backflushed through the modules. Indeed, by applying the concentrated cleaning solution directly to the oil deposits on the membrane surface a high temperature clean is not essential for an effective clean. Rinsing of hollow fibre module is also more effective by backflushing the membrane with permeate.

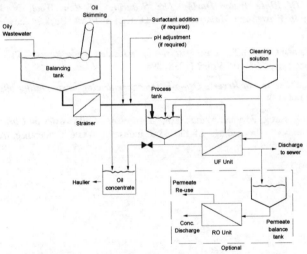

Figure 12.6 A general process flowsheet for oily waste treatment by ultrafiltration

12.7 References

1. Zaidi, A., Simms, K. and Kok, S., "*The Use of Microfiltration for the Removal of Oil and Suspended Solids from Oilfield brines*", in Membrane Technology in Wastewater Management, O.O. Hart and C.A. Buckley (eds), Permagon Press, 1992.

2. Fleischer, A., *"Separation of Oily Wastewaters - The State of The Art"*, paper presented at MARI-TECH 84 the Annual Technical Conference of the Canadian Institute of Marine Engineers, Ottawa, Canada, 1984.

3. Porter, M.C., *"Handbook of Industrial Membrane Technology"*, Noyes Publications, N.J, 1990.

4. Tanny, G.B. and Heisler, *"The application of novel Ultrafiltration Membrane to the concentration of Protein and Oil Emulsions"*, AR Cooper (ed), Ultrafiltration Membranes and Applications, Plenum Press, N.Y. 1980, pp 671-684.

5. Anderson, G.K, Saw, C.B and Le, M.S., Environmental Technology Letters 8(3) 1987 121-132.

6. Taylor, J., Larson, R.E. and Scherer, W., *"Treatment of Offshore Produced Water - An Effective Membrane Process"*, in Environment Northern Seas Conference Report, Vol. 7 (1991) 127-142.

7. Cartwright, P.S., *"The status of Ultrafiltration and Microfiltration Applications in the metal finishing industry"*, paper presented at the American Electroplaters Soc. Surface Finishing, Detroit, 1985.

8. Turpie, D.W.F., Steenkamp, C.J. and Townsend, R.B., *"Industrial Application Of Formed-In-Place Membrane Ultrafiltration And Automated Membrane Forming In The Treatment Of Rinse Water During The Scouring Of Raw Wool, in Membrane Technology in Wastewater Management"*, O.O. Hart and C.A. Buckley (eds), Permagon Press, 1992.

9. Paulson, D.J and Comb, L.F., *"Case Histories of Industrial Water Re-use"*, Proceedings of Water Re-use Symposium III, Vol. 2 (1985) 899.

10. Markind, J., *"The use of Reverse Osmosis for concentrating waste cutting oils"*, AIChE Symp. Ser. **144**, 158.

11. Senksen, M.K, Sittig, F.M and Mariaz, E.F, *"Treatment of oily wastes by Ultrafiltration / Reverse Osmosis"*, Proceedings of the 33rd Industrial Waste Conference, Ann Arbor Publishers (1979) 696.

Chapter 13

LATEX AND PAINT RECOVERY

Contents

13.1 Latex products and their uses

World wide consumption of water-based paint and latex products is estimated to be in the region of 2,000,000 tonnes per annum in 1990. This is split approximately 45% (surface coating products) and 55% (other latex products). Presently water-based paints account for about 55% of all paint products, but this number is increasing as the production of solvent-based paints is reduced through environmental concerns. The volume growth rates for paint and latex products is thought to be between 0.5 - 2% pa.

The word latex (latices for plurals) describes a milky fluid produced by plants (especially rubber) or a synthetic product resembling this. Latex is an emulsion resulting from the dispersion of one liquid in another, so that many people have claimed that these products should not be called polymer emulsions. In practice, they are frequently called polymer dispersions and the aqueous dispersions are called latices (or latexes) and the paints made from them are usually called emulsion paints. Paints are only one type of products (albeit the most important one) of a vast range of products manufactured from latices. This is because of the most remarkable range of properties that latices possess and their ability to be formulated, blended and taylor-made into any texture and body almost according to the chemist's wishes. Table 13.1.1 illustrates the diverse range of applications for different latices.

Table 13.1.1 Types of Emulsions and their uses

Manufacturers	Type of emulsions or Tradename	Applications
BASF	**Acronal** Grades 290D, S760, S360D, 168D, 295D, 567D, S300, S400, S401, 503	Trade and industrial paints. Self crosslinking type for anti-corrosion coatings. Paper and board coatings. Coatings for fibrated cement and concrete roof slabs. Architectural finishes, textured finishes, trowelling compounds, concrete paints. Thermal insularion, flexible coatings for walls and roofs. Bitumen modifers. Water-proofing slurries, mortar additives. Primers.
DSM	**Uramul** Hybrid of epoxy resin modified styrene acrylic dispersions. Grades ZB 2011 SC, CO20, CO70, CO71, 2581, TP989DF, XP328SC	Anti-corrosion maintenance paintd and metal primers. High pvc paints, textured coatings, roofcoatings. Extenor wall and concrete paints. Odourless wall paints and textured coatings for interior and exterior uses
Hoechst	**Mowilith** Grades DM760, DM762, DM765, DM766F, DM767F, DM769F, LDM767, LDM769, DM611, DM680, DM614F	Penetrative primer for porous surfaces. Wood glazes, fexible coatings. Impermeabilization. High pvc paints, ceramic tile adhesives, aluminium foil laminarion. Stiffening agent for footwear, solvent activable adhesives. Interior, exterior and textured paints, thermal insulation adhesive. Floor polishes
Rohm & Haas	**Ropaque** OP-62 **Ropaque** OP-90	Plastic pigment for paints Plastic pigment for paper
	Primal	Decorative paints, floor polish
	Multilobe	Decorarive paints
Vinamul	**Vinacryl** grades 7170, 7172, 71355, 71322, 7171252, 71264	Indusnial coatings, exterior paints, wood primers, cold seal adhesives, high gloss overprint, varnishes, heat resistant varnishs, roofing compounds, water resistant ceramic tile adhesives and glue
	Vinnapas SAF 54	Thermal and acoustic insulation. Tile adhesives
Zeneca Resins	**Neocryl** Acrylic and Styrene-acrylic copolymers	Clear and pigmented industrial coatngs for metal, wood and plastic surfaces. Anticorrosion primers, high build coatings, high gloss metal coatings Woodcoatings, decorative paints, coatings for plasrics, leather, floor polishes, printing inks, overprint varnishes, adhesives, sealants, cement additives
	Haloflex Vinyl-acrylic copolymers	Corrosion resistant coatings, overprint varnishes on fibtrboard stock.
	Neo Pac Urethane-acrylic copolymers	Paints, printing inks, overprint varnishes, floorcoatings
	Neo Tac Acrylic-urethane coploymers	Laminating, heat sealing, contact and pressure sensitive adhesives
	Neorez Polyurethane	Woodcoatings, decorative paints, coatings for plastics, leather, floor polishes, printing inks, overprint varnishes, adhesives, sealants, cement additives
	Viclan Polyvinylidenc chloride	Flexible packaging

The huge volume of latices that is required as raw materials by the many industries for the manufacture of their own products means that latex production must involve some of the world's largest chemical producers. Table 13.1.2 lists a number of major producers of resins and latex emulsions in Europe and their position as market suppliers. Some of the emulsion producers also are also emulsion users (to make end products) themselves.

Table 13.1.2 - Major European Producers of various Resins & Latex Emulsions (1990)

TABLE		
TYPES OF RESINS/EMULSIONS	**MAJOR PRODUCERS**	**ESTIMATED MARKET SHARE (%)**
Acrylic Emulsions & Dispersions	Rohm & Haas	20
	Rohm	17
	Vinamul	15
	Synthese	13
	Zeneca (previously ICI)	12
	Hoechst	12
Alkyds	DSM	25
	Cray Valley	20
	Hoechst	15
	Bayer	10
Epoxies	Ciba	30
	Dow	25
	Shell	25
Phenolics	Hoechst	30
	BP Chemicals	22
	DSM	15
Polyesters	Bayer	25
	Hoechst	22
	BASF	20
	DSM	15
Polyurethanes	Bayer	60
	DSM	20
	Rhone Poulenc	5
Vinyl Emulsions & Dispersions	Hoechst	20
	Wacker	20
	BASF	18
	Vinamul	12
	Zeneca	8

Paint products deserve a closer examination as it is the largest and single most important group of latex products. It is common practice to categorise paints according to their end-uses, notably into the decorative paints (also known as architectural coatings and as building paints) and the industrial paints. The latter can be subdivided into production-line coatings that are factory applied, and a variety of speciality types. Figure 10.1 gives a break down of the various types of paints and their annual usage. The different categories indicate the differences in the types of products involved. However, the specific category into which some classes of industrial coatings are placed may be not be completely clear, the most notable being auto-mobile refinishes which are considered to fall within the speciality coatings category. Similarly, building paints may or may not include heavy duty coatings used on steel work and industrial premises.

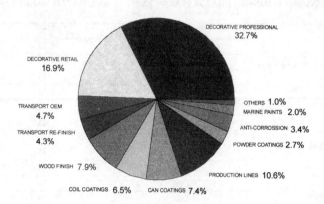

TOTAL = 605,000 tonnes pa (estimate for 1990)

Figure 13.1 Types of paints and their annual usage

13.2 Latex manufacture and wastewater generation

The manufacture of latices generally involve a variety of monomers, solvents, catalysts and organic and inorganic additives. A typical latex manufacturing facility includes office building with R&D and QC laboratories, production area, warehouse, utility building. Raw materials and cleaning solutions are stored in above and below ground tank farms and in drum storage areas. The majority of product manufacturing takes place in a series of vessels located in the process building. Most of the products are pumped to the warehouse for blending, finishing, packaging, and storage. Some products are packaged directly in the process building. Some products are stored in very large tanks ready for transfer to the road tankers for delivery to the user processes. The utility building houses water treatment and heating and cooling facilities.

Finished products are manufactured through a series of mixing and reaction stages involving a number of feed, reaction, and finishing vessels.

Emulsion manufacturing involves a feed stage, a reaction stage, and a finishing or blending stage. The reaction stage involves a number of heating and cooling phases. The reactants are heated and cooled using a vessel jacket. The vapour space above the reactants is cooled by direct contact cooling water known as "headspray". Finished emulsions are typically transferred to the warehouse for bulk storage or additional finishing.

Coalesced, iminated, and complexed latex emulsions are manufactured by making up the appropriate solution in the process building and transferring the solution to a finished latex emulsion being stored in bulk tanks. The resulting mixture is blended to make a uniform finished product. Latex is transferred from bulk to tankers for transport.

Suspension polymer manufacturing involves two stages: a reaction stage and a dewatering stage. In the reaction stage, water, monomer, and a catalyst are reacted in a single vessel to produce polymer beads suspended in a mother liquor solution or slurry. The reaction stage involves reactant and vapour space heating and cooling similar to that used in emulsion manufacturing. The slurry from the reactor vessel is transferred to the dryer area in the warehouse where the polymer beads are centrifuged from the slurry removed from suspension and dried. The dried beads are packaged directly to drums. The mother liquor solution is discharged to the wastewater system.

Two types of cleaning are associated with latex manufacturing. Dispersion and blend tanks used to make water-based products receive a water rinse before and after each batch. Road tankers and IBC cleaning operations may be carried out elsewhere by contractors. The reaction vessels and drying equipment are cleaned on a periodic basis using a water rinse or a solvent soaking followed by a water rinse. The type of cleaning used depends on the compatibility of the adjacent product runs. Wastewater generated as a result of suspension polymer manufacture includes the wasted mother liquor solution and the rinse water from cleanings. The types of wastewaters may be classified as below:

Table 13.2 Classification of principal wastewater streams in latex & paint manufacture

Type of Products		Production Event
VINYL	Emulsion feed stock	Reactor preparation blend
		Tank rinse
		Reactor cleaning and rinsing
	Coalescing Solution	Feed Tank Rinse
	Iminating Solution	Feed tank Rinse
URETHANES	Water-based Aliphatic	Dispersion Tank Rinse
	Water-based Aromatic	Dispersion Tank Rinse
	Urethane Acrylic	Dispersion Tank Rinse
	General	Reactor cleaning and rinsing
SUSPENSION POLYMERS	Styrene-based and Acrylic-based	Mother liquor overflow
		Reactor cleaning
		dryer area cleaning
EMULSION PAINTS	Water-based emulsion paint	Mixing tank & mixer cleaning
		Filter cleaning
		filling line cleaning
MISCELLANEOUS PRODUCTS	Solution polymers	Reactor cleaning
	Bulk product storage	Tank cleaning

13.3 Paint chemistry

13.3.1 Terminology

Paint is a loose term for a whole variety of materials including: lacquers, enamels, varnishes, undercoats, primers, fillers, sealers, etc. All such materials are formulated on the same principle and contain some or all of three main ingredients: pigment, binder and a vehicle (liquid). Figure 13.3 shows how paint is made up from the various component parts.

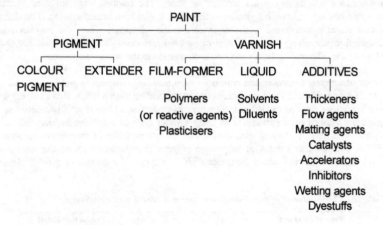

Figure *13.3 Components in paints*

Pigments have both decorative and protective properties. The ideal pigment particle size is about 0.2-0.4µm or half the wavelength of light for maximum scattering of light. Pigments employed have particle diameters from 0.01 µm (carbon black) to 50 µm (some extenders). They may be organic or inorganic. Common used pigments include Iron oxides (red, yellow, brown), TiO_2 (white & Black), lead oxides, carbon black, Zinc oxide, chalk, kaolin. Many organic pigments are organic dyes deposited on an inorganic core (e.g. aluminium hydroxide).

Binder or film-former binds the pigment particles together holding them to the surface and giving it gloss. When the coating is dry the binder is a polymer which may be formed from pre-cursors in the wet formulation.

Vehicle or liquid gives the paint the fluidity which permits paint penetration into crevices to facilitate the coating process. Liquids commonly used include water, paraffins, alcohols, esters and ketones.

Additives when added to paint in amount as little as 10 ppm and seldom over 5% can have profound effect on paint properties. Additives affecting viscosity include: silica, aluminium silicates, bentonite, metal chelates, resinous substances e.g. microgels and cellulosic thickeners. Another group of additives which may have significant influence on membrane processing are those used for modifying surface and interfacial tensions. Silicone oils are used to reduce the liquid surface tension.

Very little silicone oil is used because it finds its way almost entirely to the liquid surface. Surfactants are molecules with two ends of differing polarity and solubility. They provide a bridge between two unlike materials reducing the interfacial tension and assist dispersion.

13.3.2 Pigment dispersion

The pigment is usually in the form of a powder, in which the granules are actually aggregates of fine particles. These particles must be dispersed evenly distributed throughout the paint as a colloidal suspension. For the suspension to have the maximum stability in the solvents, the surface of each particle should be completely wetted with the varnish. Wetting alone is not enough and pigment dispersions in a solvent have poor stability. Each particle in the pigment suspension must be stabilised by polymer chains, attached to its surface by intermolecular forces, yet extending out into the varnish because of their interactions with the solvent molecules.

A resin suitable for dispersion usually contains polar groups (which provide the attraction for the pigment) and is completely soluble in the solvent mixture of the dispersion. Surfactants may be used to bridge the particle-resin interface and assist wetting. In water, ionic surfactants can provide the pigment surface with an electrical charge. Paint additives and a water-soluble thickener (e.g. hydroxyethyl cellulose) are also included in the dispersion at this stage.

Dispersion is usually carried out in a mill, a machine in which the aggregates are subjected to the shear forces and attrition. When shear is the dispersing action, the aggregates are squeezed between two surfaces moving in opposite directions, or in the same direction but at different speeds. Where attrition is part of a dispersion process, the conditions are much gentler and there is no attempt to fracture individual particles. In the viscous medium of the varnish, aggregates - not particles - are broken. Several types of mills are used. The high speed disperser is used for easily dispersed pigments and consists of a horizontal disc with a serrated edge, which rotates at high speed about a vertical axis. The ball mill is a cylinder revolving about its axis, the axis being horizontal and the cylinder partly filled with steel or ceramic balls, or pebbles. The speed of rotation is such that the balls continually rise with the motion and then cascade down again, crushing and shearing the pigment. In the sand grinder the axis of rotation is vertical, the grinding medium is coarse sand or glass beads and the charge is induced to rotate at higher speeds by revolving discs in a stationary container. The ball mill produces a batch of pigment dispersion; the sand grinder gives a continuous output of dispersed pigment .

13.3.3 Water-based Paints

Electrophoretic paints: Any type of resin can be made water dispersible. The usual procedure is to incorporate sufficient carboxyl groups into the polymer to give the resin a high acid value. These groups are then neutralised with a volatile base such as ammonia or an amine, whereupon the resin becomes a polymeric salt, soluble in water or water/ether/alcohol mixtures. If the carboxyl groups are not fully neutralised, an emulsion can be produced. The polymeric part of the salt is negatively charged and can be attracted to and discharged at the anode. This is called anodic electrodeposition. If the paint film deposited is not porous then, as the film thickness increases, so does its electrical resistance and thus the deposition rate slackens and deposition eventually stops. This has the effect of automatically controlling film thickness. Instead of acidic carboxyl groups, resins can be made to include basic, e.g. amine , groups, and these can be neutralised an acid, e.g. acetic or lactic acids, to produce resin salts soluble in water or water-solvent mixtures. Again, incomplete neutralisation produces an emulsion. In these systems, the resin is positively charged and can be discharged at a cathode. This is the basis of cathodic electrodeposition. There will be an increasing use of this type of paint as legislation, economics and world petroleum shortages encourage the use of decreasing amounts of volatile organic solvents and paints of low flammability.

Mechanical emulsions: The film-former or, if it is too viscous, a solution of the film-former, can be dispersed by vigorous agitation. The polymer may be added to the water, the water to the polymer or both may be added to the mixing vessel simultaneously. The stabilising surfactant may be introduced mixed with one component or the other. Thus the emulsion particles may be negatively or positively charged, or neutral according to the type of surfactant employed. Stability is improved by protective colloids, which are polymeric materials with highly polar and non-polar features in their molecular structures. Frequently the colloids in water paints (e.g. water-soluble cellulose derivatives) are affected by micro-organisms, so fungicides and bactericides may be included to prevent deterioration. Emulsions made by mechanical means are not popular in paints today, except for resins which are internally modified to emulsify in water, e.g. in electrodeposition paints.

Emulsion By Polymerisation: Aqueous emulsion polymerisation begins with a true solution of water-soluble monomers and initiator, or with the dispersion of insoluble monomers and initiator using surfactants. Surfactant molecules form cluster or micelles in water, with their fatty organic portions in the centre of the cluster and their water-seeking portions on the outside. The centre of the micelle is a refuge for insoluble organic monomer molecules, which are attracted by hydrophobic tails of the surfactant molecules. Reactions are started by heat, which causes the decomposition of the initiator, e.g. peroxide to free radicals. Polymerisation to insoluble, emulsified polymer occurs within some of the micelles, which then draw more monomers from other micelles. Further surfactant is adsorbed onto the growing particle surface, stabilising it.

Household emulsion paints: In Europe , the most popular paints of this type are based on vinyl acetate copolymer or acrylic latices. The former are prepared by the emulsion polymerisation of vinyl acetate, and stabilised by a combination of surfactants and protective colloids. Although polyvinylacetate (PVAc) is a relatively soft polymer, like polymethyl acrylate, its glass temperature is above room temperature. Household paints must coalesce to form a film at room temperature, so PVAc must have its glass temperature lowered, either by addition of a lacquer plasticiser, such as dibutyl phthalate, or more usually by co-polymerisation with other monomers, e.g. ethyl acrylate. Acrylic latices usually contain copolymers of acrylic or methacrylic acid as colloids and thickeners, these being solubilised by neutralisation with base. Whatever the type of latex, coalescing solvents such as alcohols, glycols, ether-alcohol esters (all with high boiling point) are also added to improve film formation. The pigment dispersion and latex are carefully blended with efficient stirring to form the paint. The additives also include fungicides to prevent mould growth and biocides to prevent bacterial degradation of the thickener.

13.4 Conventional treatment for latex & paint wastewaters

Wastewaters from latex manufacture typically contains between 1 to 5% dry weight of suspended solids and a soluble COD load between 400 to 700 mg/l. Paint wastewaters tend to have a higher solids content (2-7%) and a much higher soluble COD load (6,000 to 10,000 mg/l). The waste streams are also highly coloured. The colours may or may not be soluble depending on the type of pigments employed in the paint formulations.

Traditional wastewater treatment in this industry primarily aims at removing only the TSS (total suspended solids). Figure 13.4 shows a typical wastewater treatment scheme for paint and latex manufacture. In such a scheme the wastewater is collected in a series of settling pits where the most unstable components are settled out. The supernatant is held is a series of holding tanks prior to treatment in the sequencing batch reactors. Acid dosing help to destabilise the latex; additionally flocculants may be used to enhance the settling process. Aeration is often used to strip the surfactants to improve the settling rate. The quality of the resultant effluent tend to be very variable with TSS reduced to between 500 to 3000 mg/l and little change in the soluble COD load. The treated effluent is normally discharged into the trade sewer to receive further treatment downstream. Sludge from the settling pits and the sequencing batch reactors is further dewatered by a belt filter with lime addition before landfill or may be it may be landfilled without any further treatment.

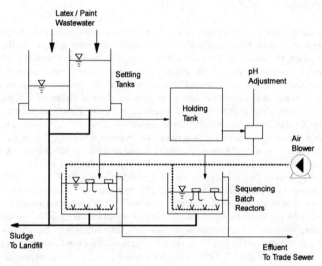

Figure 13.4 A traditional flocculation / precipitation and settling process for latex and paint wastewater treatment

13.5   Paint and latex recovery by membrane processes

13.5.1   Electrocoat paint recovery

The process of electrocoating is widely employed for many different applications[1] [2]. Virtually every car plant in the world uses it for the undercoat. Other metal electrocoat operations include the appliance manufacturing, metal furniture, coil coating, etc. The electrocoat process for primer coating involves the electrophoretic deposition of charged colloidal resinous particles in aqueous dispersion onto a conductive substrate such as a car body. The process is universally favoured because once the paint particles are deposited, they insulate the body from further deposition at that point. The electrical field current thereby directs the migration of paint particles to uncoated areas. The result is an extremely uniform, coherent and defect free coating, even on sharp edges and in recessed areas inaccessible to other methods.

Basic Principles

The aqueous paint is a dispersion of electrically charged polymer. The charges on the polymer are formed from covalently bonded acidic or basic groups. In the case of anodic paints the bound charges arise from carboxylic groups neutralised with alkali, ammonia or an amine; in cathodic paints the polymer contains nitrogen groups which are neutralised with an acid, generally acetic or lactic. The process is operated at such current densities that a rapid pH change is induced close to the metal surface, leading to precipitation of the polymers. This creates two problems. Firstly, as paint is added to make up for that lost on the metalwork the counterion concentration would steadily rise. This drift would inevitably lead to a reduction in paint efficiency. Secondly, when the coated article is removed some of the paint liquid remains on the article. This loosely adhering paint must be removed by rinsing. If left unchecked this would create a waste disposal problem. The solution to the first problem was solved by ICI by using an ion-selective membrane (see figure 13.5-1) [1]. In the anodic coating process a cation exchange membrane allows the selective removal of the potassium counter-ions, and similarly with the more modern cathodic process an anion exchange membrane is used to mediate the removal of lactate or acetate ions. The solution to the second problem was to concentrate up the paint . In the UK Pressed-Steel-Fisher company patented an RO process, while in the US, PPG industries patented an ultrafiltration process.

In a typical electrocoating process, the paint is fed continuously the electrocoat tank which is balanced only by the deposition of charged particles on the substrate. This results in the accumulation of several water-soluble ions in solution with a consequent reduction in throwing power, a lower rupture voltage, thinner films and pin-holing due to rupture. Further, excess drag-out paint from the paint tank must be rinsed off the metal parts to prevent the so-called orange-peel effect and other coating defects. The rinse water cannot be re-used and is, therefore, discarded to the sewer resulting in a severe pollution problem and a costly waste of valuable paint. UF provides a cost effective way of solving the pollution problem as well as the contaminant build-up with the added benefits of paint recovery and reuse.

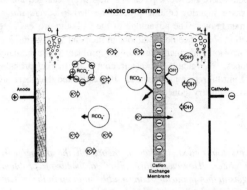

Figure 13.5-1 Schematic of anodic deposition process illustrating the use of the cation exchange membrane to remove excess potassium that would otherwise build up.

Figure 13.5-2 shows the schematic of a typical electrocoating process using UF for paint resins recovery. In such a process, a portion of the permeate is bled to the drain to control the build up of ionic contaminants in the coating tank. The bulk of the permeate is used in the rinse. Only the actual permeate is used at the last rinse station to wash any excess paint back into the recirculation tank. The dilute paint (0.05 to 0.2 %) from the recirculation rinse tank is used in the first and second rinse stations. Thus, eventually all excess paint is returned to the main paint tank. Since all excess paint is recovered a serious pollution problem is averted and, at the same time, justifying the UF unit economically. The paint savings alone typically pay for the UF unit is less than six months. Additional savings result from the reduced de-ionized water use, lowered waste treatment costs, and better control of bath composition.

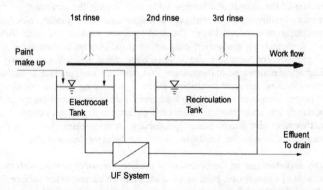

Figure 13.5.2 Schematic of a typical electrocoating process using UF for paint resins recovery.

In the nineteen-sixties most electrocoat tanks were anodic, i.e. the work piece was positively charged and served as the anode while the paint particles were negatively charged (see figure 13.5-2). These systems were ideal since many UF membranes tend to be slightly electronegative with good fouling resistance and excellent flux stability. However, in the nineteen-seventies, a cathodic paint process was developed where the work piece was negatively charged and the paint particles positively charged. This was a superior coating process but led to severe fouling problems with some of the standard UF membranes. The advantage of the cathodic process is related to the fact that there is no metal oxidation and hence no metal ions in the paint film, giving improved resistance of the finished article to corrosion. The membrane fouling problem was eventually solved by using positively charged membranes. Nowadays, the electrocoat paint market probably constitutes the largest single application of ultrafiltration in the world. It is estimated that there are over 1 ,000 UF installations in the automotive industry alone.

Further Developments

The electrocoat process looks ideal in that in theory all the permeate water can be recycled as rinsing water. However, this poses a common problem with such zero-discharge strategies in that impurities can steadily build up in the system. In this case a point would soon be reached where iron levels would effect the finished quality of the goods. For this reason some of the permeate from the UF process is purged either continuously or batch wise. During the 80's the increase environmental pressure has meant that such discharges have become an issue in certain locations. This has led some users to use reverse osmosis and nanofiltration[3] to reduce discharges. In the latter case the membrane serves to separate the iron from the organics in the permeate which can be safely returned to the process.

13.5.2   Latex concentration and recovery

The concentration of latex emulsions was one of the early applications considered for UF membranes [4,5]. However, not all latices are amenable to processing with membrane. Many latices, for example, are unstable under the high shear rates induced by pumps. The dilution of latex during cleaning and rinsing of the reactor and storage tanks may deplete the surfactant on the polymer particles or introduce multivalent ions, resulting in a decreased latex stability. Not only are they commercially worthless, de-stabilised latices also lead to severe membrane fouling. Adjustment of pH or addition of surfactants [6] can prevent coagulation of the latex on the membrane. If the feed is pre-conditioned properly, the membrane flux is often quite stable. Latex is notoriously difficult to pump. While high shear pumps tend to damage the latex, the latex itself tends to crystallise on pump seals causing heavy wearing and rapid mechanical failure. The pumping problem can be overcome to some extent by proper section of the pumping equipment. For example, lobe pumps and peristaltic pumps offer extremely low shear pumping. Peristaltic pumps and diaphragm pumps do not employ seals, however, they have very limited pumping capability. In many cases, double mechanical seals which use permeate for flushing offer a suitable solution for the pump seal problem.

Cleaning is a very important part of the operation of the latex recovery process with membrane. It is necessary, because of a progressive build up of a latex film on the membrane surface. This fouling layer causes an obstruction of the membrane pores thereby reducing the permeation rate. Consolidation of the latex layer may occur over a long period which would be accelerated with exposure to air or if the layer is allow to dry out. The types of fouling must be assessed and a suitable method for the regeneration the membrane must be designed if the long term operation of the process is to be successful. When flux decay does occur, detergent washing is usually sufficient to restore the flux. In some cases, polymer solvents may be required. Proper selection of a solvent resistant membrane and/or solvents which will disperse the latex film but not affect the membrane is crucial. For PVC latex, the solvents of choice are methyl isobutylketone (MIBK) and methyl ethyl ketone (MEK). Styrene butadiene rubber will swell in MIBK, MEK or toluene. For films from vinyl emulsions, removal will be most effective with low MW alcohols such as propyl alcohol or caustic alcohol (e.g. 20% ethanol in 20% potassium hydroxide). Generally, the membranes are first washed with the permeate or water, then detergent, followed by another permeate / water flush. The system is then drained of all water (since it will affect polymer solubility in the cleaning solution) before the cleaning solution proper is applied. Finally, the membrane is rinse with another permeate / water flush. If the module is tubular, sponge balls will enhance cleaning. The frequency of cleaning depend on the type of latex. For latex emulsion applications, a cleaning is required at least weekly. However, for suspension polymers continuous operation of several months between cleanings is not uncommon.

In the past, latex wastewaters were commonly treated by pH adjustment and / or flocculation which remove the suspended solids for landfill disposal while the supernatant which contains the bulk of the water and a high level of dissolved organic substances sent to wastewater treatment works for further treatment before discharge to a waterway. However, while discharge consent limits have become more stringent, the costs of landfill have also rocketed as land space becomes exhausted. Membrane processes become more attractive for latex wastewater treatment when there is:

- *Reduction in discharge consent limits (SS, BOD, COD)*
- *Restriction in landfill opportunities*
- *Reduction of waste volumes*
- *Re-use opportunity for the process materials*
- *Cost savings associated with above*

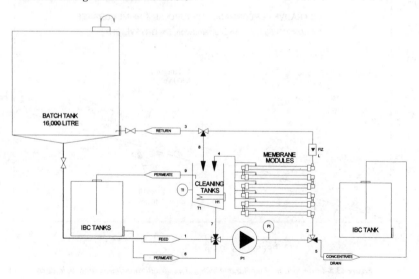

Figure *13.5-3 Schematic of a small batch process for glue (latex) recovery with UF membrane.*

Figure 13.5-3 shows the schematic diagram of a small batch process for glue (polyvinyl acetate latex emulsion) recovery in the manufacture of adhesive backed labels. In general, the dilute latex can be concentrated from 0.5% to between 30 to 50% depending on the rheology and the stability of the latex under the membrane processing condition. In many cases, the reclaimed latex can be recycled for re-formulation. The permeate can be re-used in the process or for cleaning. Where there is a significant saving in the water consumption and effluent charges, membrane process is an economical alternative even without the recovery of the latex. Both hollow-fibre (HF) and tubular membranes have been used successfully for latex concentration. Spirally wound membranes are not suited for this applications because of the high viscosity of some latices and also the feed spacers tend to generate a high shear field which causes the latex to coagulate. Figure 13.5-4 shows a comparison of tubular and HF ultrafiltration membranes in latex concentration. HF membranes are usually limited to a maximum working pressure of under 3 barg. This often means that the crossflow velocity in HF system must be significantly reduced as the viscosity of the latex increases with increasing latex concentration (see also *13.5.2 Emulsion paint recovery*). Thus, HF membrane systems tend to operate with an average flux between 10 to 20% lower than an equivalent tubular system. However, HF systems are very compact, have significantly lower capital cost and much more energy efficient than tubular systems. The advantages of both HF and tubular membranes can be combined in multi-staged systems [7]. In such a system HF modules are employed in the front stages to handle the large volume of relatively dilute materials while tubular membranes are employed in the final stage to handle the relatively small volume of very concentrated feed. The advantages of a hybrid membrane system include a higher overall flux, lower capital and running costs and compactness.

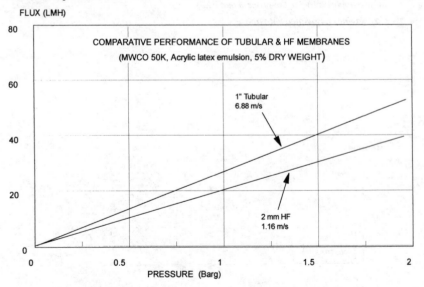

Figure 13.5-4 Comparison of tubular and hollow-fibre ultrafiltration membranes in latex concentration.

Ultrafiltration has a drawback in that the surfactants and other small molecules which stabilise the latex are not retained with the product. This gives a latex with little or no value [6] unless large volume of surfactants and other stabilising agent are added to the wastewater. Further more, the permeate water with a very high COD load often can not be discharged without further treatments. RO represents an ideal add-on process for polishing the ultrafiltration permeate. The RO permeate is of a very high quality and can be re-used in the manufacturing process. The RO concentrate containing the majority of the surfactants is re-combined with the UF concentrate to enhance the latex stability and value. Recently it has been found that tubular Reverse Osmosis membrane can be used directly with latex wastewaters to reclaim the waste latices. The latex recovered by an RO process retains virtually all the stabilising surfactants and biocides represents a much superior product compared to latex recovered by an ultrafiltration process. The RO permeate contains only traces of organics and has a typical COD (chemical oxygen demand) of less than 160 ppm. Such water is suitable for re-use in the latex manufacturing process or discharge directly to a water course with little or no further treatment.

13.5.3 Emulsion paint recovery

The wastewater is from emulsion paint manufacture is typically 2% caustic and may contain 3-5% suspended solids consisting mainly of latex and titanium pigment. This is major effluent problem as well as being a significant loss of valuable raw materials. The traditionally wastewater treatment methods are no longer acceptable as the disposal cost for the wastewater continued to soar ever higher.

In recent years UF has become an acceptable technique for emulsion paint recovery from wastewaters. Paint processing often encounters similar technical problems for latex emulsions. However, the solids in emulsion paints consist of up to 50% titanium oxides. Particles of titanium dioxide are extremely abrasive and pump wears, particularly on seals can cause frequent failures. Traditionally, air driven diaphragm pumps and peristaltic pumps have been used for paint transfer, because such pumps do not rely on any seals. However, these pumps cause severe pulsing and can shorten the membrane life considerably. Centrifugal pumps with back repellers, which ensure that abrasive particles are kept away from the pump seals, have been found the most suitable for paint recirculation duty.

Figures 13.5-5 to 13.5-8 show the effects of a number of different process variables on the performance of a HF ultrafiltration membrane system.

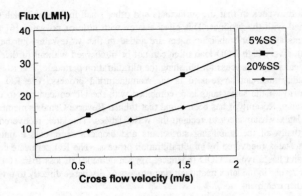

Figure 13.5-5 *Effect of crossflow velocity on flux of a HF ultrafiltration membrane in emulsion paint recovery at different solid concentrations (R25 module, 1.5 mm HF).*

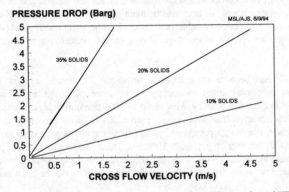

Figure 13.5-6 *Effect of crossflow velocity on pressure drop for an industrial HF module (2 mm ID) in emulsion paint recovery at different solid concentrations (R25 modules, trade emulsion paint at 42 C).*

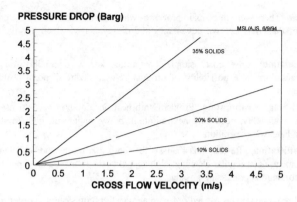

Figure 13.5-7 *Effect of crossflow velocity on pressure drop for an industrial HF module (3 mm ID) in emulsion paint recovery at different solid concentrations (R25 modules, trade emulsion paint at 39 C).*

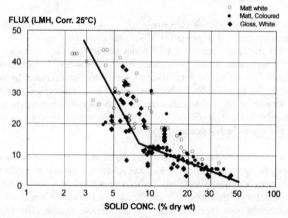

Figure 13.5-8 *Effect of solid concentration on membrane flux in emulsion paint recovery with HF ultrafiltration membranes (100k MWCO) with different types of emulsion paints.*

13.6 Paint recovery - A case study

Basis: In the course of manufacture of a white emulsion paint 10,000 L of wastewater was generated daily. The wastewater was caustic with 3-5% suspended solids consisting mainly of latex and titanium dioxide pigment. Traditionally, the wastewater was kept in a pit which was regularly emptied into road tankers for transport to a landfill site for disposal.

Process Selection: There were three main processes which might be applied to the paint wastewater problem. Each of these has its own advantages and disadvantages which may be summarised as follows:

- **Flocculation**: Low cost; simple operation; narrow operability range; need acid neutralisation; little possibility of product re-use; sludge disposal problem; batchwise operation.

- **Multi-Stage Evaporation**: Problem with heat denaturation need to use vacuum; need steam availability; need acid neutralisation; bulky equipment; high capital and running costs; batchwise operation.

- **Ultrafiltration**: Re-use possibilities for both permeate and concentrate; compact equipment; need regular cleaning; high capital costs; running cost depends on membrane life; continuous processing possible.

An experimental programme was carried out with an ultrafiltration system in order to determine the most appropriate membrane type and operating conditions for the process. Further pilot trial was then carried out on site to obtain consolidate the design data and to demonstrate the robustness of the process before the full scale membrane plant was proposed.

Membrane Process For Emulsion Paint Recovery: A continuous process was selected since the operation was to be unmanned. The plant was of a 3-stage tapered design. In operation the feed solution was fed from the stock tank by a feed pump to the first stage (which employed 100k MWCO HF) where it was concentrated from about 4% to 12% solids. The partially concentrated feed was then displaced into the second stage where it was further concentrated by the second bank of membrane modules (also HF). The feed leaving the second stage was typically 20% solids. In the final stage (tubular membrane) the paint was concentrated to over 40% solids before discharge into a paint concentrate tank. Each of the three recirculation loops was powered by a Durco centrifugal pump with repeller and water-flushed packed gland seal.

Process Economics: For membrane and evaporation the capital costs were very similar (see also example 3 in chapter 11). However, critical to the comparison was the cost of steam versus the cost of membrane replacement. Membrane was found to be more favourable provided that the membrane life was more than 6 months. Comparison with the flocculation process showed a net annual revenue expenditure, the membrane operation resulted in a net annual income. The income from the reclaimed paint would give a plant with a pay-back time of just over three years.

Operating Experience: The full size membrane plant was installed at the factory and it has been operating for several years. Experience its operation has shown that the plant has been producing reclaimed paint with excellent film forming characteristics with a solid content in the region of 40-45% dry weight. Using caustic surfactants for the cleaning procedure once a week, full flux restoration has been achieved consistently. Fouling was completely controlled using this cleaning regime. The membranes themselves have proved to be extremely resistant to the moderate level of solvents and surfactants found in the paint washings. Membrane life of 12 to 16 months was achieved and therefore the process economics have been significantly improved.

13.7 References

1. Cooke, B A, Ness, N M , and Palluel, A L L *"Industrial electrodeposition of paint"* in "Industrial Elctrochemical Processes" ed A Kuhn, 1971, Elsevier

2. Wismer, M, Dravid, A, Bosso, J F, Christenson R M, Jerbak R D,and Zwack, R R *"Cathodic Electrodeposition"* J Coatings Technology 54 (1982) 35-44

3. Davis, G J, and Weitnauer *"Reverse Osmosis Used to Treat Electrodeposition Paint Process Wastewater"* Membrane Technology Reviews 3 (1990) 61-70

4. Micheals, A.S., *"New Separation Technique for the CIP"*, in Chem. Eng. Prog., **64** (1968) 31-43.

5. Porter, M.C., *"Concentration polarisation with membrane ultrafiltration"*, Ind. Eng. Chem. Prod. Res. Dev. **11** (1972) 234-248.

6. European Patent Application EP 0512 736A2 (1990) by Rohm & Haas.

7. United Kingdom Patent Application GB 9214037.5 (1992) by Imperial Chemical Industries Plc.

APPENDIX A - POLYMERIC MEMBRANE MATERIALS

A.1   Introduction

In reading the research and patent literature it could be thought that any material can be made into a membrane. In reality the material properties limit the structural possibilities. Nevertheless, the number of materials that are commercially available is large. This variety stems from the fact that there is no such thing has a perfect membrane material. One membrane material might be the best for one application but hopeless at another. The separating properties and productivity requirements are a necessary condition but depending on the feed and operating conditions there are a whole raft of other properties that have a bearing on whether or not a material is suitable for an application:

- *Thermal stability (Glass transition temperature)*
- *Mechanical properties (brittleness, yield)*
- *Hydrophilicity*
- *Chemical stability (pH, oxidative, solvent, biological)*
- *Biocompatability*
- *Biofouling*
- *Radiation resistance*
- *Adsorption properties*
- *Contaminants (extractables, residual solvents)*

Commercial membranes are available from all basic classes of materials, i.e. polymers, ceramics, metals, inorganics, though polymers make up by far the largest group.

All membrane materials contain small quantities of catalysts, stabilisers, plasticisers and impurities. While such subtleties may not be significant to the use of the materials in other areas, even small changes can dramatically change the structure and properties. Thus an essential element of manufacturing is to secure a source with consistent set of properties.

A.2   Cellulosic Membranes (GS, RO, UF, MF)

Cellulose is the commonest naturally occurring polymer made up of anydroglucose, which have 3 hydroxyl groups. Many of the early developments in the membrane industry were founded on materials derived from it. Despite many weaknesses, cellulosic membranes are still widely used today.   While the hydroxyl groups give cellulose an extremely hydrophilic character, hydrogen bonding forces create strong intermolecular forces that make it intractable as a membrane. This problem is resolved by derivaterising the hydrogen on the hydroxyl groups, e.g. nitration, acetylation. In this way the properties of the polymer can be finely tuned. The derivaterisation makes the polymer soluble in a number of commonly available solvents. The cheap supply of cellulose, the low cost of solvents has meant that cellulose based membranes have remained popular despite a number of limitations. Most of these limitations can be traced to the biological origins of cellulose, with its limited temperature, pH, and oxidative stability.

The largest group of cellulose membranes are those based on cellulose acetate ($T_g \sim 70$ C). The degree of acetylation depends on controlling conditions. In theory up to 3 of the hydroxyl groups on each anyhydroglucose ring can be acetylated. In practice full actylation cannot be achieved and cellulose triactetate has some 2.7 groups converted. As acetylation proceeds, the level of crystallinity increases and the hydrophilicity decreases. This gives rise to a tighter membrane with higher rejection and lower flux. For this reason, cellulose triacetate is used for sea water applications.

The functional groups endow cellulose membranes with very different adsorption properties. Cellulose acetate has a low binding affinity for proteins, while cellulose nitrate has a high affinity for proteins. It is this ability to engineer materials with differing properties from a single source that has made the material a useful tool.

The biological origin of cellulose is its Achilles heal in that it is a natural food for many organisms. For this reason frequent sanitisation or continuous sanitisation is required for long term use so has to keep long term use of the membrane. While cellulosic materials have some chlorine stability its does slowly degrade the polymer by opening the polymer ring leading to loss of mechanical and other properties. Another weakness of cellulosic membranes is its limited pH stability. A process that is exacerbated by the application of temperature (see section 7.4).

A.3   Polysulphone Membranes (GS, RO, UF, MF)

Polysulphone is a widely used membrane material due to its good chemical stability, and thermal integrity. Alan Michaels developed this high performance thermoplastic into a membrane and founded Amicon on it. The uses for polysulphone membrane were extended in 1981 when Monanto launched its Prism gas separation process.

There are a number of different polysulphones. The major two that are used in the membrane industry are Udel and Victrex (polyether sulphone). Polysulphone forms the base membrane on which most composite RO membranes are made.

A.4   Polyolefins (MF, GS)

Polypropylene, and polyethylene provide good chemical, and mechanical properties. However, they are not easy to form into a microporous structure. One of the most successful methods is to stretch the partially crystallised polymer. The stretching process leads to slit like pores opening up between the fibrils. The chief weakness is their hydrophobicity that can cause problems for small sizes. These polymers have limited thermal stability. One of their chief attributes us they have excellent chemical resistance, which makes them suitable for special applications in the chemical industry.

In the mid 80's Dow launched a new membrane for gas separation. This membrane was in the form of small hollow-fibres with a 10 micron dense wall of poly(1,4-methyl penetene). This membrane is being used to concentrate nitrogen from air.

A.5   Polyamide (RO)

Polyamides were an obvious group of polymers to make membranes out of. However the aliphatic polyamides failed to meet many of the environmental demands and failed to realise a major position. With the development of aromatic polyamides the position changed, and there are a wide range of filters based on nylon 6, and nylon 66. Polyamides are naturally more hydrophilic than the polysulphones. However, they still usually require a wetting agent to make these membranes spontaneously wet.

One of the most successful polyamide membranes has been that developed by Film Tec (now owned by Dow). They used and interfacial process to condense a polyaromatic amide on the surface of a polysulphone membrane. The major weakness of membranes made from polyamide is that they have limited oxidative resistance. Another weakness if that at extreme pHs the amide group can hydrolyse.

A.6   Polyacrylonitrile (UF, MF)

Membranes made from polyacrylonitrile have been manufactured into ultrafilters. Their key advantage is that their chemical structure and partially crystalline properties endow them with good chemical stability. In particular they are widely used for treating oily water. Their use is however limited by their thermal stability.

A.7   Polycarbonate and Polyester (MF)

By applying α-radiation to thin sheets (typically 10 µm in thickness) of polycarbonate or polyester film, and then etching back along the tracks, a microporous membrane can be made with a well-defined pore size and porosity (typically 10 %). This membrane type is widely used in analytical and small scale laboratory applications requiring microfilters or loose ultrafilters.

A.8   Fluorinated Polymers (MF, ED)

Polytetrafluoroethylene (PTFE) forms the most chemically inert polymer. However, it has a low glass transition temperature and relies on crystallinity to obtain some mechanical strength. By streching partly crystallised films of PTFE fibrilation occurs and pores open up. PTFE is very hydrophobic.

PTFE is very hydrophobic. By adding fluorinated side chains on with charged end groups a polymer an ion-exchange membrane can be made with hydrophilic characteristics can be made. This is the basis of the ion-exchange membranes used in chlor-alkali production.

Polyvinidenefluoride (PVdF) is a partially fluorinated hydrocarbon. Like PTFE the polymer is hydrophobic. However, by suitable adjustments to the chemsitry the termination group can be switched from hydrogen to hydroxyl, caroxyl, amine etc. These groups are hydrophilic and thus by adjusting teh type and quantity the hydrophilicity of the polymer can be tuned.

A.9   Ion-exchange Membranes (ED, RO)

Ion-exchange membranes are made from polymers with charge groups like sulphonic acid (for cation exchanger) and quaternary ammonium groups (for anion exchanger). One way of making such polymers is to make a copolymer of styrene and divnylbenzene, and then sulphonating this with hot concentrated sulphuric acid. This material can be converted to an anion-exchanger by chloromethylation and amination with a triamine. The membrane is prepared by pouring the monomers with a binding polymer onto a mesh of PVC or polypropylene, followed by copolymerising and sulphonation.

Another group of ion-exchange membranes is based on polyethylene. The charged groups are introduced through sulphochlorination.

The good mechanical, thermal (Udel - Tg =195), and chemical properties of polysulphones suggest that it might be a good candidate for a reverse osmosis membrane. However, its lacks the hydrophilicity required. One way of remedying this is to use extremely aggressive chemicals to add sulphonic acid groups. The difficulty of this approach is control of the extent of sulphonation, and chemical degradation. Anothere approach that avoids this is to create a random copolymer that includes a monomer that is readily sulphonatable. The key advantage of this is the exact control of the number of sulphonic acid groups through the physical formulation. Reverse osmosis membranes made from these polymers have good chlorine resistance.

A.10   Polyimides (UF, GS)

The polyimides form a chemically inert group of polymers with very high glass transition temperature. Ube developed a polyimide membrane for gas separation, and Nitto-Denko developed a polyimide membrane for ultrafiltration.

A.11   Silicone Rubber (GS)

The most widely used class of silicon polymers consist of a backbone of alternating silicon and oxygen atoms, with methyl side chains (polydimethylsiloxane - PDMS). This polymer is used in gas separation for enriching oxygen, and in forming a seal coating on the Prism polysulphone membrane used for extracting hydrogen. The selectivity of silicon rubber is low but it has a high permeability. There have been many attempts to increase its selectivity by varying the side chain groups with little success.

A.12   Elastomers/ Rubbers

Sometimes small and insignificant the seal is a vital part of most membrane equipment ensuring that the integrity of the membrane module. These seals are largely made from elastomers. These highly elastic materials readily return to their original form when the stress is released. Elastomers are polymers that are above their glass transition temperature, but are constrained from flowing by cross-links. A wide variety of elastomers are available (see table A.1)

Table A.1 Summary of various widely available elastomers

Elastomer	Abbreviation	Cost ratio	Comment
Polyisoprene rubber (natural rubber)	NR	1	Suitable for oils
Chloroprene (Neoprene)	CR	2	
Hypalon	CSM	3	A super version of neoprene
Nitrile Rubber	NBR	2	Low permeability. Suitable for acids
Ethylene Propylene rubbers	EPDM, EPM, EPT, EPR	1	Suitable for oils. Water resistant very high
Fluorinated propylene rubber	FPM		Trade Name is Viton
Full fluorinated rubber	FFR		Trade Name is Kalrez. Has similar chemical and heat resistance to PTFE.
Dual laminate elastomers			PTFE, PFA, or FEP covering a resilient inexpensive base such as EPDM
Silicon rubber	SiR	5	
Fluorinated silicon rubber	FSi	40	

These materials are used to provide seals between the permeate and the feed, to seal elements inside pressure vessels, to avoid by-passing (e.g. brine seals in reverse osmosis elements). As ever there is a trade-off between properties, and cost. The low temperature of membrane operation means that some of the lower cost elastomers can be used. However, even within this group there is a significant variety of properties. The rubber industry provides a lot of information on the performance of these materials and there is a wide level of common standardisation. Some of the key properties are listed in table A.2.

Table A.2 Properties used to characterise elastomers

• Tensile Strength	• Thermal Stability
• Elastic Modulus	• Hardness
• Tear Strength	• Oxidative stability (ozone)
• Resilience	• Chemical compatability
• Elongation	• Leachables
• Compression Set	• Biological growth
• Permanent Set	• Density

While elastomers are thermally stable at room temperatures, problems can arise if sustained high temperatures are required or they are subjected to short term exposure (e.g. for cleaning). Figure A.1 shows the temperature stability of various elastomers.

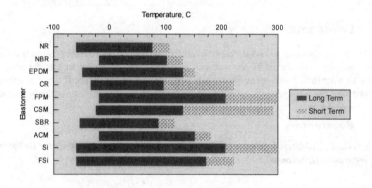

Figure A.1 Shows short and long term temperature stability of various elastomers

The common use of chlorine in system can lead to rubber perishing, and thermal effects can increase the rate of this degradation (see table A.3 for oxidative stability). Rubbers have a significant affinity with certain organic chemicals. The resulting swelling can present difficulty. Uptake of some component from a cleaning chemical can lead to slow release of the material during subsequent operation. Also it should be noted that seals contain a rich cocktail of chemicals, which can leach out during use. Rubber left in water rapidly obtains a slimy feel. This is the result of organic debris (usually from cells) plating out the surface. This organic slime, low flow conditions are an ideal location for the development of a micro-ecosystem.

Table *A.3. Some chemical compatibility data*

Elastomer	Ozone (2 ppm)	5 % H$_2$O$_2$	Chlorine (5 ppm)
NR	Poor	Good	Poor
NBR	Poor	Good	Poor
EPDM	Fair-Good	Excellent	Fair
CR	Fair	Excellent	Fair
FPM	Good	Excellent	Good

Table *A.4. Physical property data of some elastomers*

Elastomer	Specific Gravity	Tensile Strength psig @ 70 F	Elongation Max % @ 70 F	Hardness Range Shore
NR	0.93	4000 10-17	750	30-100
NBR	1	3000 7-14	600	30-100
EPDM	0.86	3000 7-14	600	30-90
CR	1.23	3000 10-17	600	33-95
FPM	1.8-1.9	3,000	450	55-95
CSM	1.12-1.28	3000 14	600	40-95
FFR	2.12-2.22	3,500	350	65-100

A.14   Lubricants

Lubrication is widely used to assist assembly and reduce wear on components such as pumps. However, such materials should be selected with care. Just 1 mL of a lubricant can cover 100 m^2 of membrane to a depth of 100 Å. In addition many lubricants contain silicones. If these materials come into contact with the membrane they change the hydrophilic properties of the membrane surface.

A.15   References

Staude, E, Breitbach, L "*Polysulphone and Their Derivatives: Materials for Different Separation Operations*" J Applied Polymer Science 43 (1991) 559-566

ANNEX B - Trade Names & Acronyms

Abbreviations, trade names abound in the membrane industry. The following list collects together names used by companies to identify their product. There is no common convention in their use. Some companies use the name to describe a particular membrane, a range of membranes, a system design. An additional complexity is that the large number of sales of companies, parts of companies, agreements mean that the name is often associated with more than one company.

Name	Company	Technology	Description
Abcor®	Koch		Spiral wound Tubular products
Accurrel	Enka	MF	Hollow-fibre/tubular polypropylene membranes
Aciplex	Asahi Chemical	ED	Ion-exchange membrane
ACM™	DuPont	RO	Advanced Composite Membrane - polyamide on polysulphone support
Acroflux	Gelman	MF	MF cartridge
Amicon®	Grace		Name of membrane division
Aminidyne	Pall	AM	Affinity membrane with functional -NH2 groups
Anopore	Whatman Scientific	MF,UF	Alumina ceramic membrane
Aqualytics	Graver	ED	Company division supplying ED technology including bipolar membranes
Aquapore	Ionics	MF	Tubular membrane
Aquasource	Degremont	UF	Hollow-fibre membrane for production of potable water in municipal sector
Aquatech	Allied-Signal	ED	Bipolar membrane - water splitting membrane (see Aqualytics)
Aries	Osmonics	MF	Point of use multi-cartridge system
Asypor	Domnick Hunter	CF	mixed cellulose ester
Auto SDI™	King Lee	RO	Automated SDI measuring device
Avir	A/G Technology	GS	Air separation systems
Bekipor®	NV Bekaert SA		Stainless steel fibre matrix >1 µm
Bekopore	Beko Kondest Tech	UF	Tubular polysulphone membranes
Betaklean	Cuno	CF	Absolute rated industrial cartridge filter for high temperature applications
Betapure™	Cuno	CF	classification type rigid cartridge separating particles by sieving
Bioclean™	Argo Scientific		Cleaning chemical for RO membranes
Biofil™	Nanosearch	NF/UF	Tight UF membrane made from crystallised bacterial membrane materials
Bioguard	Pacific-Aquatech		Biocontrol pre-treatment for RO
Biodyne	Pall	AM	Affinity membrane with DNA/RNA specific groups
BTS	Memtec (Aus)	CF	
Carbosep®	Tech-Sep (Rhone-Poulenc)	UF	Inorganic tubular module(ZrO_2/TiO_2/carbon)
Carboxydyne	Pall	AM	Affinity membranes with functional -COOH groups
Carre	Graver Separations	MF	Tubular stainless steel inorganic membranes and systems
Celgard	Hoechst (Celanese)	MF/UF	Microporous polypropylene film and hollow-fibres
Cel-tan	Hoechst	UF	Cassette modules for ultrafiltration
Centistart	Sartorius	UF, MF	System for small samples
Ceraflo®	SCT(US Filter)	MF	Ceramic cross-flow alumina membrane
CeraMem®	Membrain	MF, UF	ceramic membrane module
Ceramesh	USF Acumem	MF	Flexible ceramic membrane
Ceratrex	Osmonics	MF	ceramic membrane
Ceraver	Alcoa	MF	Tubular alumina membrane
Character-U	Kuraray	UF	Cartridge filter
CodeLine™	Advanced Structures	RO	FRP pressure vessels for spiral wound RO elements
CTG-Klean	Cuno	CF	sealed, clean change filter system
Curophan	Enka	HD	A cellulose regenerated by means of the cuprammonium process
Cyclopore™	Whatman	MF	polycarbonate microfilters - track etch for analysis purposes
Diaflo	Amicon Grace	UF,MF	Range of hollow-fibre cartridges using polysulphone membranes
Diamite Series™	King Lee	RO	Cleaner for RO membranes
Durapore	Millipore	MF	Polyvinylidene difluoride membrane filter discs
Duo-Fine	Memtec(Aus)	CF	
Durapore	Millipore	MF	PVdF membrane
Durosan™	Desalination Systems	RO	Special outer wrap for RO
Durotherm™	Desalination Systems	RO	Range of high temperature RO elements
Dynaceram	TDK	UF	Alpha-alumina
Dynaperm	Dynaflow	MF	Cross-flow microfiltration tubes
Eliminator™	Flocovery Systems	RO	light industrial commercial RO systems
Emflon	Pall		PVDF (gas and vent filtration)
Epocel®	Pall	CF	Epoxy resin impregnated cellulose fibres
ESPA™	Hydranautics	RO	Energy saving polyamide
Eval	Kuraray	HD	ethylene-vinyl alcohol copolymer
Fastek	Osmonics	RO	Spiral wound RO elements

FiberCor™	Minntech Corp	CF	Membrane division of the company
FiberFlo™	Minntech Corp	MF	Hollow-fibre filter
Filinert	Nuclepore	MF	PTFE membrane discs
FilmTec®	Dow	RO, NF	Range of elements with interfacial RO membrane
Filtryzer	Toray	HD	Hollow fibre made from polymethylmethacrylate
Flemion	Asahi Glass	ES	Fluorinated ion-exchange polymer
Flocon®	FMC	RO	Cleaning chemical
Floclean™	FMC	RO	Membrane cleaner
Flocide®	FMC	RO, UF	Biocide for sanitizing RO/UF membrane and distribution systems
Flosep™	Tech-Sep (Rhone-Poulenc)	UF	Hollow-fibre modules
Flotrex	Osmonics	CF	Pleated cartridge filtration
Fluoropore	Millipore	MF?	PTFE membrane
Fluorodyne	Pall		PVDF
Fyne	PCI Membrane Systems	NF	Process for dealing with coloured upland waters using tubular membranes
Gambrane	Gambro	HD	An experimental polycarbonate copolymer
Generon™	Dow	GS	N2 from air
Gortex	Gore	MF	PTFE membrane
HGC	Pall	CF	High Dirt Capacity all polypropylene pleated filters
Hemophan	Enka	HD	A modified cuprophan with improved biocompatability
High*Flux Series™	King Lee	RO	Cleaner for RO membranes
Hisep	Asahi Glass	GS	Membrane for oxygen enrichment
Hi-VII	Memtec(Aus)	CF	
Hollosep	Toyobo	RO	Hollow fibre RO membrane
Hydrophilic Nadir®	Hoechest	UF	Regenerated cellulose or polaramid membranes
Hyflow Tetpor	Domnick Hunter	CF	Air filtration using PTFE
Hypersperse™	Argo Scientific	RO	Anti-scalant and dispersant for RO
Hytrex	Osmonics	CF	polypropylene microfibre cartridge filter
Immobilon	Millipore	AM	Affinity membrane
Ionac	Ionac Chemical Company	ED	Ion-exchange membrane
Kerasep™	Tech-Sep (Rhone-Poulenc)		Monolithic ceramic membrane
KL Series™	King Lee	RO	Cleaner for RO membranes
Lavisol	Pacific-Aquatech		Membrane cleaner for RO
Liqui-Cel	Hoechst-celanese	GC	
Magiblock	Osmonics	RO	Fluid Control System
Magnum	Fluid Systems	RO	60" long spiral wound RO elements
MaxCell™	A/G Technology	CF	
MegaModule	USF Acumem	MF	Large hollow-fibre membrane element
MemBio	Memcor(Australia)	MF	Membrane/ Biological Reactor System
Membralox®	SCT(US Filter)	UF	Alumina/Zirconia membrane
Membrana	Enka		Tubular microfiltration module
Mempore	Porex Technology	AM	Affinity membrane
Memsep	Domnick Hunter	AM	Affinity membrane
Mem*Recon™	King-Lee	RO,UF	Process of cleaning spirals by disassembly
Memtrex	Osmonics	CF	Pleated polypropylene polycarbonate, nylon elements
Memstor™	King-Lee	RO,UF	A micobiological growth inhibitor for storage of elements
Metaguard	Pacific-Aquatech		Powder form antiscalants/dispersants
Metricel	Gelman Sciences		Cellulose membranes
Micro-Klean	Cuno	CF	Graded density filter constructed from resin bonded agglomerated fibres
Microclean UHP™	King Lee	RO	For removing biofilms in DI systems
Microdyne	Akzo/Frings	MF	Large membrane systems based on Accurel
Microfluor	Cuno	MF	PTFE
MicroTreat Bio™	King Lee	RO	Biocontrol agent for RO membranes
MicroTreat TF™	King Lee	RO	Biocontrol agent for RO membranes
Microza	Asahi Chemical	MF	
Microza SR-205	Asahi Chemical	MF	Polypropylene hollow fibres
Microza SR-309	Asahi Chemical	MF	Fluoropolymer hollow fibres
Microza LGV	Pall	UF	Pre-treatment for process water applications
Minnclean™	Minntech		Membrane cleaner and preservative
Minncare®	Minntech		Disinfectant/sannitant for membranes
Microdyne	Enka	MF	Polypropylene/nylon 6 capillary tube cartridges
Microfiltrex	Fairey	F	Stainless steel
Mini-tan	Millipore	UF,MF	A small flat sheet membrane stack
Mitex		MF	PTFE membrane
Molsep	Hoechst	UF	Multitubular and Hollow fibre modules UF
N₆₆™	Pall	CF	Nylon 66 filters
Nadir	Hoechst		Flat sheet/tubular membranes
Nafion	DuPont	ES	Fluorinated ion-exchange membrane
Neosepta®	Tokuyama Soda	ED	Ion-exchange membrane
Neosepta-F®	Tokuyama Soda	ED	Ion-exchange membrane

Nepton	Ionics	ED	Ion-exchange membrane
Novasette	Filtron Technology		Flat sheet membrane module
Nuclepore®	Nuclepore Corp	MF	Polycarbonate track-etch membranes
Nypor	Domnick Hunter	MF	Nylon
OptiClean	Pacific-Aquatech		Membrane cleaner for RO
Optimem	Acumem	RO, UF, MF	Cl2 resistant membranes
Organoguard	Pacific-Aquatech		Organic pre-treatment controller for RO membranes
Orion	Osmonics	RO, UF	Point of use system
Osmo®	Osmonics	RO,NF,UF	
Osmotik®	Osmosis Technology	RO	Domestic point of use RO element
PallSep VMF™	Pall	MF	Vibratory membrane filter system using PTFE membrane on stainless support
PEC-1000	Toray	RO	Polyether composite designed for sea water RO
Pellicon	Millipore		Mini cassette flat sheet module (RO,UF, MF)
PermaCare	Houseman	RO	Scale inhinbitoir/antiscalant for RO membranes
PermaTreat	Houseman	RO	Scale inhinbitoir/antiscalant for RO membranes
PermaClean	Houseman	RO	Cleaning chemical for RO membranes
Permasep	DuPont	RO	Hollow-fibre polyaramid module for sea/brackish water applications
Persep®	Tech-Sep (Rhone-Poulenc)		Spiral modules
Plasmax	Toray	PS	
Pleaide®	Tech-Sep (Rhone-Poulenc)	UF	Pplate and frame ultrafiltration/microfiltration system
PMM™	Pall	CF	Metal membranes made of sintered stainless steel
Poly-Fine	Memtec(Aus)	CF	
Posidyne™	Pall	CF	Positively charged membrane in pleated cartridge
Powderguard	Pacific-Aquatech		Powder form antiscalants/dispersants
Pretreat Plus™	King Lee	RO	Anti-scalant/dispersant for RO membranes
Preservol	Pacific-Aquatech		Preservative for RO
Prism	Air Products	GS	Polysulphone gas separation elements developed by Monsanto
Profile™	Pall	CF	Media filter with varying fibre size but constant density
Progard	Argo Scientific		Preservative for RO membranes
Prosep®	Kurita	MF	Hollow-finre PTFE module
Propor-PES	Domnick Hunter	CF	Polyethersulphone
Proppor	Domnick Hunter	CF	Polyproylene
ProTec RO™	King-Lee	RO,UF	Antifoulant for RO UF
PSS®	Pall	CF	Metal membranes made of sintered stainless steel
QuixStand	A/G Technology Corp		Benchtop hollow-fibre system
Rectan	Ube	MF	Polypropylene
ROdesign	Hydranautics	RO	Program for simulating and projecting performance of a system
Romicon	Koch Membrane Systems	MF, UF	Hollow-fibre cartridges
ROPRO	Fluid Systems	RO	Program for simulating and projecting performance of a system
ROSA	Dow	RO	Program for simulating and projecting performance of a system
ROSDA	USF Acumem	RO	Program for simulating and projecting performance of a system
Roga®	Fluid Systems	RO	reverse osmosis spiral wound elements
Rogun™	Argo Scientific	RO	Biocontrol agent
Romembra	Toray	RO	
Sartolon	Sartorius		Polyamide membrane discs
Sartocon	Sartorius	MF	Flat sheet filtration system
Sartopro	Sartorius		
Sartofluor	Sartorius		
Scepter™	Graver Separations	MF,UF	Stainless steel tubular membranes
Schumasiv	Schumacher	MF,UF	Ceramic membranes (MF- α-Al_2O_3, UF-ZrO_2,TiO_2)
Selemion	Asahi Glass	ED	Ion-exchange membrane
SelRO	Kiryat-Weizmann	NF/UF	Range of tubular and spiral wound membranes
Sepa	Osmonics	UF	UF flat sheet membranes
Separex	Hoechst-Celanese	GS	
SepraTech	Ionics	RO-MF	Division of Ionics dealing with filtration processes
Silicium	Atech Innovations	UF	Silicon carbide membrane
SP-MF®	Selecto Scientific	CF	Small element designed to remove cryptosporidium, cysts, turbidity
Spira-Cel®	Hoechst	UF	Spiral wound module -treating oil-water emulsions uses Hydrophilic Nadir
Spiragas	UOP	GS	
Super-cor™	Koch	UF	Tubular membranes
TLC™	Osmonics	RO	Fastek thin layer composite RO elements for domestic use
TFC®	Fluid Systems	RO	Spiral wound RO elements
TFCHR	Fluid Systems	RO	High rejection thin film composite membrane
TFCL	Fluid Systems	RO	Spiral wound RO elements
TFM®	Desalination Systems	RO	Thin-Film Membrane used in Desal3 products
TonkaFlo	Osmonics	RO	High pressure pumps designed for reverse osmosis
Tuiffryn	Gelman Science	UF	Low protein binding polysulphone
Ucarsep	Carre	UF	Zirconia membrane
Ultimem	USF Acumem	RO,UF,MF	Membrane systems

App B - 3

Ultipor®	Pall	MF	
UltraBar	PCI Membrane Systems	UF	Hollow-fibre system
Ultra-cor®	Koch	UF	Tubular membranes
Upilex	Ube		
Ventrex	Osmonics		Polypropylene and PTFE for gases
Veracel	Gelman Sciences	MF	PVDF
Versapor	Gelman Sciences	MF	acrylic polymers
Vinostart	Sartorius		CA for beverage market
VirA/Gard	A/G technology	UV	Hollow-fibre membrane cartridge for viral retention
V-Sep	New Logic	UF,RO	Vibratory shear-enhanced processing
Wafergard PF	Millipore		PTFE
Xampler™	A/G Technology	CF	
Xenoguard	Pacific-Aquatech		Organic pretreatment controller for RO membranes
Xiga	X-Flow	UF,MF	Process using cross-flow UF hollow-fibre element for backwash recovery
Xpress®	Hydac Technology	UF	Capillary
Xpress®	A/G Technology	UF	High Pressure ultrafiltration membranes
Zetaplus™	Cuno	CF	charge modified depth filter with electrokinetic adsorption
Zetapore™	Cuno	CF	charge modified Nylon 66 - 0.4 microns and upwards

ACRONYMS

Analytical Measures & Techniques

EDXS	Electron Dispersive X-ray Scattering
FTIR	Fourier Transform Infra-Red
FTU	Formazin Turbidity Units - This is based on Formazin.
GC	Gas Chromatatography
GPC	Gel Permeation Chromatography
IR	Infra-red
JTU	Jackson Turbidity Units - derived from measurement on the Jackson Candle Turbidemeter (the first type developed)
LSI	Langelier Stability Index
MWCO	Molecular Weight Cut-off
MRI	Magnetic Resonance Imaging
NIR	Near infra red
NTU	Nephelometric - this terminology is used with instruments using 90 scattering.
POU	Point of Use
POE	Point of Entry
SDI	Silt Density Index
SEM	Scanning Electron Microscopy
SIMS	Secondary Ion Mass Spectroscopy
SFM	Scanning Force Microscopy
STM	Scanning Tunnelling Microscopy
TEM	Transmission Electron Microscopy
TIC	Total Inrganic Carbon
TOC	Total Organic Carbon
TOX	Total Organic Halogens
UV	Ultraviolet

Chemical

AE	Acid extractables (class of DPBs)
ANC	Acid Neutralising Capacity
AOC	Assimable Organic Carbon
BE	Base extractables (class of DPBs)
BNC	Base Neutralising Capacity
BOD	Biological Oxygen Demand
BOM	Background Organic Matter
CBOD	Carbonaceous Biochemical Oxygen Demand
CK	Chlorinated ketones
CP	Chlorinated phenols
COD	Chemical Oxygen Demand
DBP	Disinfection by-products
DOC	Dissolved Organic Carbon
HAA	Halo Acetic Acid
HAN	Haloacetonitriles
HS	Halogenated solvents
IOCs	Inorganic chemicals
NBOD	Nitrogenous Biochemical Oxygen Demand
NOM	Natural organic matter
PAHs	Polyaromatic Hydrocarbons
SOC	Synthetic Organic Chemicals

THM	Trihalomethane
THMFP	Trihalomethane Formation Potential
TTHMs	Total Trihalomethanes
VOCs	Volatile Organic Chemicals

Legislation

GL	Guide Level (GL)
IPDWR	Interim primary drinking water regulation
MAC	Maximum Admissible Concentration (EC)
MCG	Maximum Concentration Goal (US)
MCL	Maximum Contaminant Level (US Water Stds)
MRC	Minimum Required Concentration (EC)
MRDLs	Maximum Residual Disinfectant Levels
NPDWR	national primary drinking water regulations
pcv	Prescribed Concentration Values
RMCLs	Recommended maximum concentration levels
SMCL	Secondary Maximum Contaminant Level (US Water Stds)
SWDA	Safe Drinking Water Act
SWTR	Surface Water Treatment Rule

Membrane Technology

DI	Dialysis
DD	Donnan dialysis
ED	Electrodialysis
GC	Gas Contactor
GS	Gas Separation
HD	Haemodialysis
HF	Haemofiltration
MF	Microfiltration
NF	Nanofiltration
PE	Perstraction
PS	Plasma Separation
PV	Pervaopration
RO	Reverse Osmosis
UF	Ultrafiltration

Societies & Organisations

AIDE	
ASTM	American Society for Testing & Materials
AWWA	American Water Works Association
BEWA	British Effluent and Water Association
BP	British Pharmacopoeia
BS	British Standards
EA	Environment Agency of the UK
EPA	Environmental Protection Agency
EP	European Pharmacopoeia
ESMST	European Society of Membrane Science & Technology
EU	European Union
IDA	International Desalination Association

IWSA	International Water Services Association
NAMS	North American Membrane Society
NCCLS	The National Committee for Clinical Laboratory Standards
PHS	US Public Health Service
NSF	National Sanitation Foundation
UKWIR	United Kingdom Water Industry Research
USEPA	United States Environmental Protection Agency
USP	United States Pharmacopeia
WHO	World Health Organisation
WQA	Wayer Quality Association
WRc	Water Research Centre (UK research organisation for water industry)

Polymers

ABS	Acrylobutyl styrene
CA	Cellulose acetate
CTA	Cellulose triacetate
cPVC	Chlorinated poly vinylchloride
EPDM	Ethylene propyldiene monomer (rubber)
PA	Poly amide
PEG	Poly ethylene glycol
PES	Poly ether sulphone (Victrex)
PS	Poly sulphone (Udel)
PSt	Poly styrene
PTFE	Poly tetrafluoroethylene
PVC	Poly vinylchloride
PVDC	Poly vinylidene chloride
PVDF	Poly vinylidene fluoride
uPVC	Unplasticsed Polyvinylchloride

Water Treatment Processes

BAF	Biologically Aerated Filter
BAFF	Biological Aerated Filter
BFB	Biological Fluidised Bed
DAF	Dissolved Air Flotation
GAC	Granular Activated Carbon
PAC	Powdered Activated Carbon

APPENDIX C - MEMBRANE PROCESSES GLOSSARY

BIPOLAR membranes consist of a laminate of a cationic and anionic ion-exchange membrane. In the presence of an electric field they produce acid on one side and alkali on the other. Allied-Signal developed a membrane and membrane system in the early 80's(Aquatech). So far the process has not seen much commercial success.

DIAFILTRATION is a process in which water is added at the same rate as water is removed through the membrane.

DIALYSIS is the transfer of molecules between two streams separated by a membrane. The largest use is in blood treatment where on one side of the membrane is blood and on the other an isotonic solution. The low molecular weight species diffuse across the membrane and thereby removed while maintaining the osmotic environment of the blood constant.

DONNAN DIALYSIS is a variant of dialysis in which the membrane carries a bound charge. Consequently charge as well as size plays a role in selectivity.

ELECTRODIALYSIS is a process for removing ions from water using an electric field to drive ions through a membrane. The membranes used are either cation or anion selective. This membrane is used on a large scale in Japan for the production of salt from sea water. In other parts of the world it is used for producing potable water, e.g. Malta.

ELECTRODEIONISATION is a variant of electrodialysis in which ion-exchange beads are placed between ion-exchange membranes. This provides a conducting route for the current, and extends the range to which electrodialysis can be used to low ionic strength systems, e.g. high purity water applications.

ELECTROSYNTHESIS is using an ion-exchange membrane in an electrolysis cell. The key application is the production of chlorine and caustic from salt.

GAS SEPARATION is the separation of gas molecules using pressure. Membranes fall into two types. The first is those that exploit the physical properties of the material. The second relies on the pore structure being less than mean free path of the molecules. An example of Knudsen separation is the use of microporous alumina membranes to separate different isotopes of uranium. An example of a solution-diffusion type membrane is polydimthylsiloxane to concentrate oxygen from air. Another example is the Prism membrane that is based on polysulphone. This is used to recover hydrogen in the ammonia process.

HAEMODIALYSIS is the application of dialysis to blood. Typically it uses hollow-fibre elements where blood passes down the lumen and isotonic fluid passes on the shell side of the element

HAEMOFILTRATION is an application of ultrafiltration to blood whereby water and low molecular weight solutes (typically < 20,000 Daltons) are removed by the application of pressure using an ultrafiltration membrane.

HAEMOXIDATION uses a membrane to introduce air into blood and thereby reduce problems like foaming

LIQUID MEMBRANES use a liquid that has a high affinity for the component to be extracted. The liquid is either supported on a membrane or is emulsified in the fluid. For effectiveness the liquid extractant must have low solubility in the other media.

MEMBRANE DISTILLATION uses a number of hydrophobic microporous membranes. The membrane acts as a single effect distillation stage.

MICROFILTRATION is a filtration application that rejects bacterial size materials (> 0.1 microns).

NANOFILTRATION occupies no-man's land between reverse osmosis and ultrafiltration. Too many nanofilters are regarded as loose RO membranes, while to others they are a tight ultrafiltration membrane. The rejection behaviour is more complex than the typical reverse osmosis application, and charge of the various solution species can play a significant role in determining the actual separation achieved. In recent years this a number of large applications have been installed in Florida for membrane softening.

PERTRACTION is a membrane based solvent extraction process for the removal of organic components from industrial waste water. The membrane does not exert any selectivity.

PERVAPORATION was term coined by Kobler in 1906. The essential idea is that liquid is on one side of the membrane and gas on the other. A temperature gradient and concentration gradient are generated across the membrane. The major membrane type is made from polyvinylalcohol. The major use of the membrane is in dehydration of alcohol, and as an alternative to azeotropic distillation.

PIEZODIALYSIS has not been commercialised. The basic idea is that a membrane which consist of an ion-exchange membrane with cation and anion components. This permits the ions to pass through the membrane but preventing the passage of water.

PLASMAPHERESIS is an ultrafiltration/microfiltration process for blood which allows the removal of large biological molecules such as imuno-complexes

REVERSE OSMOSIS is the use of pressure to remove water against the osmotic pressure of feed. Its principal use is to remove inorganic salts from water. The first commercial membranes were developed in the early 60's as a result of research by Loeb and

Sourirajan. A major use is the production of potable water from sea or brackish water, but there are a large number of other applications. Initially the key membrane was cellulose acetate, but increasingly interfacial membranes based on a cross-linked polyamide have had a dominating position in the market.

ULTRAFILTRATION is the use of pressure to concentrate colloidal materials by removal of water. Typically it is used to separate molecules in the size range 1,000 to 250,000 Daltons. The first commercial membranes were sold by Sartorius in 1927. As a process ultrafiltration grew into prominence on the back of reverse osmosis in the early 60's. Some of its largest applications are in the car industry where it is used in the electrocoat process.

APPENDIX D - UNITS & CONVERSIONS

D.1 Introduction

The large US market for membranes ensures that US units are widely used in the membrane business. The wide variation in size of plants means that different units are used depening on scale. Also it should be noted that the actual units can vary with the context, e.g. permeability for a fibre can be defined in terms of internal, external, or mean diameter.

D.2 Basic Conversion Factors

1 US gallon	= 231 in³	1 psi	= 6894.8 Pa
1 inch	= 2.54 cm	1 bar	=100000 Pa
1 foot	= 30.48 cm		

It should be noted that the US gallon (3.7854 L) is smaller than the UK gallon (4.5461 L).

D.3 Flux

FLUX		L/m²/day	L/m²/hr	Gal/ft²/day (GFD)	m³/m²/day	mL/cm²/s	m³/m²/s
1	L/m²/day	1	4.167E-02	2.4542E-02	1.000E-03	1.1574E-06	1.1574E-08
1	L/m²/hr	24	1	0.59	2.400E-02	2.7778E-05	2.7778E-07
1	GFD	40.75	1.7	1	4.075E-02	4.7160E-05	4.7160E-07
1	m³/m²/day	1,000	41.67	24.54	1	1.1574E-03	1.1574E-05
1	mL/cm²/s	8.64E+05	3.60E+04	2.1205E+04	864	1	0.01
1	m³/m²/s	8.64E+07	3.60E+06	2.1205E+06	8.64E+04	100	1

D.4 Flow Rate

		gal/day (GPD)	m³/day	L/min	m³/s	GPM	m³/hr	ML/day
1	gal/day	1	3.7854E-03	2.6288E-03	4.3813E-08	6.9444E-04	1.5773E-04	3.7854E-06
1	m³/day	2.6417E+02	1	6.9444E-01	1.1574E-05	1.8345E-01	4.1667E-02	1.0000E-03
1	L/min	3.8041E+02	1.44	1	1.6667E-05	2.6417E-01	6.0000E-02	1.4400E-03
1	m³/s	2.2824E+07	8.64E+04	6.00E+04	1	1.5850E+04	3,600	86.4
1	GPM	1.4400E+03	5.45E+00	3.79E+00	6.3090E-05	1	0.23	0.01
1	m³/hr	6.3401E+03	2.40E+01	1.67E+01	2.7778E-04	4.4	1	0.02
1	ML/day	2.6417E+05	1.00E+03	6.94E+02	1.1574E-02	183.45	41.67	1

D.5 Permeability

	m³/m²/s/Pa	cm³/cm²/s/atm	L/m²/hr/atm	L/m²/hr/bar	"A"-value	GFD/atm
1 m³/m²/s/Pa	1	1.0133E+07	3.6477E+11	3.6000E+11	1.0133E+12	2.1486E+11
1 cm³/cm²/s/atm	9.8692E-08	1	3.6000E+04	3.5529E+04	1.0000E+05	21,204.62
1 L/m²/hr/atm	2.7415E-12	2.7778E-05	1	9.8692E-01	2.7778E+00	5.8902E-01
1 L/m²/hr/bar	2.7778E-12	2.8146E-05	1.0133E+00	1	2.8146E+00	5.9682E-01
1 "A"-value	9.8692E-13	1.0000E-05	3.6000E-01	3.5529E-01	1	0.21
1 GFD/atm	4.6543E-12	4.7160E-05	1.6977E+00	1.6755E+00	4.7160E+00	1

D.6 Permeability (Mass)

	Kg/m²/s/Pa	g/cm²/s/atm	10^{-5}gm/cm²/s/atm
1 Kg/m²/s/Pa	1	1.0133E+04	1.0133E+09
1 g/cm²/s/atm	9.8692E-05	1	1.00E+05
1 1e-5gm/cm²/s/atm	9.8692E-10	1.00E-05	1

D.7 Permeability

The permeability of a material is defined in a variety of ways. Darcy's law states that the flow rate, Q, through an homogeneous microporous film is proportional to surface area, S, and the pressure gradient. The constant of proportionality, A, is known as the permeability and in a planar system .

$$Q = -A\, S\, \frac{\partial P}{\partial x}.$$
(D.1)

If there is no consumption or reaction of material conservation of matter implies that the pressure field varies linearly across the film of thickness, L, and thus

$$Q = A\, S\, \frac{\Delta P}{L},$$
(D.2)

where ΔP is the pressure drop. An assumption that underlies this equation is that the flow within the material is laminar. If the flow is laminar and the fluid is Newtonian, then the flow-rate is expected to be inversely proportional to viscosity. Thus the relationship is written as

$$Q = -\frac{A_1\, S}{\eta}\, \frac{\partial P}{\partial x},$$
(D.3)

where A_1 is known as the specific permeability and has the dimesions of area. The specific permeability is a function of the geometry of the porous structure. If the structure was composed of parallel pores then

$$A = \frac{1}{8}\, \varepsilon\, r^2,$$
(D.4)

where r is the radius of the pore and ε is the porosity of the material.

Most membranes cannot be treated as a homogeneous material. For a structure which shows a simple gradation in the porous structure

$$Q = A'\, S\, \Delta P,$$
(D.5)

where

$$\frac{1}{A'} \equiv \int dx \frac{1}{A(x)} .$$

(D.6)

For simple planar pieces of membrane a surface is readily defined and readily measured. However, for hollow-fibres the surface area for the inside and outside of the fibre are different. This means it is essential to define the permeability with respect to a particular fibre diameter (eg inside, ouside, geometric mean). It is most common to use the diameter which corresponds to the surface of filtration.

References

P C Carman "*Flow of Gases Through Porous Media*" Publ. Butterworths, 1956

APPENDIX E - MASS BALANCE EQUATIONS

E.1   Membrane Performance Measures

In discussing the performance of membrane systems its performance is defined in terms of the observable flows and concentrations.

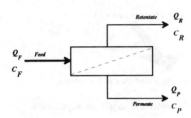

Figure E.1 Schematic outline of membrane separation process.

Three terms which characterise the separation are the

Rejection, $\qquad R \equiv 1 - \frac{c_P}{c_F}$,

Recovery $\qquad Y = \frac{Q_P}{Q_F}$,

Concentration Factor $\qquad CF \equiv \frac{c_R}{c_F}$.

These terms are not independent, and a mass balance on the system gives

$$CF = 1 + R\left(\frac{Y}{1-Y}\right).$$

These equations can be applied to an element, a stage, or the total system. However, the rejection for an element or system will invariably larger than the microscopic rejection since the solute is concentrated up within the system, i.e.

$$R_m > R_e > R_s > R_{sy}$$

where the subscripts m, e, s, sy stand for membrane, element, stage, and system respectively

E.2   Relationship between Macroscopic and Microscopic Rejection

The relationship between the element rejection and that of the membrane is dependent on a large number of factors. A simple relationship can be made if some gross assumptions are made. These include

- *negligible pressure drop in the element,*

App E - 1

- *intrinsic rejection of the membrane is independent of concentration,*
- *the flux is uniform along the duct.*

Consider the flow of feed along a duct lined with membrane (see figure E.2)

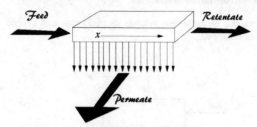

Figure *E.2 Schematic of flow along a duct lines with membrane.*

The mass balance equations for the water and solute are respectively

$$\frac{\partial}{\partial x}(J_f) = -\frac{4}{d_h}J_p \qquad (E.2.1)$$

$$\frac{\partial}{\partial x}(c_f J_f) = -\frac{4}{d_h}c_p J_p \qquad (E.2.2)$$

where d_h is the hydraulic diameter of the feed channel (see Appendix H). Assuming that the pressure and concentration changes along the duct are such that the permeate flow rate is constant equation E.2.1 can be integrated to give

$$J_f = J_F - J_p\left(\frac{4x}{d_h}\right) \qquad (E.2.3)$$

Equations E.2.2 and equation E.2.1 can be combined to give

$$\frac{\partial}{\partial x}(c_f) = \frac{4}{d_h}(c_f - c_p)J_p, \qquad (E.2.4)$$

The permeate concentration is related to the feed concentration through the rejection viz. $c_p = c_f(1-r)$. Using the latter definition in equation E.2.4 gives

$$\frac{\partial}{\partial x}[\ln(c_f)] = \frac{4r}{d_h}\frac{J_p}{J_f}, \qquad (E.2.5)$$

which is readily integrated, viz.

$$c_f(x) = c_F\left[1 - \frac{J_p}{J_F}\left(\frac{4x}{d_h}\right)\right]^{-r} . \qquad (E.2.6)$$

Now the feed flux and permeate flux are related to the feed and permeate flow rates by

$$Q_P = J_p S, \qquad (E.2.7)$$

$$Q_F = J_F A \qquad (E.2.8)$$

where S is the membrane surface area and A is the cross-sectional area. Now

$$d_h = V/S = AL/S,$$ (E.2.9)

where L is the length of the duct. Hence

$$c_f(x) = c_F\left[1 - \frac{Q_P}{Q_F}\left(\frac{x}{L}\right)\right]^{-r},$$ (E.2.10)

from which is follows that

$$CF \equiv c_R / c_F = 1 / \left[1 - \frac{Q_P}{Q_F}\right]^r = \left(\frac{1}{1-Y}\right)^r.$$ (E.2.11)

Combing this equation with the mass balance equation E.2.4 a relationship between the element rejection and that of the membrane can be obtained. This effect is shown in figure E.3a. As can be seen the effect of element recovery is only significant if the element recovery is large, when the rejections are high.

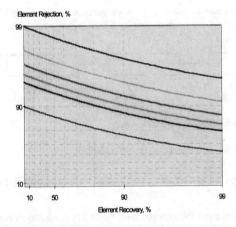

Figure E.3a *Effect of element recovery on element rejection for membrane with intrinsic rejection of 90, 95, 98,99 %. The calculation is made assuming the membrane rejection and flux is constant throughout element.*

However, for solutes which have low or negative rejections the effect of recovery is dramatic at low recoveries (see figure E.3b).

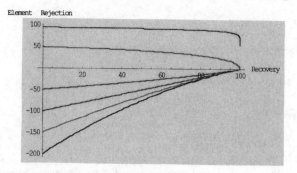

Figure *E.3b Effect of element recovery on element rejection for membrane with intrinsic rejection of -200, -150, -100, -50, 50, 95 %. The calculation is made assuming the membrane rejection and flux is constant throughout element.*

E.3 Series of Membrane Elements

In reverse osmosis it is common to have up to 6 elements in series. Each element creates a concentration enhancement. Thus, the permeate quality falls along the sequence.

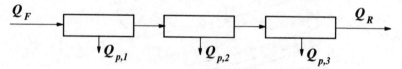

Figure *E.4 Schematic of a series of membrane elements*

A mass balance on the water for each element gives for n elements

$$Q_R = (1 - Y)Q_F = \prod_{j=1}^{n}(1 - Y_j)Q_F, \qquad \Rightarrow \qquad (1 - Y) = \prod_{j=1}^{n}(1 - Y_j),$$

If pressure losses are negligible and the concentration factor effect on flux is negligible then the permeate flows in each case will be approximately the same. In this case the system recovery is simply n times the recovery of the first element

$$Y = nY_1 .$$

Also it is readily shown under the same assumptions that

$$Y_j = \frac{Y_1}{1 - jY_1} .$$

and this can be used to estimate the recovery being used for the last element. In practice pressure losses and concentration factors mean that the productivity rate falls through the sequence.

App E - 4

E.4   Reject Recycle

These continuous designs however are confined to relatively large systems. For small systems there maybe insufficient elements to allow staging. This can be overcome by introducing a recycle in which reject is fed back to the feed (see figure E.5). Recycle like this saves adding an extra stage e.g. if there are only enough elements to construct an a two stage design, but a three stage recovery is required, then introducing recycle allows on to do this.

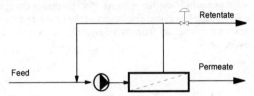

Figure E.5 Schematic of a continuous system utilising a recycle to maintain cross-flow conditions (feed and bleed process)

Whilst recycle allows one to maintain better cross-flow conditions within the elements, there is both a quality and energy penalty.

Mass balance equations for the system are readily established. The recycle allows on to operate the system recovery, Y, at a reasonable value while operating the stage recovery at a more modest value, Y_s. A simple relationship can be established [1] between the system and stage recoveries:

$$Y = Y_s \left(\frac{1+\Theta}{1+Y_s\Theta} \right)$$

where Θ is the recycle ratio which is defined by

$$\Theta \equiv \frac{\text{Internal Recycle Flow Rate}}{\text{System Reject Flow Rate}}$$

Recycling some of the reject raises the concentration being fed by the pump by

$$\frac{1+Y_s\Theta}{1+Y_{sy}\Theta(1-R_s)} \sim 1 + Y_{sy}\Theta \qquad for \quad R_s \sim 1$$

The consequence for this is that for the system rejection, R_{sy}, decrease from the stage rejection, R_s, as the system recovery is increased by increasing the recycle

$$R_{sy} = \frac{R_s}{1+\Theta Y_{sy}(1-R_s)}$$

Inspection of this equation shows that the higher the stage rejection the less significant is the recycle. However, for nanofilters the effect is very significant. A measure of recycle is the percentage of water that passes through the pump more than once. It can be shown by

$$n = \frac{Y}{Y_{sy}} = \left(\frac{1+\Theta}{1+Y_{sy}\Theta} \right)$$

E.5 Permeate Recycle

In some applications the key issue is permeate quality e.g. landfill leachate, high purity water for electronics, high purity for pharmaceuticals, potable water from sea water. Unfortunately, elements are not always available that can meet the quality requirement. This problem can be overcome by using a *multi-pass* system in which the permeate from one bank of membranes is passed through a second bank. This can be done by back-pressurising the first stage (see figure 2.6.6) or introducing an inter-stage pump to boost the pressure. In this design the reject from the second pass is returned to the feed of the first stage. The quality of the permeate from the first pass is such that this secondary membrane system can frequently operate at much higher recoveries and fluxes than the primary separation stage.

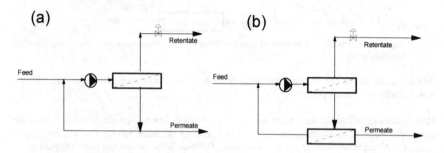

Figure E.6 Schematic of a permeate recycle system (a) with direct recycle (b) with additional membrane stage on permeate line.

The size of some systems might be too small to have a second pass as in figure E.6b. However, the effect can sill be achieved, though less efficiently, by feeding some of the permeate from the first pass directly back to the feed to the first pass (figure E.6a). The effect of this is to reduce the feed concentration to the element, and thereby improve the system rejection. While twin pass systems can achieve higher quality they do at the expense of higher energy and more membrane area for a given permeate flow. For a given rejection target it can be shown that for every halving in transmittance a doubling in membrane area is required.

A similar analysis can be carried out to that for recycle on the reject. In this case though the recycle reduces the system recovery.

$$Y = \frac{Y_{sy}}{1 - \Theta(1 - Y_{sy})}$$

where

$$\Theta = \frac{\text{Recycle Flow Rate}}{\text{Permeate Flow Rate from System}}$$

The average number of passes is given by

$$n = \left(\frac{1 - Y}{1 - Y_{sy}}\right) = \frac{1 - \Theta}{1 - \Theta(1 - Y_{sy})}$$

E.6   Feed & Bleed

A design which is more popular with UF and MF processes than RO is the feed and bleed processes shown in figure E.7 .

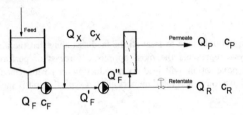

Figure *E.7 Schematic of feed and bleed process*

The relationship between the system performance and the element performance is given by

$$Y = \Theta\left(\frac{Y_e}{1-Y_e}\right)$$

$$R = R_e\left[\frac{1-Y}{1-R_eY}\right]$$

where

$$\Theta = \frac{\text{Recycle Flow Rate}}{\text{Permeate Flow Rate into System}} = \frac{Q_X}{Q_F}$$

E.7   Batch and Semi-batch Processing

Batch processes are widely used in applications which have a variable and small demand. The mass balance equations for the process are readily written down, viz.

$$\frac{d}{dt}(V) = -Q_P \qquad\qquad (E.7.1)$$

$$\frac{d}{dt}(M_s) = -c_pQ_p \qquad\qquad (E.7.2)$$

where V is the total volume of water on the feed side, M_s is the total mass of solute on the feed side, Q_p is the permeate flow rate, and c_p is the concentration of solute in the permeate[1]. These equations can be combined to give

$$V\frac{d}{dt}(c_f) = c_f R\, Q_p \qquad\qquad (E.7.3)$$

If the permeate flow rate and rejection are constants i.e. independent of concentration and time, then it follows from these equations that

$$V(t) = V(0)\,(1 - \tfrac{t}{\tau}) \qquad\qquad (E.7.4)$$

---

[1]   For simplicity, the analysis ignores variation in the feed concentration (i.e. between the return line and the feed tank).

$$c_f(t) = c_f(0)/(1 - \tfrac{t}{\tau})^R \qquad (E.7.5)$$

where the time constant, τ, is defined by

$$\tau \equiv \frac{V(0)}{Q_P} \qquad (E.7.6)$$

and represents the time it would take to treat all the feed assuming that the initial feed rate can be maintained. This analysis represents the most optimistic forecast of performance. Increases in osmotic pressure and polarisation will lead to a reduction in performance. For such analysis equations E.8.1 and 2 have to be integrated numerically. A typical result is shown in figure E.8

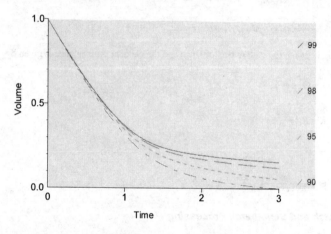

Figure *E.8 Graph showing how the feed-side volume falls with time for membrane with initial rejections ranging from 90 to 99 % (scales are normalised - for conditions see table E.8)*

In many batch applications the key issue is recovery of solute. From the above equations it follows that the amount of material recovered, is related to the concentration factor by the formul

$$Y_s \equiv \frac{M_s(t)}{M_s(t)} = \left[\frac{c_f(t)}{c_f(0)} \right]^{-\left(\frac{1-R}{R} \right)}, \qquad (E.7.7)$$

which is of more general validity.

The equation shows, as expected, that the higher the concentration enhancement the lower the solute recover. Also the looser the membrane the faster is the fall off with concentration factor. What is less obvious is that with a looser membrane (i.e. one with lower rejection) one can achieve higher concentration factors at the expense of lower solute recoveries in the same time. For example a membrane with 90 % rejection might yield 10 % less recovery but achieve the same concentration factor in about half the time that a 95 % rejecting membrane takes (see figure E.9).

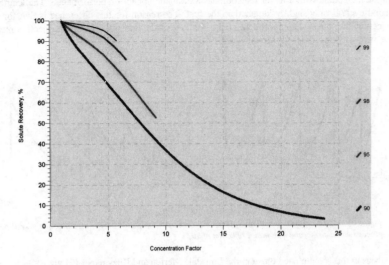

***Figure** E.9 Graph showing the variation of solute recovery with concentration factor for membranes with different rejections (see legend) operating for the same process time (3 times τ). See table E.8 for conditions.*

It is particularly important to consider the implications of the osmotic pressure when using a reverse osmosis membrane. The following provides the results of solving the mass balance equations in a semi-batch reverse osmosis process. The most common way this process is operated is to set a high and low level trip in the feeder tank. When the low level trip is hit new feed is added quickly returning the volume to the high level, and in so doing diluting the concentrated feed. If a high rejecting membrane is used in such applications a problem can arise

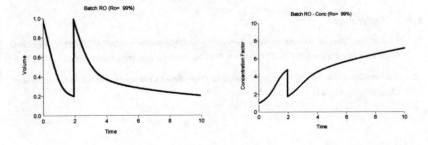

***Figure** E.10 These graphs show how volume and concentration factor evolve with time for a semi-batch process with a membrane with initial rejection of 99 %.*

App E - 9

Selecting a membrane with a lower rejection has a dramatic impact on performance. The length of the second cycle is only slightly longer than the first. The reason for this difference is that for the high rejecting case the system approaches the osmotic pressure limit. Lowering the rejection means that the osmotic pressure is substantially less.

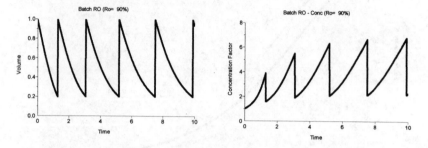

Figure E.11 *These graphs show how volume and concentration factor evolve with time for a semi-batch process with a membrane with initial rejection of 90 %.*

The analysis in this section used those in the Lonsdale Merten and Riley paper [2] viz.

$$Q = SA(\Delta P - R\Pi_f)$$

$$R = 1 / \left(1 + \frac{P_c}{\Delta P - R\Pi_f}\right)$$

The characteristic pressures were taken to be those that gave the initial rejection figures quoted at the start of the batch operation. The surface area and permeability was normalised out using the time constant

$$\tau \equiv \frac{V_o}{SA\Delta P}$$

In the case of the batch recycle the lower volume set-point was set at 20 % of the initial load.

E.8   References

1   P T Cardew, *"Recycle Reverse Osmosis Systens: Some Performance Guides"* in *"Effective Membrane Processes - New Perspectives"* Ed D R Paterson Publ. Mechnical Engineering Publications Ltd (1993).

2   H K Lonsdale, U Merten and R L Riley, *"Transport Properties of Cellulose Acetate Osmotic Membranes"* J Appl Poly Sci 9 (1965) 1341-1362

APPENDIX F - WATER

F.1 Introduction

Water is the most important process fluid for industry, due to its low cost and availability. However, this is not the only use (see figure F.1). The most common source of this water is from municipal suppliers. Despite the exacting standards their process treatment is one which is dictated by the requirements of potability, which are related to health and aesthetic issues, rather than suitability for industrial use. The water quality will vary with the seasons if a surface supply, or change as sourcing switches for distribution reasons. Parameters such as silica which is not a determinand for the municipal supplier can be critical for those operating high pressure boilers. Thus, an important element in any industrial application is to recognise the qualities of the water required for all its uses. Failure to recognise this differentiation will mean sooner or later process problems will arise.

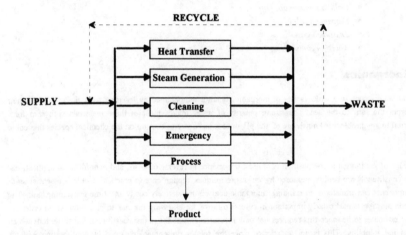

***Figure** F.1 Industrial users have a variety of uses for water. Each use has different requirements and leads to different wastes.*

Some 70 % of the Earth's surface is covered with water. 95 % of all water is in oceans and salty seas, and of this 4 % is locked away in the frozen ice-caps. Of the 1% remainder, 0.01 % is in streams, lakes and rivers. The majority of water used in industry derives from this latter supply. Only in arid areas or when on board ship the source of water maybe derive from saline sources such as sea water.

Table F.1 Salinity classification of waters

Type	Total Dissolved Solids, mg/L
Fresh	< 1000
Brackish Water	1,000-5,000
Highly Brackish	5,000 - 15,000
Saline	15,000 - 30,000
Sea Water	30,000-40,000
Connate Brine	> 40,000

The impurities found in water reflect where it has been, in the air, over hard rocks, through hard rocks, in marshes, in rivers, etc. Three major sources of water are

- surface water (that from rivers, lakes etc.)
- ground water (from wells)
- sea water

and a classification as simple as this can provide some idea of the issues likely to be encountered. One of the key questions for reverse osmosis is just how much dissolved salts are present. However, within this classification there is a large range of compositions which impact on its performance

In terms of chemistry most natural waters exist over a fairly limited range of pH and redox potential. Despite this there is a tremendous richness which is further complicated by microbes which pervade all waters.

- *Dissolved inorganics*
- *Dissolved organics*
- *Colloidal materials, Oils*
- *Suspended solids*
- *Dissolved gases*
- *Micro-organisms*

F.2   Inorganics

All natural waters contain a wide variety of inorganic constituents, obtained by leaching from the ground in their passage from the rain to the sea. The strong polarity of water makes most of these materials split into ions. Fundamental to any problem is knowledge of the pH since it has a strong bearing on the chemical species that could be present.

A knowledge of the inorganic composition is of most importance in reverse osmosis, and nanofiltration applications. Inorganic constituents are readily measured by various techniques, though care to practical details is essential since some components can transform on standing. Inorganic analysis is relatively costly and time consuming task. The detailed composition is particularly important in reverse osmosis for assessing the risk to precipitation. In order for reasonable estimates to be made this requires not only measurements of the major constituents but those ions which because of the solubility. This is not unexpected since the solutes present are obtained by equilibration with its environment.

One of the major uses of reverse osmosis is to reduce TDS of potable water. Most suppliers provide computer programmes to predict the performance. These programmes are usually limited to those ions that occur in natural water (see table F.2.1). For many process applications these programmes do not provide a basis of prediction, and designs are invariably derived from pilot testing. For any design a fundamental starting point is a good water analysis.

Table F.2.1. *Solution species which are considered in RO projection programmes. The focus is on ions that are common in most natural waters, and those that can create precipitation problems. Ammonium and phosphate are features which occur in waste waters and occur in some programmes. With the increasing practice of phosphate dosing to reduce Pb solubilisation and corrosion. While strontium, and barium occur only in low quantities in water they can create troublesome precipitates.*

Anions	Cations	Neutral	Other
Chloride	Calcium	Silica	pH
Sulphate	Sodium	Iron	Alkalinity
Bicarbonate/Carbonate	Magnesium	(Manganese)	Hardness
Fluoride	Potassium		Temperature
Nitrate	Strontium		Turbidity
(Phosphate)	Barium		
	(Ammonium)		

For cleaning applications the hardness of water is critical, and membranes can be used to reduce them. However, just as they can lead to troublesome deposits in boilers etc. the concentration that can occur in a reverse osmosis process can also lead to precipitation of hard salts. Hardness principally comes from calcium and magnesium, though in some processes other ions can contribute to hardness e.g. strontium, barium.

Table F.2.3. *Hardness classification of water*

Classification	Hardness	
	grains per gallon	mg as CaCO3/L
Very Soft	< 1	< 17
Soft	1 - 3.5	17 - 60
Slightly Hard	3.5 - 7.0	60 - 120
Hard	7.0 - 10.5	120 - 180
Very Hard	10.5 -	180 +

Hardness alone does not tell us whether a precipitate might form. For that information on the counterion is required. The two anions of principal interest are sulphate and carbonate. Carbonate levels depend on the pH because of the bicarbonate/carbonate equilibrium etc. Carbonate concentrations are not usually measured. Instead a solution is titrated with acid until a set pH is reached. The quantity of acid required is known as the alkalinity (M-alkalinity). From the starting pH and alkalinity the quantity of carbonate species in water can be calculated using known equilibrium data., and hence the possible dangers from calcium carbonate precipitation can be assessed .

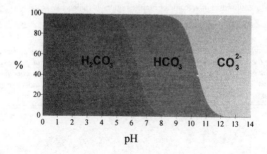

Figure F.2 *Calculated speciation of carbonic acid at 25 C as a function of pH*

Appendix F-3

A common problem with water analyses are errors from sampling, calibration, and interferences etc. A number of simple checks can help verify

- the mass of the ions adds up to TDS
- the moles of anions and cations agree within 5 %
- the TDS and conductivity are in reasonable agreement
- alkalinity is slightly greater than the hardness

These latter two are based on observations on natural waters.

F.3 Gases

As the water passes through the air it adsorbs the gases that make it. Thus, most natural waters contain small quantities of various gases unless steps have been taken to remove them. The level of oxygen is particularly important in that it determines what type of organism will bread in the water.

Table F.6. Gases in natural waters

Common	Contaminants	Additives
· Oxygen	· Hydrogen sulphide	· Chlorine
· Nitrogen	· Methane	· Choloramines
· Carbon dioxide	· Sulphur dioxide	· Ozone
	· Ammonia	· THMs
	· Radon	

In waters which have had a high oxygen demand and have had the oxygen stripped out (e.g. marshes) gives rise to anaerobic conditions as organisms switch to sulphur as a source and the production of hydrogen sulphide. Methane and sulphur dioxide are associated with volcanic waters.

Another group of gases are those associated with disinfection in water treatment e.g. chlorine. In recent years some municipalities have introduced chloroamines. Typical concentrations for total chlorine is 0.7 ppm, while chloramines are usually present at levels of 3 ppm on account of their lower biocidal activity. Ozone is a common disinfectant in waters with high humics or for bottling plants but has a short life due to its high reactivity. This material rapidly decomposes and is rarely of significance for membranes, unless being considered for water recovery.

One gas that is particular to ground waters in volcanic or granite regions is radioactive radon.

> **Henry's Law**
> Many gases distribute themselves between the gas and liquid phase according to Henry's law which states that the concentration in the water is proportional to the partial pressure in the gaseous phase. For gases like carbon dioxide hydrolysis of the dissolved species complicates the issue, and in this case there is an additional uptake to account for that lost through disproportionation to bicarbonate, and carbonate.

F.4   General Measures of Water Quality

Conductivity

Electrical conductivity is a widely used in-situ global measure of the inorganic constituents of water. When conductivity is plotted against TDS for natural water an approximately linear correlation is found (see Nalco Handbook) viz. $TDS \approx 0.6 * \kappa$, where the TDS is in mg/L and the conductivity is in µS/cm (and corrected to 25 C). For natural waters there is also a good correlation between ionic strength and conductivity, viz. $I \approx 1.5 * 10^{-5} * \kappa$

Suspended Solids

Most potable supplies contain a small amount of fine suspended solids. This can vary from fine silts to corrosion products, natural colloids. While such material might be thought to be extremely small e.g. 1 mg/L. If such a feed was operated on a reverse osmosis plant this material could end up at the membrane surface. At a permeate flow rate of 100 L/m²/hr such a feed could lead to a film which is 10 microns thick after ? hrs of operation. Fortunately some of this material is eluted. For this reason membrane plant was operated with

Colour

When we think of clean water our mind conjures up a picture of blue tranquillity. This picture derives from the reflection of the blue sky on a sunny day. When viewed in a test tube "pure" water is colourless. Unfortunately, the water exiting from a tap has a brown hue. This colour can have a number of origins. This colour usually come from inorganic residues of which iron and manganese are the most common, and humics. This colour content has implications for its uses. Operations like laundry washing or dying pastel shades will be upset by these factors. If an area is vulnerable to such problems then suitable treatment becomes a requirement.

In order to quantify the colour scale an arbitrary scale has been developed for measuring colour intensity by comparison with a standard. The most common standard is based on potassium chloroplatinate. In this a solution is said to have a colour of 5 mg/L if the colour intensity is the same as that of a solution of 5 mg/L potassium chloroplatinate.

The colour of water cam be a useful indicator of the suitability of water for certain processes. The presence of colour indicates the potentiality for deposition, taste and odour issues. A very common causes of colour is from inorganic precipitates of iron (red water) and manganese (black water), or organic components (humics). If the source is iron or manganese residues then microfiltration can be used to remove these. If the source is the organic components then tight ultrafiltration or even reverse osmosis might be required.

Ultra-Violet

Many organic molecules show no colour in the visible light band. However in the ultra-violet range all organic molecules which contain conjugated carbon molecules will show strong adsorption signals.

Refractive Index (Brix)

Membranes are widely used in food applications. One method that is widely used to characterise the feed and performance of a membrane unit is that for fruit juices is to measure the refractive index of water.

Turbidity

Turbidity is a measure of the light absorbed in water by suspended and colloidal matter. Turbidity provides a measure of the lack of clarity or brilliance of water. Practically, turbidity is easy and reliable technique to use. However, it only provides a gross measure and has limited sensitivity. For potability it is desirable to have less than 1 and is usually required to be below 5. suspended matter are not synonymous terms. Suspended matter is that material which can be removed through conventional filtration.

Appendix F-5

Particle Sizing	Particle size analysis is becoming more readily available. While industrial devices do exist there is still some questions as to their reliability and interpretation of events.
BOD	Biochemical Oxygen Demand (BOD) is one of the three common ways of characterising organic carbon. BOD is the mass of dissolved oxygen required by a specific volume of liquid for the process of biochemical oxidation under prescribed conditions over five days at 20 C in the dark. The resulting value is expressed in milligrams of oxygen per litre of sample. In waste water industries BOD is widely used. Its origin dates back to the turn of the century where a Royal Commission on sewage disposal proposed its use as a means of assessing the rate of biochemical oxidation that would occur in a stream to which a polluting effluent was discharged. It was devised as a way of assessing the degree of degradation of biological matter that would occur in the maximum travel time in rivers in the UK (i.e. in 5 days). Despite its shortcomings it remains a widely used method of quantifying the required treatment for a waste water though the temperature and time vary from country to country. Unless steps are taken to inhibit ammonia oxidation to nitrate the BOD is a measure of both carbonaceous biochemical oxygen demand (CBOD), and nitrogenous biochemical oxygen demand (NBOD). Typically BOD measures 50 to 60 % of the total oxidisable carbon. Another complication of BOD is that heavy metal impurities can act as a poison for the organisms and this can create problems
COD	Another measure of organic material in water is the Chemical Oxygen Demand (COD). COD is the amount of oxygen required to cause chemical oxidation of the organic materials in water. The test is based on treating the water with a known amount of dichromate, digesting it at an elevated temperature to oxidise the organic matter and titrating the unconsummed or residual dichromate. The oxygen equivalent of the dichromate destroyed is reported as the COD in mg/L. In contrast with BOD chemical oxygen demand is a relatively quick technique involving measuring the extent to which a solution can be oxidised by chromate solution. Typically the COD of water is some 2 to 3 times greater than the BOD. Both COD and BOD are indicators of the environmental health of a surface water supply, and are very commonly used in wastewater treatment.
TOC	The third measure of organics is TOC. This is determined by oxidising all carbon to CO_2. From the membrane standpoint microfiltration will have a reasonable rejection of BOD but will provide little impact on TOC since this frequently is dominated by the small molecules.

F.5 Organics

Just as water leaches inorganics from the rocks that it passes through, so does it extract organics from the residues of life and industry. Table F.4 shows the quantity of organics that are typically present in different waters.

Table F.7   *Typical values for TOC in various waters*

Water Type	Typical TOC range
Ground	0.1 - 2 mg/L
Sea	0.5 - 5 mg/L
Surface	1.0 - 20 mg/L
Biologically treated	8.0 - 20 mg /L
Waste	50 -1000 mg/L
Swap	80 - 300 mg/L

The naturally organic compounds that occur are classified as humic or fulvic acids. In modern times pesticides have penetrated our ecosystem and are typically present at levels of 1-10 ppb. The wide variety of organic constituents

Appendix F-6

present mean that with the exception of specific determinants of interest. One of the most important classifications is volatility. Organics with a high volatility can usually be addressed by such process as air/steam stripping. Another characteristic is the polarity.

Table F.8 *Classification of organic compounds found in water (from JAWWA 70 (11) by permission. Copyright 1978. The American Water Works Association)*

		VOLATILITY		
		Volatile	**Semivolatile**	**Non-volatile**
POLARITY	**Polar**	Alcohols, Ketones Carboxylic Acids	Alcohols, Ketones Carboxylic Acids, Phenols	Polyelectrolytes Carbohydrates, Fulvic acids
	Semipolar	Ethers ,Esters Aldehydes	Ethers,Esters Aldehydes, Hetrocylics	Proteins, Carbohydrates Humic Acids
	Nonpolar	Aliphatic hydrocarbons Aromatic hydrocarbons	Aliphatic, Aromatics Alicyclics, Arenes	Nonionic polymers, Lignins Hymatomelanic Acid
		Low	**Medium**	**High**

MOLECULAR WEIGHT

F.6 Oils and Detergents

Dilute dispersions of oils stabilised with detergents occur in some waste streams. Some are easy to separate due to their density difference with the surrounding media. Other though are difficult to separate on account of their density, size, and concentration. Membrane processes have been designed to concentrate some of these dilute waste streams.

Oils can be particularly troublesome in some membrane processes on account of their tendency to plate out on the membrane surface, and thereby changing the nature of the membrane and hence its performance.

F.7 Microbiological Constituents

The variety of life forms in water is legion. At the top end of the range are the oocysts of protozoa species such giardia, and cryptosporidium. These cysts give rise to opportunistic infections in people who are weakened by other illness', giving rise to giardiasis (acute gastro-enteritis), cryptosporidiosis. This is an area of current concern for water utilities which extract surface water with little treatment. Fungal spores are pervasive. While they rarely cause health risks they can lead to product spoilage. In the case of market gardens which attempt to husband the water by recycling, spore removal is important to keeping the levels of spores and fungus in control. The smallest living organisms are the bacteria and these pervade all water systems. Surfaces such as membranes provide an ideal place for bacteria to grow in that the hydrodynamic conditions are steady, and there is a steady supply of new nutrients. If membranes are run under fairly constant conditions membranes will go through various stages of wildlife development. If oxidants such as chlorine the rate of this development can be curtailed significantly. The alternative strategy is to frequently clean the membrane. This means more than kill the wildlife but remove the residue, otherwise.

F.8 References

BS 2486 - Treatment of water for land boilers

Stumm and Morgan "*Aquatic Water*"

Scheonick "*Water Chemistry*"

"Water Treatment Principles and Design",

P W Atkins "*Physical Chemistry*" Oxford University Press

BS 2690 - Methods of testing water used in industry

Methods for the Examination of Water and Associated Materials (HMSO)

ASTM Stds 1983 Vol 1101 and 1102

American Public Health Association Standard Methods for the examination of Water and Waste Water 19th Ed

APPENDIX G - WORLD WIDE WEB

In the last year the world wide web has become a key method for companies to promote their products. Equally it provides users with rapid information on products and services. The following is a list of world wide web address of companies serving the membrane market.

A/G Technology Corporation
hhtp://www..agtech.com/

AEA Technology: Separation Processes Services (SPS)
hhtp://www..aeat.co.uk/pes/sps.html

Ametek Inc
hhtp://www..ametek.com/

Amicon Inc
hhtp://www..amicon.com/

Anglian Water plc
hhtp://www..anglianwater.co.uk

Aqualytics
hhtp://www..aqualytics.com/

Bekaert
hhtp://www..bekaert.com/

Degremont
hhtp://www..degremont.fr/

Domnick Hunter
hhtp://www..domnickhunter.com/

The Dow Chemical Company
hhtp://www..dow.com/

DuPont
hhtp://www..dupont.com/

Elga
hhtp://www..elga.co.uk/

Gelman Sciences Inc
hhtp://www..argus-inc.com/Gelman/Gelman.html

W L Gore and Associates
hhtp://www..gorefabrics.com/

Graver
hhtp://www..graver.com/

Hoechst
hhtp://www..hoechst.com/

Ionics Inc
hhtp://www..ionics.com/

Koch Membrane Systems Inc
hhtp://www..kochmembrane.com/

Membrex
hhtp://www..membrex.com/

Memtec America Corp
hhtp://www..memtec.com/

Millipore Corporation
hhtp://www..millipore.com/

Osmonics
hhtp://www..osmonics.com/

Osmotik Technology Inc
hhtp://www..osmotik.com/

Pall Corporation
hhtp://www..pall.com/

Porex Technologies Corp
hhtp://www..porex.com/

Rhône-Poulenc
hhtp://www..rhone-poulenc.com/

Schuller Corporation
hhtp://www..schuller.com/

Sulzer
hhtp://www..sulzer.com/

US Filter
hhtp://www..usfilter.com

Whatman International
hhtp://www..whatman.co.uk

APPENDIX H - FLOW IN DUCTS

Membrane technology is largely about the flow of fluids in ducts. Some of the basic definitions and results are collected together. The two commonest geometries are the parallel plate and hollow-fibre configurations (see figure H.1).

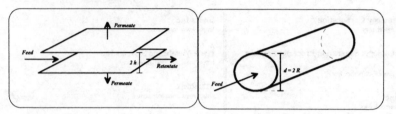

Hydraulic Diameter, Radius

In order to generalise and provide a definition which is geometry independent

$$d_h \equiv 4 * \left(\frac{\text{Enclosed Volume}}{\text{Wetted Surface Area}} \right) \tag{H.1}$$

$$r_h \equiv \left(\frac{\text{Enclosed Volume}}{\text{Wetted Surface Area}} \right) = \frac{d_h}{4} \tag{H.2}$$

Table *H.1   Hydraulic factors for slit and cylindrical ducts*

	Hydraulic Radius	Hydraulic Diameter	k_H	k_S	
Cylinder	d/4	d	$\frac{1}{32}$	8	d = internal diameter
Slit	h	4h	$\frac{1}{48}$	12	2h = slit width

Hagen-Pouseiuille's Equation

In laminar flow the velocity profile adopts a parabolic profile when flowing down a slit or hollow-fibre viz.

$$u(z) = \tfrac{3}{2}\bar{u}(1 - \beta^2) \qquad where \qquad \beta \equiv \tfrac{z}{h}$$

$$u(r) = 2\bar{u}(1 - \beta^2) \qquad where \qquad \beta \equiv \tfrac{r}{R}$$

and $\bar{u}$ is the average flow-rate. The flow-rate is related to the pressure drop by a formula of the type

$$\bar{u} = k_H \left(\frac{d_h^2}{\eta} \right) \frac{\Delta P}{L}$$

where k_H is a geometric factor (see table H1)

Shear

The shear at the membrane surface wall for a cylindrical duct is given by

$$\tau = -\eta \left(\frac{\partial u}{\partial r} \right)_{r \to R} \qquad \text{for hollow-fibre}$$

For hollow-fibre this gives

$$\tau = 4\eta \frac{\bar{u}}{R} = 4\eta \frac{\bar{u}}{R} = 8\eta \frac{\bar{u}}{d_h}$$

an in general

$$\tau = k_S \, \eta \, \frac{\bar{u}}{d_h}$$

where k_S is a geometric factor

Porous Walls

The flow through the membrane walls alters the velocity profile. Using the continuity equation, and the assumption of separation of variables then an approximate solution for the axial and radial velocities is for a cylinderical duct given by

Duct type	Axial	Normal
Cyndrical	$u(r,z) \cong 2\bar{u}(z)\left[1 - \left(\frac{r}{R}\right)^2\right]$	$v(r,z) \cong \frac{3}{2}J\frac{r}{R}\left[1 - \frac{1}{3}\left(\frac{r}{R}\right)^2\right]$
Slit	$u(x,z) \cong \frac{3}{2}\bar{u}(z)\left[1 - \left(\frac{r}{R}\right)^2\right]$	$v(x,z) \cong \frac{3}{2}J\frac{x}{h}\left[1 - \frac{1}{3}\left(\frac{x}{h}\right)^2\right]$

The variation of the mean flow rate $\bar{u}(z)$ can be obtained by a mass balance and gives

$$\bar{u}(z) = \bar{u}(0) - \left(\frac{4J}{d_H}\right)z$$

These formulae provide useful approximations at low wall Reynolds numbers. Higher order corrections have been obtained by Berman[ ]. These equations providing a useful starting point but in practice they have to considerably extended to allow for pressure losses down the fibre etc.

Index

Index

Index